METHODS IN MOLECULAR BIOLOGY

Series Editor
John M. Walker
School of Life and Medical Sciences
University of Hertfordshire
Hatfield, Hertfordshire, UK

For further volumes:
http://www.springer.com/series/7651

For over 35 years, biological scientists have come to rely on the research protocols and methodologies in the critically acclaimed *Methods in Molecular Biology* series. The series was the first to introduce the step-by-step protocols approach that has become the standard in all biomedical protocol publishing. Each protocol is provided in readily-reproducible step-by-step fashion, opening with an introductory overview, a list of the materials and reagents needed to complete the experiment, and followed by a detailed procedure that is supported with a helpful notes section offering tips and tricks of the trade as well as troubleshooting advice. These hallmark features were introduced by series editor Dr. John Walker and constitute the key ingredient in, each and every volume of the *Methods in Molecular Biology* series. Tested and trusted, comprehensive and reliable, all protocols from the series are indexed in PubMed.

Duchenne Muscular Dystrophy

Methods and Protocols

Edited by

Kevin Flanigan and Stefan Nicolau

Jerry R. Mendell Center for Gene Therapy, Abigail Wexner Research Institute, Nationwide Children's Hospital, Columbus, OH, USA

Humana Press

Editors
Kevin Flanigan
Jerry R. Mendell Center for Gene Therapy
Abigail Wexner Research Institute
Nationwide Children's Hospital
Columbus, OH, USA

Stefan Nicolau
Jerry R. Mendell Center for Gene Therapy
Abigail Wexner Research Institute
Nationwide Children's Hospital
Columbus, OH, USA

ISSN 1064-3745 ISSN 1940-6029 (electronic)
Methods in Molecular Biology
ISBN 978-1-0716-4810-0 ISBN 978-1-0716-4811-7 (eBook)
https://doi.org/10.1007/978-1-0716-4811-7

This Humana imprint is published by the registered company Springer Science+Business Media, LLC, part of Springer Nature.
The registered company address is: 1 New York Plaza, New York, NY 10004, U.S.A.

Preface: Introduction and Overview

Duchenne muscular dystrophy (DMD) occupies a special place in the histories of neurology and pathology. Its name refers to Guillaume Duchenne de Boulogne, as the first description of the disease is commonly attributed to his brief account of a patient in 1861, subsequently expanded upon in 1868, [1, 2] although it has become quite clear that other accounts (including those of Edward Meryon and Gaetano Conte) preceded those of Duchenne [3]. Regardless of eponymous claims, these clinical and histopathologic descriptions were important in establishing degeneration of myofibers as a separate process, distinct from degeneration of motor neurons or elements of the central nervous system, laying the groundwork for modern myology.

A similar primacy of place is occupied by DMD in the field of human genetics. The *DMD* gene, encoding the dystrophin protein, was among the first two disease genes (along with that of X-linked chronic granulomatous disease) mapped by positional cloning [4–6]. The subsequent characterization of the protein from the cDNA sequence was perhaps the earliest example of what was formerly referred to as "reverse genetics" [7]. This quickly led to the explanation for the distinction between the severe DMD and less severe Becker muscular dystrophy phenotype linked to the same locus: mutations maintaining an open reading frame result in expression of a partially functional protein and the milder BMD, whereas complete truncation of the reading frame results in no expression and the more severe BMD [8].

In molecular therapeutics, DMD once again is leading the field. Several therapies based upon application of the reading frame rule have now received either accelerated or complete approvals from the US Food and Drug Administration (FDA). Four phosphorodiamidate morpholino oligomers directed toward skipping of specific exons (exons 45, 51, or 53) to restore an open reading frame have received accelerated approval [9–12]. Recently, the FDA approved delandistrogene moxeparvovec, the first adeno-associated viral (AAV) gene therapy for DMD; due to AAV genome packaging limits, this therapy delivers a microdystrophin lacking portions of the central rod domain of the dystrophin protein, but maintaining a reading frame encoding critical N-terminal and C-terminal protein interaction domains [13, 14].

Despite the identification of the gene and protein early in the modern molecular age, and the development of a first-generation of molecular therapies, studying the *DMD* gene and the dystrophin protein continues to present challenges. The locus on the X chromosome comprises 79 exons and 8 promoters over 2.4 megabases that in muscle are spliced into the most prevalent 427 kilodalton isoform (Dp427m). Although advances in sequencing technology have improved diagnostic capabilities, the size of the genomic locus provides a large target for a variety of unusual mutations, some of which are not detectable by standard clinical diagnostic testing. The size of the protein presents challenges to reproducibility in quantification of expression by immunohistochemical, immunoblotting, and other methods, and the lack of standardized methodology across labs makes meaningful comparisons for both prognostic and therapeutic trial outcome measures difficult.

The dystrophinopathies, and in particular DMD, present a significant emotional, social, and economic burden to patients and to society, and intense efforts aimed at understanding underlying pathogenic mechanisms and developing molecular therapies are underway. It is with these considerations in mind that we have undertaken the preparation of this volume. We have aimed to provide protocols for key laboratory techniques employed in the study of the dystrophin protein in the context of Duchenne and Becker muscular dystrophies. We have organized the volume by what we hope are useful categories. These protocols span a wide gamut, from biochemical assays to dystrophin quantification methods to measurements of muscle physiology. We begin with special methods of *DMD* sequence analysis, followed by methods of studying expression of dystrophin and related proteins. This is followed by methods to culture and study patient-derived cells and tissues, and finally methods to analyze animal models of DMD.

We are exceedingly grateful to each of the authors who graciously gave of their time and expertise to participate in this volume, and particularly appreciate their willingness to share their laboratory's best practices for studying the dystrophin protein and models in which it is absent or altered. We hope that that the results prove profitable to our colleagues and will contribute to the standardization of practices in the study of dystrophin and dystrophin-related disease.

Columbus, OH, USA ***Kevin Flanigan***
Columbus, OH, USA ***Stefan Nicolau***

References

1. Duchenne GBA (1861) De l'électrisation localisée et son application à la pathologie et à la thérapeutique, 2nd edn. Ballière, Paris
2. Duchenne GBA (1868) Récherches sur la paralysie musculaire pseudohypertrophique, ou paralysie myosclérosique. Arch Gén Méd 11(5)
3. Tyler KL (2003) Origins and early descriptions of "Duchenne muscular dystrophy". Muscle Nerve 28(4):402–422
4. Koenig M, Hoffman EP, Bertelson CJ, Monaco AP, Feener C, Kunkel LM (1987) Complete cloning of the Duchenne muscular dystrophy (DMD) cDNA and preliminary genomic organization of the DMD gene in normal and affected individuals. Cell 50(3):509–517
5. Antonarakis SE (2021) History of the methodology of disease gene identification. Am J Med Genet A 185(11):3266–3275
6. Kunkel LM (2005) 2004 William Allan Award address. Cloning of the DMD gene. Am J Hum Genet 76(2):205–214
7. Koenig M, Monaco AP, Kunkel LM (1988) The complete sequence of dystrophin predicts a rod-shaped cytoskeletal protein. Cell 53(2):219–228
8. Monaco AP, Bertelson CJ, Liechti Gallati S, Moser H, Kunkel LM (1988) An explanation for the phenotypic differences between patients bearing partial deletions of the DMD locus. Genomics 2(1): 90–95
9. Clemens PR, Rao VK, Connolly AM et al (2023) Efficacy and safety of viltolarsen in boys with duchenne muscular dystrophy: results from the phase 2, open-label, 4-year extension study. J Neuromuscul Dis 10(3):439–447
10. Servais L, Mercuri E, Straub V et al (2022) Long-term safety and efficacy data of Golodirsen in ambulatory patients with Duchenne muscular dystrophy amenable to exon 53 skipping: a first-in-human, multicenter, two-part, open-label, phase 1/2 trial. Nucleic Acid Ther 32(1):29–39

11. Wagner KR, Kuntz NL, Koenig E et al (2021) Safety, tolerability, and pharmacokinetics of casimersen in patients with Duchenne muscular dystrophy amenable to exon 45 skipping: a randomized, double-blind, placebo-controlled, dose-titration trial. Muscle Nerve 64(3):285–292
12. Charleston JS, Schnell FJ, Dworzak J et al (2018) Eteplirsen treatment for Duchenne muscular dystrophy: Exon skipping and dystrophin production. Neurology. 90(24):e2146–e2154
13. Zaidman CM, Proud CM, McDonald CM et al (2023) Delandistrogene moxeparvovec gene therapy in ambulatory patients (aged >/=4 to <8 years) with Duchenne muscular dystrophy: 1-year interim results from study SRP-9001-103 (ENDEAVOR). Ann Neurol 94(5):955–968
14. D'Ambrosio ES, Mendell JR (2023) Evolving therapeutic options for the treatment of duchenne muscular dystrophy. Neurotherapeutics. 20(6):1669–1681

Contents

Contributors

CAMILA F. ALMEIDA • *Jerry R. Mendell Center for Gene Therapy, Abigail Wexner Research Institute, Nationwide Children's Hospital, Columbus, OH, USA*

MARGARET J. BEATKA • *Diverge Translational Science Laboratory, Milwaukee, WI, USA*

MATTHEW J. BURKE • *Department of Molecular Microbiology and Immunology, School of Medicine, The University of Missouri, Columbia, MO, USA*

EMILY H. CANESSA • *School of Pharmacy and Pharmaceutical Sciences, Binghamton University, Johnson City, NY, USA*

FELIPE DE SOUZA LEITE • *Division of Genetics and Genomics, Boston Children's Hospital, Boston, MA, USA; Department of Pediatrics, Harvard Medical School, Boston, MA, USA*

DONGSHENG DUAN • *Department of Molecular Microbiology and Immunology, School of Medicine, The University of Missouri, Columbia, MO, USA; Department of Chemical and Biomedical Engineering, College of Engineering, The University of Missouri, Columbia, MO, USA; Department of Neurology, School of Medicine, The University of Missouri, Columbia, MO, USA; Department of Biomedical Sciences, College of Veterinary Medicine, The University of Missouri, Columbia, MO, USA*

DIANE M. DUNN • *Department of Human Genetics, University of Utah, Salt Lake City, UT, USA*

EMMA C. FRAIR • *Jerry R. Mendell Center for Gene Therapy, Abigail Wexner Research Institute, Nationwide Children's Hospital, Columbus, OH, USA*

BAYARDO I. GARAY • *Department of Medicine, Lillehei Heart Institute, University of Minnesota, Minneapolis, MN, USA*

GIANNI N. GIARRANO • *Department of Physiology and Cell Biology, The Dorothy M Davis Heart and Lung Research Institute, College of Medicine, The Ohio State University, Columbus, OH, USA*

J. PATRICK GONZALEZ • *Solid Biosciences Inc., Charlestown, MA, USA*

LIUBOV V. GUSHCHINA • *Jerry R. Mendell Center for Gene Therapy, Abigail Wexner Research Institute at Nationwide Children's Hospital, Columbus, OH, USA; Department of Pediatrics, The Ohio State University, Columbus, OH, USA*

EMANUELA GUSSONI • *Division of Genetics and Genomics, Boston Children's Hospital, Boston, MA, USA; Department of Pediatrics, Harvard Medical School, Boston, MA, USA; The Stem Cell Program at Boston Children's Hospital, Boston, MA, USA*

MARGARET HABERMAN • *Diverge Translational Science Laboratory, Milwaukee, WI, USA*

YETRIB HATHOUT • *School of Pharmacy and Pharmaceutical Sciences, Binghamton University, Johnson City, NY, USA*

ROLAND W. HERZOG • *Department of Pediatrics, Herman B Wells Center for Pediatric Research, Indiana University, Indianapolis, IN, USA*

ZACHARY M. HOWARD • *Department of Physiology and Cell Biology, College of Medicine, The Ohio State University, Columbus, OH, USA*

PAUL M. L. JANSSEN • *Department of Physiology and Cell Biology, College of Medicine, The Ohio State University, Columbus, OH, USA; Dorothy M. Davis Heart and Lung Research Institute, The Ohio State University, Columbus, OH, USA*

Patricia Lam • *Jerry R. Mendell Center for Gene Therapy, Abigail Wexner Research Institute, Nationwide Children's Hospital, Columbus, OH, USA*
Michael W. Lawlor • *Diverge Translational Science Laboratory, Milwaukee, WI, USA; Medical College of Wisconsin, Milwaukee, WI, USA*
Miguel A. Lopez Perez • *Department of Physiology and Cell Biology, The Dorothy M Davis Heart and Lung Research Institute, College of Medicine, The Ohio State University, Columbus, OH, USA*
Jeovanna Lowe • *Department of Physiology and Cell Biology, College of Medicine, The Ohio State University, Columbus, OH, USA*
Paul T. Martin • *Jerry R. Mendell Center for Gene Therapy, Abigail Wexner Research Institute, Nationwide Children's Hospital, Columbus, OH, USA; Department of Pediatrics, The Ohio State University College of Medicine, Columbus, OH, USA*
Mohammed A. Mashali • *Department of Physiology and Cell Biology, College of Medicine, The Ohio State University, Columbus, OH, USA; Dorothy M. Davis Heart and Lung Research Institute, The Ohio State University, Columbus, OH, USA; Faculty of Veterinary Medicine, Department of Surgery, Damanhour University, Damanhour, Egypt; Center for Cardiovascular Research, Abigail Wexner Research Institute, Nationwide Children's Hospital, Columbus, OH, USA*
Hui Meng • *Diverge Translational Science Laboratory, Milwaukee, WI, USA*
Stefan Nicolau • *Jerry R. Mendell Center for Gene Therapy, Abigail Wexner Research Institute, Nationwide Children's Hospital, Columbus, OH, USA; Department of Pediatrics, The Ohio State University, Columbus, OH, USA*
Emily M. Ott • *Diverge Translational Science Laboratory, Milwaukee, WI, USA*
Rita C. R. Perlingeiro • *Department of Medicine, Lillehei Heart Institute, University of Minnesota, Minneapolis, MN, USA*
Mariah J. Prom • *Diverge Translational Science Laboratory, Milwaukee, WI, USA*
Jill A. Rafael-Fortney • *Department of Physiology and Cell Biology, College of Medicine, The Ohio State University, Columbus, OH, USA*
Neha Rastogi • *Department of Physiology and Cell Biology, College of Medicine, The Ohio State University, Columbus, OH, USA; Dorothy M. Davis Heart and Lung Research Institute, The Ohio State University, Columbus, OH, USA*
Nancy S. Saad • *Department of Physiology and Cell Biology, College of Medicine, The Ohio State University, Columbus, OH, USA; Dorothy M. Davis Heart and Lung Research Institute, The Ohio State University, Columbus, OH, USA; Faculty of Pharmacy, Department of Pharmacology and Toxicology, Helwan University, Cairo, Egypt*
Jessica Sutton • *Diverge Translational Science Laboratory, Milwaukee, WI, USA*
James Teixeira • *Department of Molecular Microbiology and Immunology, School of Medicine, The University of Missouri, Columbia, MO, USA*
Tatyana A. Vetter • *Jerry R. Mendell Center for Gene Therapy, Abigail Wexner Research Institute, Nationwide Children's Hospital, Columbus, OH, USA; Department of Pediatrics, Ohio State University College of Medicine, Columbus, OH, USA; Diverge Translational Science Laboratory, Milwaukee, WI, USA*
Nicolas Wein • *Jerry R. Mendell Center for Gene Therapy, Abigail Wexner Research Institute, Nationwide Children's Hospital, Columbus, OH, USA; Nantes Université, CHU de Nantes, INSERM, TaRGeT - Translational Research in Gene Therapy, UMR 1089, F-44200, Nantes, France; Department of Pediatrics, The Ohio State University, Columbus, OH, USA*

NOAH L. WEISLEDER • *Department of Physiology and Cell Biology, The Dorothy M Davis Heart and Lung Research Institute, College of Medicine, The Ohio State University, Columbus, OH, USA; Department of Molecular and Cellular Biochemistry, College of Medicine, University of Kentucky, Lexington, KY, USA*
ROBERT B. WEISS • *Department of Human Genetics, University of Utah, Salt Lake City, UT, USA*
DEBORAH ZYGMUNT • *Jerry R. Mendell Center for Gene Therapy, Abigail Wexner Research Institute, Nationwide Children's Hospital, Columbus, OH, USA*

Chapter 1

Multiplex Ligation-Dependent Probe Amplification (MLPA) for the Detection of Copy Number Mutations in the *DMD* Gene

Diane M. Dunn and Robert B. Weiss

Abstract

The multiplex ligation-dependent probe amplification (MLPA) assay is designed to simultaneously screen all 79 *DMD* gene exons for copy number variation in Duchenne and Becker muscular dystrophy (DMD/BMD) patients. Large deletions or duplications are found as mutations in >65% of DMD/BMD patients, making the MLPA assay a leading genetic diagnostic tool for patients and their families. This assay requires only standard molecular biology, PCR amplification equipment and access to a capillary electrophoresis instrument for sizing fluorescently labeled amplicons. MLPA is in wide use for copy number analysis of the dystrophin gene, with key reagents and software available through MRC Holland, the group that originated this quantitative multiplex PCR technique.

Key words Copy number mutations, DMD, MLPA, CNV, Molecular diagnosis, Dystrophinopathy, Genetic disease

1 Introduction

The human dystrophin (*DMD*) gene spans approximately 2.4 million base pairs along the X chromosome, making it ideal for developing assays to detect copy number variation. Due to its enormous size, the dystrophin gene exhibits a high frequency of structural mutations, primarily large deletions and duplications, that alter one or more of the 79 exons encoding the muscle and cardiac-specific Dp427m transcript isoform. Together, these two classes of structural mutations account for over 65% of all pathogenic DMD/BMD mutations [1–3]. Although there are "hot spots" for particular deletions and duplications, all exons must be examined for the molecular diagnosis of de novo mutations, which account for one third of all mutations observed in families [4]. Since the dystrophin gene was among the first human disease genes to be mapped and sequenced and given the high prevalence

Kevin Flanigan and Stefan Nicolau (eds.), *Duchenne Muscular Dystrophy: Methods and Protocols*, Methods in Molecular Biology, vol. 2975, https://doi.org/10.1007/978-1-0716-4811-7_1,

of DMD/BMD across many human populations [5], numerous molecular assays have been developed and tested for genetic diagnosis of *DMD* structural mutations in these families. The multiplex ligation-dependent probe amplification (MLPA) technique for detecting copy number variants (CNVs) was first introduced in 2002 by Dr. Jan Schouten and his colleagues at MRC Holland [6], an Amsterdam-based biotechnology company specializing in molecular biology kits and reagents for genetic testing. The technique remains a workhorse for clinical genetic testing with over 350 MLPA assays available for CNV testing of medically relevant human disease genes.

MLPA is a quantitative multiplex PCR technique that uses a single primer pair to simultaneously amplify up to 60 probe pairs, each targeting a unique genomic sequence. The MLPA technique provided added robustness to two early quantitative multiplex PCR assays developed for CNV testing of DMD/BMD patients. In 1996, Stephen Abbs and colleagues introduced 2 fluorescent multiplex PCR assays to amplify 25 exons of the dystrophin gene [7]. The labeled amplicons were separated and detected on a then state-of-the art, slab gel-based automated DNA sequencer from Applied Biosystems, which quantitated the amount of PCR product from each targeted exon. They also introduced "dosage quotient" (DQ) analysis for quantifying copy number differences between a test and reference sample, a technique still used to analyze MLPA products. A second route to multiplex PCR for CNV detection was the introduction of multiplex amplifiable probe hybridization (MAPH) which overcame the challenge of controlling the variability in annealing efficiency and amplification among exon-specific PCR primers [8]. This MAPH technique used DNA hybridization to immobilized patient DNA with probes prepared from cloned PCR products followed by amplification of the captured probes with a common set of PCR primers. This relatively simple and quick technique was optimized for the dystrophin gene by Dr. Johan den Dunnen and colleagues [9] at Leiden University Medical Center (LUMC). Finally, the two groups of researchers from LUMC and MRC Holland applied the MLPA technique to detect CNVs in the dystrophin gene and showed the assay achieved high accuracy, robustness, and simplicity [10].

For detecting CNVs in the *DMD* gene, each MLPA probe pair is designed to hybridize to adjacent genomic sequence in one of the 79 Dp427m exons (Fig. 1). Specificity is achieved from the annealing of two adjacent oligonucleotides followed by the DNA ligation of the "nick" separating the 3′-hydroxyl group of the "left" probe synthetic oligonucleotide from the 5′-phosphate group of the "right" probe M13-derived oligonucleotide. This ligation step uses a thermostable ligase to enhance specificity, and the resultant ligation product is the target sequence for the multiplex PCR step (Fig. 1). Since MLPA probemixes typically have up to 60 probe

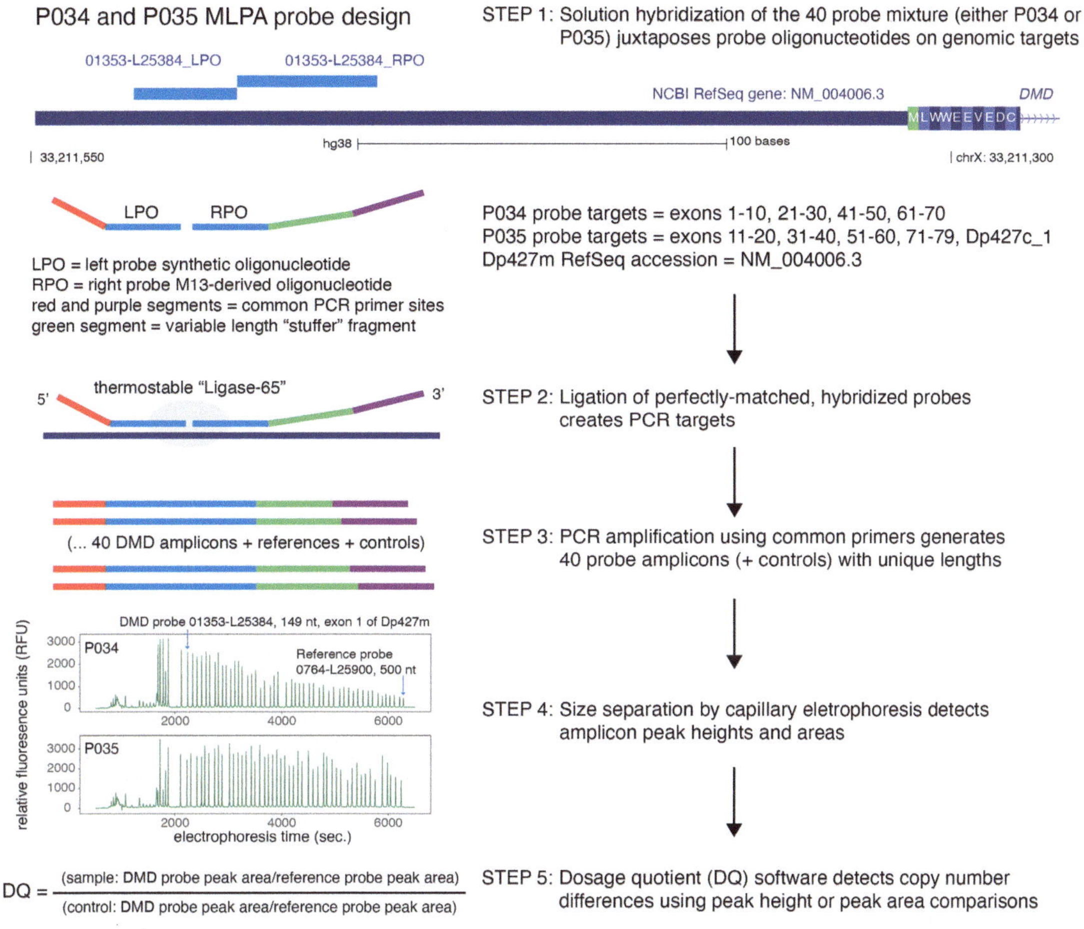

Fig. 1 Flow chart depicting each step of the MPLA technique for *DMD* whole exon deletions and duplications. The genomic location of the left probe (LPO) and right probe (RPO) oligonucleotides used to detect Dp427m exon 1 are shown schematically with a hg38 UCSC genome browser view. The exon composition of the 80 probe mixtures (P034 and P035) are shown beside a diagram of common elements found in the left probe synthetic oligonucleotides and the right probe M13-derived oligonucleotides. The MLPA procedure is shown sequentially from top to bottom for the five main steps: hybridization, ligation, PCR amplification, fragment separation by capillary electrophoresis, and dosage quotient analysis

pairs, the 80 DMD exons (79 exons from Dp427m plus the Dp427c alternate exon_1) are split evenly between 2 probemixes: P034 and P035. Additionally, the probemixes contain nine probe pairs that detect different regions on the X chromosome that are used for DQ analysis, plus quality probe pairs that serve as control for DNA concentration (Q-fragments), denaturation (D-fragments), and the X/Y chromosomes. These controls generate shorter amplicons that precede the 40 DMD amplicons and 9 reference amplicons, which are all jointly analyzed after separation by capillary electrophoresis (Figs. 1 and 2). With the rapid advance in genome technology since the first draft sequence of the human

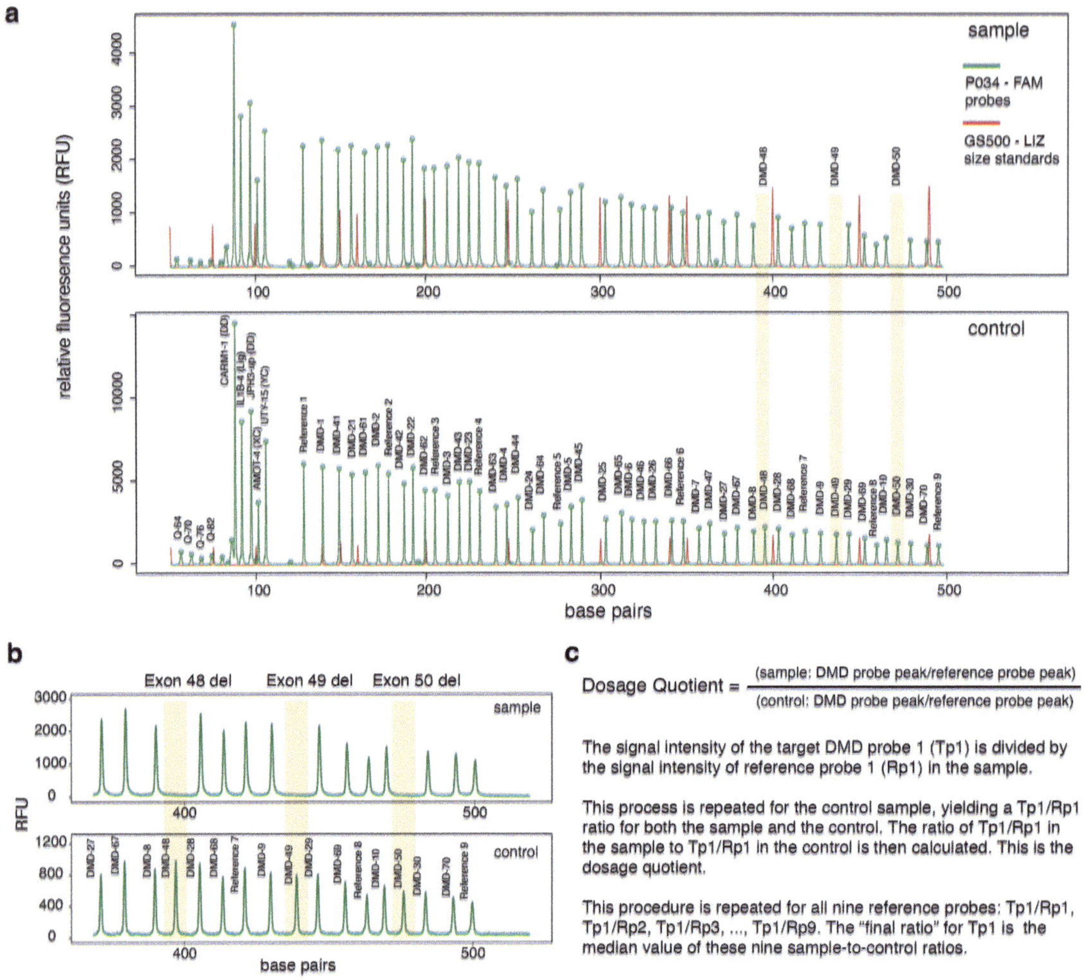

Fig. 2 Dosage quotient analysis of MLPA electropherograms using Coffalyser. NET™ (**a**) Capillary electrophoretic separations of the P034 MLPA amplicons from a DMD patient with a *DMD* exon 48–50 deletion (sample, upper panel) and a reference without *DMD* copy number variation (reference, lower panel). The 40 DMD probe fragments, the 9 reference probes, and the small (<120 bp) control fragments are annotated and shown in green (FAM dye). The GS500 sizing ladder (LIZ dye) is shown in red. (**b**) Zoomed view of the region containing amplicons for exons 48, 49, and 50, which are highlighted in yellow. (**c**) Overview of the dosage quotient calculation used by Coffalyser.Net™ software to detect copy number variation using the P034 and P035 probesets

genome, many genome-wide approaches to CNV analysis, such as microarrays and short/long read next-generation sequencing [11], are now in wider use, but the MLPA method described here still allows a simple and direct approach for analyzing CNVs in the dystrophin gene [12]. The preferred method requires access to a capillary electrophoresis instrument for fragment analysis of the fluorescent amplicons, but access to these instruments through core facilities or commercial vendors is still widespread, and the MLPA reagents and analysis software are available from MRC Holland.

2 Materials

2.1 Genomic DNA Purification from Whole Blood

1. Puregene® RBC Lysis Solution.
2. Puregene® Cell Lysis Solution.
3. Puregene® Protein Precipitation Solution.
4. RNase A (10 mg/mL).
5. Isopropanol (2-propanol).
6. 70% ethanol.
7. DNA Hydration Solution.
8. 50 mL conical tubes.
9. Serological pipettes: 25 mL, 10 mL, 5 mL.
10. Sterile aerosol filter tips: 200 μL, 1000 μL.
11. Paper towels, Kimwipes.
12. Disposable gloves.
13. 1.8 mL cryotube.
14. Vortex mixer.
15. Bulb pipette, pipette-aid, or electronic pipettor.
16. Micropipettes: P200, P1000.
17. Benchtop centrifuge: one which is comparable to an Eppendorf 5810 with swinging bucket rotor A-4-62, including adapters for 15 mL/50 mL tubes.
18. Water baths: 37 °C and 65 °C.
19. NanoDrop Spectrophotometer ND-1000, or a comparable instrument for quantitatively measuring genomic DNA.

2.2 MLPA Protocol

1. SALSA® MLPA® Reagent Kit (MRC-Holland, *see* **Note 1**).
2. SALSA® MLPA® Probemixes: P034 DMD-1 B2 and P035 DMD-2 B1 (MRC-Holland).
3. TE: 10 mM Tris, pH 7.0, 1 mM EDTA.
4. Sterile double-distilled water (ddH_2O).
5. Microcentrifuge configured for 1.5 mL microfuge tubes.
6. 12-well strip tubes with caps.
7. Sterile aerosol filter tips: 2 μL, 20 μL, 200 μL, 1000 μL.
8. Micropipettes: P2, P20, P200, P1000.
9. Microfuge tubes.
10. 96-well plastic PCR tube racks.
11. Thermocycler.

12. Benchtop centrifuge: one which is comparable to an Eppendorf 5810 with swinging bucket rotor A-4-62, including plate holders.

2.3 Capillary Electrophoresis and Dosage Quotient Analysis

1. POP-7™ Polymer.
2. 1× 3730 Running Buffer: 100 mL 10× 3730 Running Buffer + 900 mL ddH_2O.
3. Hi-Di™ Formamide.
4. Microcentrifuge tubes.
5. HotSeal™ sealing film.
6. GeneScan™ 500 LIZ™ Size Standard.
7. Skirted 96-well PCR plate.
8. ABI 3730xl DNA Genetic Analyzer or equivalent capillary electrophoresis instrument. Any of the current series of ABI Genetic Analyzers (SeqStudio, SeqStudio Flex, or ABI 3500 series) can be used for separation and fluorescent detection. Electrophoretic conditions for the SeqStudio instrument may have to be adjusted since these capillaries do not use POP-7™ Polymer.
9. Coffalyser.Net™ Data analysis software (*see* **Note 2**).

3 Methods

3.1 Genomic DNA Purification from Whole Blood

There are a number of methods for purifying genomic DNA, and this protocol describes a standard liquid-phase DNA isolation procedure starting with whole blood collected from patients. Other sources of DNA from cultured cells or tissue biopsy are also suitable, and protocols for DNA isolation from these starting materials (Qiagen) can be substituted. The most important consideration is to have the isolated DNA at high concentration (>10 ng/μL) and free of contaminants that might inhibit the enzymatic reactions.

Whole blood should be collected in an anticoagulant tube (a lavender top EDTA tube is preferred). Disposable gloves should be worn throughout the entire protocol. 150–500 μg of genomic DNA is the expected yield from 10 mL of whole blood. The yield is mainly dependent on the number of leukocytes in the 10 mL sample and the length of time from when the sample was drawn and the storage time (*see* **Note 3**).

1. Combine 10 mL of whole blood and 30 mL of Puregene® RBC Lysis Solution in a sterile 50 mL conical tube (*see* **Note 4**). Invert the tube 15× to mix thoroughly, and incubate for 5 min at room temperature; invert again at least once during the incubation.

2. Centrifuge the tube in the benchtop centrifuge for 5 min at 2000 × *g* to pellet the leukocytes. Pour the supernatant into another 50 mL conical tube, and discard the tube in a biological hazard waste bin, leaving 200–400 μL of liquid in the original tube.
3. Vortex the tube vigorously on high speed for 15 s to resuspend the cells in the residual liquid.
4. Add 10 mL of Puregene® Cell Lysis Solution with a serological pipette and pipet up and down 15 times to lyse the cells (*see* **Notes 5** and **6**).
5. Add 50 μL of RNase A solution to the cell lysate. Mix the sample by inverting the tube 25 times, and incubate in a 37 ° C water bath for 15 min.
6. Equilibrate the sample at room temperature for 15 min.
7. Add 3.33 mL of Puregene® Protein Precipitation Solution using a 5 mL serological pipette to the cell lysate.
8. Vortex the tube vigorously on high speed for 20 s to mix the lysed cells with the solution.
9. Centrifuge at 2000 × *g* for 5 min. The precipitated proteins will form a tight dark brown pellet (*see* **Note** 7).
10. Pour the supernatant containing the DNA (leaving behind the precipitated protein pellet) into a clean 50 mL conical tube containing 10 mL of 100% isopropanol.
11. Mix the sample by inverting gently 50 times until the white threads of DNA form a visible clump (*see* **Note 8**).
12. Centrifuge at 2000 × *g* for 5 min; the DNA will be visible as a white pellet.
13. Gently pour off the supernatant making sure the white pellet remains at the bottom of the tube. Drain the tube on a paper towel for 20 s.
14. Add 10 mL of 70% ethanol using a 10 mL serological pipette, and invert the tube ten times to wash the DNA pellet.
15. Centrifuge at 2000 × *g* for 5 min. Gently pour off the ethanol making sure the white pellet stays near the bottom of the tube. Drain the tube on a paper towel for 5 min.
16. Invert the tube so it is right side up and allow to air dry for 10–15 min.
17. Resuspend the DNA pellet in 600 μL of DNA Hydration Solution (*see* **Note 9**), vortex gently for 3 s, and rehydrate the DNA by incubating in a 65 °C water bath for 1 h.
18. Measure the concentration of the DNA sample using a Nano-Drop Spectrophotometer or equivalent instrument.

19. For storage, centrifuge the tube briefly and transfer the sample to a 1.8 mL cryotube. The DNA should be stored at 4 °C, or for long-term storage at −20 °C or −80 °C.

3.2 MLPA Protocol

Two sets of reactions are performed independently to amplify the 79 exons that correspond to the Dp472m muscle isoform (RefSeq NM_004006.3) and the alternate exon_1 promoter from the Dp427c isoform. The primer pairs are contained in the SALSA MLPA P034 DMD-1 B2 and P035 DMD-2 B1 Probemixes available from MRC Holland. This protocol is written for processing 1–12 genomic DNA samples using 12-well strip tubes held in 96-well PCR tube racks. Volumes of the master mixes should be adjusted depending on how many DNA samples are being processed. Master mixes are increased by 10% to allow for pipette inconsistencies. The overnight incubation for the hybridization step makes this a 2-day protocol, and the reactions can be stored frozen at the completion of the *Day 2* steps if there is a wait period to get access to a capillary electrophoresis instrument.

1. Configure the required programs on the thermocycler (Table 1).

Day 1

2. Label one 12-well strip tube with P034 and the sample names and a second 12-well strip tube with P035 and the sample names.
3. Add 5 μL of genomic DNA (50–150 ng total) sample #1 to the first well of the P034 strip tube and to the first well of the P035 strip tube and repeat for the number of samples being processed (*see* **Note 10**). If the stock DNA concentration is greater

Table 1
Thermal cycling programs

Program	Temperature	Time	Cycles
MLPA1 (DNA denaturation)	98 °C 25 °C	5 min Hold	1
MLPA2 (hybridization reaction)	95 °C 60 °C	1 min Hold	1
MLPA3 (ligation reaction)	54 °C 98 °C	15 min 5 min	1
MLPA4 (PCR reaction)	95 °C 60 °C 72 °C 72 °C 25 °C	30 sec 30 sec 1 min 20 min Hold	35 1 1

Table 2
Preparation of P034 or P035 hybridization master mix

Component	1× Master mix (μL)	12× Master mix (μL)
MLPA buffer	1.6	19.2
Probemix P034 or P035	1.6	19.2
Total volume	3.2	38.4

than 30 ng/μL, add the appropriate amount of TE to bring the volume up to 5 μL.

4. Seal each of the strip tubes with a 12-well strip cap, and place the tubes in the thermocycler. Run program: *MLPA1.*
5. Thaw the MLPA buffer and MLPA P034 and P035 Probemixes. Vortex the tubes briefly to mix.
6. Prepare hybridization master mixes according to Table 2.
7. Remove the strip tubes from the thermocycler and centrifuge at 1200 rpm for 1 min in the benchtop centrifuge in a microplate holder.
8. Slowly pipette each of the hybridization mixes 8×. Add 3 μL of P034 hybridization mix to each of the P034 samples in the strip tube. Add 3 μL of P035 hybridization mix to each of the P035 samples in the second strip tube.
9. Seal each of the strip tubes with a fresh 12-well strip cap, and place the tubes in the thermocycler. Run program: *MLPA2* for 14–24 h (*see* **Note 11**).

Day 2

10. Remove the strip tubes from the thermocycler and centrifuge at 1200 rpm for 1 min in a microplate holder.
11. Thaw Ligase Buffer A and Ligase Buffer B. Vortex the tubes briefly to mix.
12. Prepare a single Ligation master mix for both sets of reactions according to Table 3.
13. Add the Ligase-65 enzyme to the Ligation master mix last. Slowly pipette the Ligation master mix 8×. Dispense 32 μL of the Ligation master mix to each of the P034 and P035 samples.
14. Seal each of the strip tubes with a fresh 12-well strip cap, and place the tubes in the thermocycler. Run program: *MLPA3.*
15. Thaw the SALSA® PCR primer mix. Vortex the tube briefly to mix.
16. Prepare a single PCR master mix for both sets of reactions according to Table 4.

Table 3
Preparation of ligation master mix

Component	1× Master mix (μL)	12× Master mix (μL)
Ligase buffer A	3.3	39.6
Ligase buffer B	3.3	39.6
ddH2O	27.5	330
Ligase-65 enzyme	1.1	13.2
Total volume	35.2	422.4

Table 4
Preparation of PCR master mix

Component	1× Master mix (μL)	12× Master mix (μL)
SALSA® PCR primer mix	2.2	26.4
ddH2O	8.25	99
SALSA® polymerase	0.55	6.6
Total volume	11	132

17. Add the SALSA® Polymerase to the PCR master mix last. Slowly pipette the PCR master mix 8×. Dispense 10 μL of the PCR master mix to each of the P034 and P035 samples.
18. Seal each of the strip tubes with a fresh 12-well strip cap, and place the tubes in the thermocycler. Run program: *MLPA4.*
19. Freeze the PCR reactions at -20 °C until the reactions will be run on the ABI 3730 Capillary Electrophoresis instrument. Samples are stable when stored at −20 °C up to 1 week.

3.3 Capillary Electrophoresis

An aliquot of each MLPA reaction is suspended in formamide containing a fluorescent dye standard which is used for sizing the DMD PCR products. MLPA reactions are run on a capillary electrophoresis instrument for data collection. The electropherograms are analyzed using Coffalyser.Net™ Data analysis software.

1. Prepare a stock mix of Hi-Di™ Formamide + GeneScan™ 500 LIZ™ Size Standard: 600 μL Hi-Di™ Formamide +15 μL GeneScan™ 500 LIZ™ (*see* **Note 12**).
2. Thaw the frozen PCR reactions at room temperature.
3. Aliquot 13 μL of the size standard stock mix to 2 wells of a 96-well skirted PCR plate for every sample, and add 1 μL of the PCR reactions (*see* **Note 13**).

4. Cover the 96-well plate with HotSeal™ sealing film and heat the reaction products for 5 min at 95 °C.
5. Load the plate onto the ABI3730xl DNA Analyzer Capillary Electrophoresis instrument.
6. Perform electrophoresis using the ABI instrument Data Collection software to generate .fsa (or .ab1) files. The .fsa file is a file type produced by Data Collection software for fragment analysis data and is a binary file that uses the ABIF file format specification.

3.4 Dosage Quotient Analysis

The fluorescent signals from MLPA amplicons detected and measured by the capillary electrophoresis instrument generates a fragment profile. To determine exon copy numbers, the MLPA signals of an unknown sample must be compared to a control sample (or samples) with normal dystrophin exon copy numbers. MRC Holland provides a free software package, the Coffalyser.Net™ Data analysis software [13], that is a robust solution for processing unknown and control .fsa files through quality control checks and dosage quotient analysis of P034 and P035 electropherograms. A distinct advantage of this software package is a pre-configured Microsoft SQL Server database with P034 and P035 fragment size and probe nomenclature specifications. One limitation of the software is that it requires a computer running an up-to-date Microsoft Windows version (64-bit, x64 Windows 10 or 11) and administrator privileges to install the software; however, this is a fairly standard configuration found on most personal or lab PC computers.

Extensive on-line documentation for installing and using Coffalyser.Net™ can be found on the MRC Holland website at https://www.mrcholland.com/technology/software/coffalyser-net and https://support.mrcholland.com/downloads/files/coffalyser-net-reference-manual/download. The installation and reference manual are well-organized and detailed, so instead of reiterating those instructions, Fig. 2 shows a high-level view of dosage quotient analysis comparing an exon 48–50 deletion patient sample with a control sample. The ABI electropherograms for P034 fragments are shown in Fig. 2a, with the control sample in the lower panel displaying annotations for the 40 DMD exon probes, 9 reference probes, and the quality control probes. This figure is similar to standard plots produced by Coffalyser.Net™, but this analysis used R code and the RStudio integrated development environment for R to load, analyze, and plot the .fsa data. The "binner" R package can read .fsa fragment files from an ABI Genetic Analyzer and perform peak finding, normalization, plotting, and formatting of output files for detected peaks with fragment size and peak height/area measurements. To reproduce the normalized and peak-detected plots shown in Fig. 2a, use RStudio and install the "binner" package by running install.packages("devtools") and install_github("plantarum/binner"). The "binner" function used in Fig. 2a was "readFSA" that reads raw .fsa files and returns an object of class fsa with sized and peak-detected data "fsaNormalize" that normalizes the RFU values across each sample in the fsa object, "plot(fsa object, ...)", and "fsa2PeakTab" that converts the fsa object into a peaks table. The peaks from the Q-fragments (64–82 nt), the D-fragments (88–96 nt), and the X- and Y-fragments (100 and 105 nt) are seen at the beginning of the electropherogram (Fig. 2a). These are quality control fragments with the Q-fragments increasing in signal with lower amounts of starting genomic DNA (<100 ng/μL) and the D-fragments decreasing in signal with incomplete denaturation. An excellent guide to troubleshooting these electropherograms can be found in a flow chart on page 11 of the MRC-Holland MLPA® protocol:

https://www.mrcholland.com/products/4838/MLPA%20General%20Protocol%20MDP-v008.pdf

The final step to determine exon copy number is calculation of the metric termed "final ratio" by the Coffalyser.Net™ software. The signal intensity of each DMD fragment (target probe1 or Tp1) is divided by the signal intensity of one of the nine reference fragments (reference probe 1 or Rp1) in the unknown sample. This procedure is repeated for the control sample, resulting in a Tp1/Rp1 ratio for both the sample and the control. The ratio of Tp1/Rp1 in the sample to Tp1/Rp1 in the control is then calculated, yielding the dosage quotient. This process is repeated for all nine reference probes: Tp1/Rp1, Tp1/Rp2, Tp1/Rp3, ..., Tp1/Rp9. The final dosage quotient or "final ratio" for Tp1 is the median value of these nine sample-to-control ratios (Fig. 2c). The targets for all nine reference probes are on the X chromosome and thus can be used for uniform copy number determination for both male and female samples. The sex of the test and control samples does not need to be matched, but for more cautious interpretations, matching the sex is recommended as well as using multiple control samples (typically three) for dosage quotient analysis. The Coffalyser.Net™ software facilitates comparisons of a single test sample versus multiple controls and outputs confidence intervals on the final ratios. Deletion mutations are clearly the most unambiguous with final ratios at or near zero in male samples. Duplications, and complex mutations such as mixed duplications/triplications, can be more challenging to interpret, requiring careful inspection of the final ratios and confidence intervals. Somatic mosaicism can also contribute to ambiguous results, along with false positives due to polymorphisms within the target sequence that destabilize hybridization of the probe to its target. Fortunately, the P034 and P035 targets were carefully designed to avoid common variants, and currently in the gnomAD v4 release (https://gnomad.broadinstitute.org/news/2023-11-gnomad-v4-0) which includes genotyping data from >800,000 individuals, no variant with an allele frequency > 0.005 is found within these DMD MLPA probe targets.

4 Notes

1. The SALSA® MLPA reagents are available from MRC-Holland (https://www.mrcholland.com) as the sole source supplier. As a side note, the acronym "SALSA" stands for Selective Adaptor Ligation, Selective Amplification. It refers to a different technique developed by MRC Holland, unrelated to MLPA, but underscores that these are products of MRC Holland.
2. This software [13] is freely available from https://www.mrcholland.com/technology/software. The Windows-only computer requirements include a genuine copy of Microsoft Windows (64 bit, x64) with the latest updates, including Microsoft .NET Framework 4.8, an account with administrative rights during installation, and a system that meets the requirements of Microsoft SQL Server 2022. Coffalyser.Net™ stores its data in a Microsoft SQL Server database, which needs to be hosted on your computer or somewhere within your network. If you choose the MRC Holland-supplied preconfigured SQL Server during installation, then Coffalyser.Net™ will install the pre-configured SQL Server 2022 Express on a single (personal or shared lab) computer. See the "Software" section on the MRC Holland website for detailed download and installation instructions.
3. Blood draws should be stored at 4 °C until the genomic DNA is purified. This will help to keep the leukocytes intact. The yield of DNA generally decreases with time before purification. For longer term storage, immediate freezing at −80 °C is recommended, followed by thawing at 37 °C for 15 min and gentle mixing before DNA purification.
4. 5 mL of blood can also be the input volume for this protocol with the following change: in **step 1**, decrease the volume of Puregene® RBC Lysis Solution to 15 mL. All other steps should be followed as written.
5. If the cells remain clumped after mixing, incubate the sample at 37 °C for 15 min and then proceed with **step 5**.
6. Samples in Puregene® Cell Lysis Solution are stable for at least 2 years at room temperature.
7. If the protein pellet is not tight, repeat **step 8** followed by incubation on ice for 5 min and then repeat **step 9**.

8. When the DNA quantity is low, you may not see white threads. Continue with the protocol as stated, but the yield may be less than anticipated.
9. The volume of DNA Hydration Solution may be adjusted depending on the size of the DNA pellet. If the pellet is large, resuspend in 1 mL; if the pellet is not clearly visible, resuspend in 300 μL.
10. Include at least one control sample when setting up the reactions. These controls with normal copy numbers for *DMD* exons are required for the standard analysis pipeline using Coffalyser.Net™ software.
11. There isn't a notable difference in data recovery when the ligation reaction is incubated from 14–24 h. However, incubating the reaction for longer than 24 h resulted in poor data recovery.
12. The Hi-Di™ Formamide + GeneScan™ 500 LIZ™ stock mix is stable at 4 °C up to 3 months.
13. Loading each sample in two wells instead of a single well increases data recovery from the redundant use of two capillaries, since individual capillaries can sometimes fail due to temporary clogs or other electrophoretic anomalies.

Acknowledgments

This work was supported in part by a grant from the National Institutes of Health (R01NS085238).

References

1. Den Dunnen JT, Grootscholten RM, Bakker E, Blonden LAJ, Ginjaar HB, Wapenaar MC, Van Paassen HMB, Van Broeckhoven C, Pearson RL, Van Ommen GJB (1989) Topography of the Duchenne muscular dystrophy (DMD) gene: FIGE and cDNA analysis of 194 cases reveals 115 deletions and 13 duplications. Am J Hum Genet 45:835–847
2. Aartsma-Rus A, Ginjaar IB, Bushby K (2016) The importance of genetic diagnosis for Duchenne muscular dystrophy. J Med Genet 53: 145–151
3. Chamberlain JS, Gibbs RA, Ranier JE, Nguyen PN, Caskey CT (1988) Deletion screening of the Duchenne muscular dystrophy locus via multiplex DNA amplification. Nucleic Acids Res 16(23):11141–11156
4. van Essen AJ, Kneppers AL, van der Hout, Scheffer H, Ginjaar IB, ten Kate LP, van Ommen GJ, Buys CH, Bakker E (1997) The clinical and molecular genetic approach to Duchenne and Becker muscular dystrophy: an updated protocol. J Med Genet 34:805–812
5. Wilson LA, MacKen WL, Perry LD, Record CJ, Schon KR, Frezatti RSS, Raga S, Naidu K, Köken ÖY, Polat I, Kapapa MM, Dominik N, Efthymiou S, Morsy H, Nel M, Fassad MR, Gao F, Patel K, Schoonen M, Bisschoff M, Vorster A, Jonvik H, Human R, Lubbe E, Nonyane M, Vengalil S, Nashi S, Srivastava K, Lemmers RJLF, Reyaz A, Mishra R, Töpf A, Trainor CI, Steyn EC, Mahungu AC, Van Der Vliet PJ, Ceylan AC, Hiz AS, Çavdarll B, Semerci Gündüz CN, Ceylan GG, Nagappa M, Tallapaka KB,

Govindaraj P, Van Der Maarel SM, Narayanappa G, Nandeesh BN, Wa Somwe S, Bearden DR, Kvalsund MP, Ramdharry GM, Oktay Y, Yiş U, Topaloğlu H, Sarkozy A, Bugiardini E, Henning F, Wilmshurst JM, Heckmann JM, McFarland R, Taylor RW, Smuts I, Van Der Westhuizen FH, Sobreira CFDR, Tomaselli PJ, Marques W, Bhatia R, Dalal A, Srivastava MVP, Yareeda S, Nalini A, Vishnu VY, Thangaraj K, Straub V, Horvath R, Chinnery PF, Pitceathly RDS, Muntoni F, Houlden H, Vandrovcova J, Reilly MM, Hanna MG (2023) Neuromuscular disease genetics in under-represented populations: increasing data diversity. Brain 146:5098–5109
6. Schouten JP, Mcelgunn CJ, Waaijer R, Zwijnenburg D, Diepvens F, Pals G (2002) Relative quantification of 40 nucleic acid sequences by multiplex ligation-dependent probe amplification. Nucleic Acids Res 30(12):e57
7. Yau SG, Bobrow M, Mathew CG, Abbs SJ (1996) Accurate diagnosis of carriers of deletions and duplications in Duchenne/Becker muscular dystrophy by fluorescent dosage analysis. J Med Genet 33:550–558
8. Armour JAL, Sismani C, Patsalis PC, Cross G (2000) Measurement of locus copy number by hybridisation with amplifiable probes. Nucleic Acids Res 28(2):605–609
9. White S, Kalf M, Liu Q, Villerius M, Engelsma D, Kriek M, Vollebregt E, Bakker B, Van Ommen G-JB, Breuning MH, Den Dunnen JT (2002) Comprehensive detection of genomic duplications and deletions in the DMD gene, by use of multiplex amplifiable probe hybridization. Am J Hum Genet 71(2): 365–374
10. Lalic T, Vossen RHAM, Coffa J, Schouten JP, Guc-Scekic M, Radivojevic D, Djurisic M, Breuning MH, White SJ, den Dunnen JT (2005) Deletion and duplication screening in the DMD gene using MLPA. Eur J Hum Genet 13:1231–1234
11. Porubsky D, Eichler EE (2024) A 25-year odyssey of genomic technology advances and structural variant discovery. Cell 187:1024–1037
12. Abbs S, Tuffery-Giraud S, Bakker E, Ferlini A, Sejersen T, Mueller CR (2010) Best practice guidelines on molecular diagnostics in Duchenne/Becker muscular dystrophies. Neuromuscul Disord 20:422–427
13. Coffa J, Van den Berg J (2011) Analysis of MLPA data using novel software Coffalyser. NET by MRC-Holland. In: Eldin AB (ed) Modern approaches to quality control. IntechOpen

Chapter 2

Total RNA-seq as a Tool to Study *DMD* Splicing and Transcriptional Dynamics

Diane M. Dunn and Robert B. Weiss

Abstract

Advances in RNA analysis of dystrophin (*DMD*) mutations have been significantly influenced by improvements in methods to study mRNA transcription and splicing. Early breakthroughs, such as the 1988 cDNA sequencing of the 14-kb muscle transcript isoform of the dystrophin gene, paved the way for RNA-based mutation detection via RT-PCR. These studies revealed that point mutations often cause premature polypeptide chain termination, and RNA polymerase requires approximately 16 h to transcribe the 2.1 Mb Dp427m transcription unit. Next-generation technologies for RNA sequencing (RNA-seq) have enabled broader and more quantitative transcriptome analysis of the *DMD* locus. This protocol outlines an RNA-seq approach using rRNA depletion to examine muscle biopsy total RNA from DMD/BMD patients, specifically focusing on resolving pseudoexon and other intronic mutations. While not achieving the ultra-deep read depths of targeted RNA-seq methods, this cost-effective strategy provides sufficient coverage to robustly evaluate *DMD* exon and intron mRNA levels, generating data with comparable read depths for both.

Key words RNA-seq, Complex *DMD* mutations, Pseudoexons, Splicing, Transcription termination, Dystrophinopathy

1 Introduction

Advances in RNA analysis of dystrophin (*DMD*) mutations have repeatedly benefitted from improved methodologies for studying mammalian transcription. The landmark cDNA cloning and sequencing of the 14 kb muscle transcript isoform of the dystrophin gene in 1988 [1] enabled the use of reverse transcription-polymerase chain reaction (RT-PCR) as a sensitive method for RNA-based dystrophin mutation detection and analysis [2, 3]. Early mutation studies demonstrated that most point mutations lead to polypeptide chain termination (stop, frameshift, and splice site mutations) rather than missense mutations [4–6]. RT-PCR also identified aberrant splice products, including exon skipping due to intraexonic deletion [7] and pseudoexon

Kevin Flanigan and Stefan Nicolau (eds.), *Duchenne Muscular Dystrophy: Methods and Protocols*, Methods in Molecular Biology, vol. 2975, https://doi.org/10.1007/978-1-0716-4811-7_2,

inclusion caused by intronic mutations [8]. Early experiments in myogenic cultures demonstrated that RNA polymerase requires ~16 h to transcribe the 2.1 Mb dystrophin gene, during which most of the 79 exons in the Dp427m mRNA isoform undergo co-transcriptional splicing. Notably, steady-state mRNA levels of Dp427m exons follow a 5–3′ gradient, with proximal 5′ exons expressed at roughly twice the levels of distal 3′ exons. This gradient is explained by a transcription model in which Dp427m exon 1 is attached to a transcribing RNA polymerase for ~16 h before exon 79 is synthesized. After transcription is completed, the mature Dp427m mRNA has a relatively short half-life of ~4 h in skeletal muscle [9]. These early insights into dystrophin mRNA dynamics have been corroborated in muscle biopsies using in situ RNA hybridization techniques, such as RNAscope, which have further illuminated the effects of dystrophin mutations on nuclear and cytoplasmic mRNA distribution [10, 11].

The adoption of automated DNA sequencing methods facilitated larger surveys of DMD/BMD patient cohorts [12, 13], while RT-PCR remained essential for confirming or elucidating the specific effects of mutations on transcript isoforms and their levels. The advent of next-generation sequencing (NGS) technologies enabled transcriptome-wide analysis through RNA sequencing (RNA-seq). Key steps in RNA-seq workflows include RNA extraction, mRNA enrichment or rRNA depletion, cDNA synthesis, and library preparation, followed by high-throughput sequencing and computational analysis to map reads and quantify transcript abundance. Initial RNA-seq applications successfully identified intronic mutations causing pseudoexon inclusion [14] but faced challenges with uniform exon coverage when using mRNA enrichment based on oligo-dT priming. Targeted mRNA capture techniques have also been used to increase the coverage depth across the entire DMD locus [15].

In this protocol, we describe an rRNA depletion strategy applied to total RNA from DMD/BMD patients lacking exonic mutations, enabling a broader search for intronic mutations associated with pseudoexon inclusion [16]. While this approach does not achieve the ultra-deep read depths of targeted methods, it provides a cost-efficient, genome-wide assay capable of generating sufficient data to evaluate steady-state levels of *DMD* intronic and exonic mRNA (Fig. 1a). This protocol achieves *DMD* intronic read depths comparable to those of transcripts encoding abundant skeletal muscle proteins, reflecting the proportional relationship between RNA-seq read counts and transcription unit size. In the example shown in Fig. 1a, from a patient with a *DMD* intron 61 c.9163 + 2510G > A mutation that creates a pseudo 3′-terminal exon, a decrease in intronic read depth downstream of the mutation is clearly detectable. Aggregate read counts for Dp427m exons ranks within the top 100 protein-coding transcripts (excluding

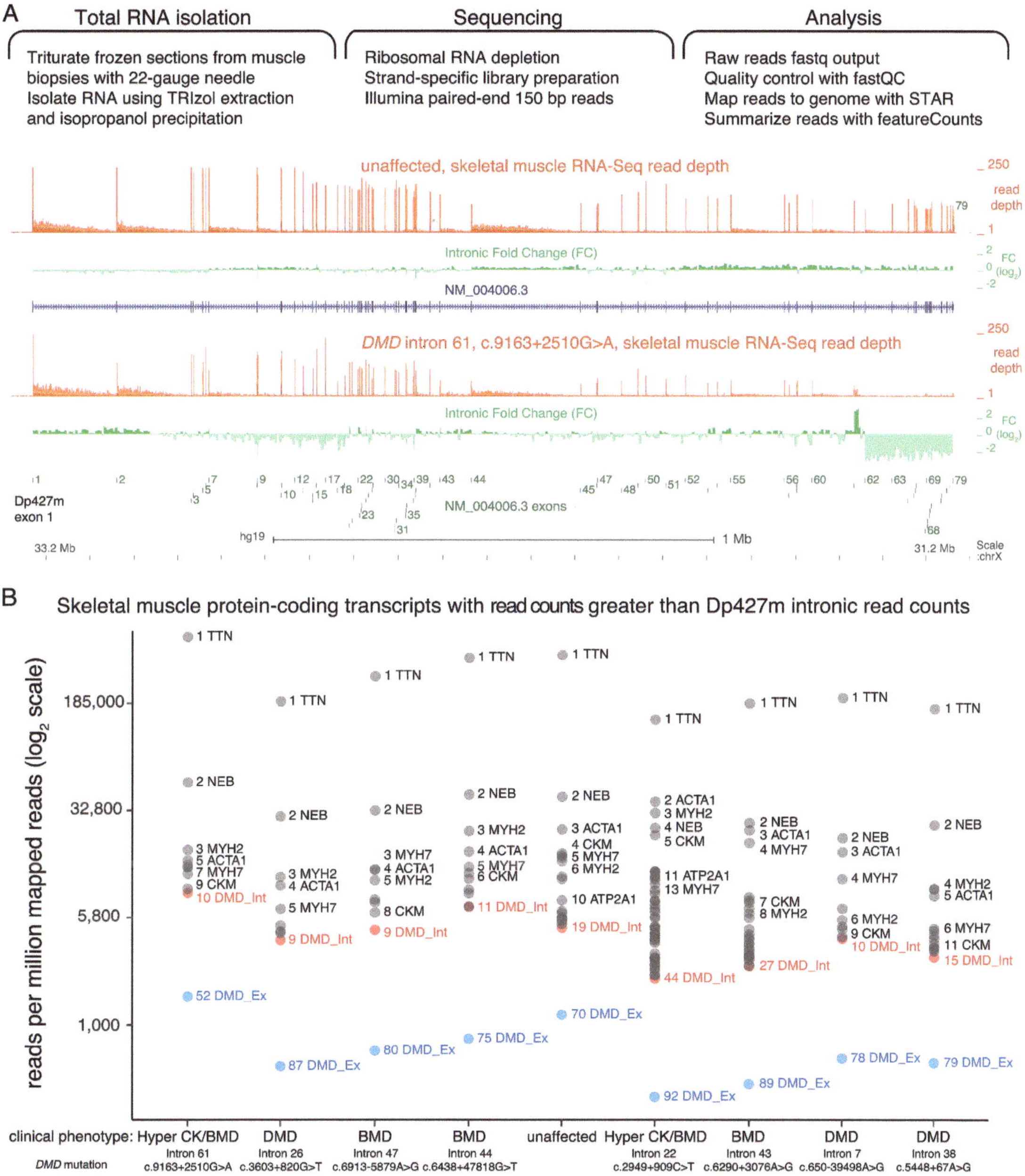

Fig. 1 Overview of RNA isolation, RNA-seq read mapping and read summarization for the human *DMD* locus. (**a**) The flowchart outlines the three main phases of the protocol. Below, two RNA-seq examples mapped to the *DMD* region (hg19) are shown. Read depth is represented by the line height, indicating the number of reads aligned at each position, visualized using bigWig or bedGraph files displayed in the UCSC Genome Browser. An intronic "fold-change" track shows intron coverage differences relative to an unaffected control, plotted in 5-kb windows. (**b**) Normalized RNA-seq data from eight different D/BMD mutations and one unaffected sample (see Waldrop et al. [16] for sample details) are plotted on the x-axis, with the top protein-coding transcripts expressed in skeletal muscle shown on the y-axis. Each point represents the normalized reads per million mapped reads (RPM) value for a transcript, annotated by its abundance rank and gene symbol. Transcripts with RPM values higher than the aggregate value for Dp427m *DMD* introns (red) are displayed for each sample, along with the RPM value for the Dp427m exons (blue)

highly variable mitochondrial transcripts), while Dp427m intron read counts rank in the top 20 (Fig. 1b). For every one million reads mapped to the skeletal muscle protein-coding transcriptome, this protocol yields approximately 1000 reads for Dp427m exons and 4500 reads for Dp427m introns. Targeting 50–200 million reads per sample in an RNA-seq experiment ensures substantial coverage across the *DMD* locus.

2 Materials

2.1 RNA Purification from Muscle Tissue

1. RNase-free lysis buffer: 20 mM Tris-HCl pH 7.4, 2.5 mM $MgCl_2$, 150 mM NaCl, 1 mM DTT, 1% Triton X100.
2. DEPC-treated water.
3. TRIzol™ LS Reagent.
4. Isopropanol (2-propanol).
5. Chloroform.
6. 75% ethanol (*see* **Note 1**).
7. 70% ethanol in a spray bottle.
8. RNase AWAY™.
9. Disposable gloves.
10. Sterile needles: 18 gauge, 21 gauge, 22 gauge.
11. Sterile 1 mL syringes.
12. RNase-free 1.5 mL microfuge tubes.
13. Microfuge tube rack.
14. Sterile aerosol filter tips: 200 μL, 1000 μL.
15. Kimwipes™.
16. Microcentrifuge (4 °C).
17. Micropipetters: P200, P1000.
18. Picofuge™ (Stratagene).
19. Vortex mixer.
20. −20 °C freezer.
21. −80 °C freezer.
22. NanoDrop ND-1000 Spectrophotometer.

2.2 rRNA Depletion

1. An rRNA depletion kit such as NEBNext® rRNA Depletion Kit (Human/Mouse/Rat).
2. Ice bucket with shaved ice.
3. Disposable gloves.
4. 12-well strip tubes with caps.
5. 96-well plastic PCR tube racks.

6. Benchtop Centrifuge: one which is comparable to an Eppendorf 5810 with swinging bucket rotor A-4-62 including plate holders.
7. Sterile aerosol filter tips: 2 μL, 20 μL, 200 μL, 1000 μL.
8. Micropipetters: P2, P20, P200, P1000.
9. Picofuge™.
10. Thermocycler.
11. 70% ethanol in a spray bottle.
12. RNase AWAY™.

2.3 RNA Clean-Up with RNAClean Beads

1. RNAClean XP beads and AMPure® XP beads (Beckman Coulter).
2. 80% ethanol (*see* **Note 1**).
3. DEPC-treated water or equivalent nuclease-free water.
4. 12-well strip tubes with caps.
5. 96-well plastic PCR tube racks.
6. Microfuge tubes.
7. Magnetic stand for 1.5 mL microfuge tubes.
8. Micropipetters: P20, P200.
9. Sterile aerosol filter tips: 20 μL, 200 μL.
10. Microfuge tube rack.
11. Ice bucket with shaved ice.
12. −80 °C freezer.
13. 70% ethanol in a spray bottle.
14. RNase AWAY™.

2.4 RNA Fragmentation and Illumina Library Preparation

1. A strand-specific RNA library construction kit such as NEBNext® Ultra™ II Directional RNA Library Prep Kit for Illumina or equivalent.
2. SuperScript™ II Reverse Transcriptase.
3. RNAClean XP beads or equivalent RNase-free SPRI paramagnetic beads.
4. 80% ethanol (*see* **Note 1**).
5. Micropipetters: P2, P20, P200, P1000.
6. Sterile aerosol filter tips: 2 μL, 20 μL, 200 μL, 1000 μL.
7. 12-well strip tubes with caps.
8. 96-well plastic PCR tube racks.
9. Microfuge tubes.
10. Magnetic stand for 1.5 mL microfuge tubes.
11. Microfuge tube rack.

12. Picofuge™.
13. Thermocycler.
14. Ice bucket with shaved ice.
15. −20 °C freezer.
16. 70% ethanol in a spray bottle.
17. RNase AWAY™.

2.5 Library Validation and Sequencing

1. Agilent TapeStation system or equivalent automated electrophoresis instrument.
2. Illumina NovaSeq 6000 Sequencing system or equivalent.

3 Methods

Purification of RNA from muscle tissue and all subsequent protocols should be performed on a dedicated RNA workbench, routinely cleaned with RNase AWAY™ and 70% ethanol before use. Always wear disposable gloves and use RNA-dedicated micropipetters and RNase-free reagents when working with RNA. Reagents that are stored at −20 °C should be thawed at room temperature, briefly centrifuged, and kept on ice prior to and during use, unless otherwise specified. Return reagents to −20 °C immediately after being used in a protocol to maintain their integrity. This protocol begins with frozen muscle biopsy tissue that has been cut into thin sections using a cryostat while maintaining the frozen state. Approximately 40 sections from a biopsy diameter of 2–4 mM with a section thickness of 10–20 microns is a reasonable starting quantity. The RNA yield will depend on the diameter of the frozen muscle biopsy and the presence of adipose tissue in the sample. Additional slices should be cut and included in the purification if the muscle diameter is less than 2 mM or there is an abundance of fat tissue.

3.1 Total RNA Purification from Muscle Tissue

1. Frozen muscle biopsies sectioned on a cryostat are resuspended in 450 μL of RNase-free lysis buffer (*see* **Note 2**) in a 1.5 mL microfuge tube. The sample is triturated by passing the liquified sections through a 22 gauge needle attached to a disposable 1 mL syringe for ten full strokes. If the needle clogs, switch to an 18 gauge or 21 gauge needle before returning to the 22 gauge needle to ensure the tissue is thoroughly homogenized. Close the lid on the microfuge tube.
2. Vortex the microfuge tube on high speed for 10 s and centrifuge the homogenized tissue briefly in a Picofuge™. Split the homogenate between two 1.5 mL microfuge tubes, with 200 μL homogenate added to 600 μL TRIzol™ LS each (*see* **Note 3**).

3. Vortex vigorously on high speed for 15 s. Incubate for 15 min at room temperature.
4. Add 160 μL chloroform to each tube for phase separation (160 μL chloroform for 600 μL TRIzol™ LS), close the tubes, and shake each tube vigorously for 10 s.
5. Incubate samples for 5 min at room temperature.
6. Centrifuge at 12,000 × *g* for 15 min at 4 °C.
7. Carefully transfer the upper aqueous phase to a fresh microfuge tube using a P200 pipettor. Use caution to avoid the interface.
8. Add 400 μL of isopropanol to the aqueous phase (400 μL for every 600 μL TRIzol™ LS used) and mix well by shaking the tubes vigorously for 20 s.
9. Incubate at −20 °C overnight to precipitate the RNA (*see* **Note 4**).
10. Centrifuge at 12,000 × *g* for 15 min at 4 °C in a microfuge.
11. Pour off the supernatant as soon as centrifugation is completed, being careful not to dislodge the small white RNA pellet.
12. Wash the RNA pellet by adding 500 μL of freshly prepared 75% ethanol and gently inverting the tube ten times.
13. Centrifuge at 12,000 × *g* for 15 min at 4 °C.
14. Pour off the supernatant as soon as centrifugation is completed, being careful not to dislodge the small white RNA pellet.
15. Place the tube upside down on a Kimwipe™ with the lid open and allow the remaining liquid to drain for 5–15 min. Once the liquid is completely drained, turn the microfuge tube upright, and air-dry the RNA pellet for an additional 5–10 min.
16. Resuspend each pellet in 25 μL DEPC-treated water.
17. Combine the RNA from the two tubes and measure the concentration by OD260 using a NanoDrop ND-1000 Spectrophotometer (or equivalent) (*see* **Note 5**).
18. Store the RNA in a −80 °C freezer. The RNA will remain stable at −80 °C for several years.

3.2 rRNA Depletion

The starting material for this part of the protocol should be 50 ng to 1 μg of purified total RNA. Ideally, 1 μg of RNA is used, but satisfactory libraries can be prepared with as little as 50 ng of starting material. The NEBNext® rRNA Depletion Kit provides the necessary reagents for this process. This rRNA depletion procedure removes cytoplasmic and mitochondrial rRNA through the use of tiling oligodeoxynucleotides. Evaluate the quality of the input RNA by analyzing the sample with an Agilent TapeStation and RNA ScreenTape to determine the RNA Integrity Number (RIN). For intact RNA (RIN > 7), set the fragmentation time in the thermocycler program "*RNA_Frag* (Fragments the RNA)" to 10 min, and for partially degraded RNA (RIN 2–7), set the fragmentation time to 5 min. The protocol steps assume that NEBNext® rRNA Depletion kits and NEBNext® Ultra II Directional RNA Library Prep Kit for Illumina are used for the kit reagents. An Illumina TruSeq Stranded Total RNA Library Prep Ribo-Zero Gold (Human/Mouse/Rat) kit can be substituted for the NEBNext® kits. Before starting, configure the following programs on a thermocycler:

A. ***rRNA_Hyb* (Hybridization of the Probes to the RNA).** 95 °C for 2 min; ramp down to 22 °C at 0.1 °C/s; 22 °C hold.

B. ***RNaseH_Dig* (Thermostable RNase H Digestion).** 50 °C for 30 min.

C. ***DNaseI_Dig* (DNase I Digestion).** 37 °C for 30 min.

1. Place a 12-well strip tube into a 96-well plastic PCR tube rack. Dilute 50 ng to 1 μg of each RNA with nuclease-free water to a final volume of 11 μL in the appropriate number of wells of a 12-well PCR strip tube.
2. Assemble the RNA/probe hybridization master mix according to Table 1. Adjust the volumes of the master mix depending on the number of samples being processed.
3. Pipette 4 μL of the hybridization master mix into each RNA sample, mix thoroughly by pipetting up and down ten times (*see* **Note 6**), and seal the tubes with a strip cap.
4. Centrifuge the strip tube at 1200 rpm for 1 min in a benchtop centrifuge.
5. Preheat the thermocycler to 95 °C and place the strip tube into the thermocycler and run the ***rRNA_Hyb*** program. (This will take ~15–20 min to complete).
6. Once the samples have equilibrated to 22 °C, centrifuge the strip tube at 1200 rpm for 1 min in a benchtop centrifuge.
7. Assemble the RNase H Digestion master mix according to Table 2. Adjust the volumes of the master mix depending on the number of samples being processed.
8. Pipette 5 μL of the RNase H Digestion master mix into each RNA sample, mix thoroughly by pipetting up and down ten times, and seal the tubes with a fresh strip cap.
9. Centrifuge the strip tube at 1200 rpm for 1 min in a benchtop centrifuge.

Table 1
RNA/probe hybridization master mix

Component	1× Master mix (μL)	12× Master mix (μL)
NEBNext v2 rRNA depletion solution	2.1	25.2
NEBNext probe hybridization buffer	2.1	25.2
Total volume	4.2	50.4

Table 2
RNase H digestion master mix

Component	1× Master mix (μL)	12× Master mix (μL)
RNase H reaction buffer	2.1	25.2
NEBNext thermostable RNase H	2.1	25.2
Nuclease-free water	1.1	13.2
Total volume	5.3	63.6

Table 3
DNase I digestion master mix

Component	1× Master mix (μL)	12× Master mix (μL)
DNase I reaction buffer	5.1	61.2
DNase I	2.6	31.2
Nuclease-free water	23.5	282.0
Total volume	31.2	374.4

10. Place the strip tube into the thermocycler and run the ***RNaseH_Dig*** program.
11. Centrifuge the strip tube at 1200 rpm for 1 min in a benchtop centrifuge.
12. Assemble the DNase I Digestion master mix according to Table 3. Adjust the volumes of the master mix depending on the number of samples being processed.
13. Pipette 30 μL of the DNase I master mix into each RNA sample; mix thoroughly by pipetting up and down ten times, and seal the tubes with a fresh strip cap.
14. Centrifuge the strip tube at 1200 rpm for 1 min in a benchtop centrifuge.

15. Place the strip tube into the thermocycler and run the ***DNaseI_Dig*** program.
16. Centrifuge the strip tube at 1200 rpm for 1 min in a benchtop centrifuge and proceed immediately to Subheading 3.3.

3.3 RNA Clean-Up with RNAClean XP Beads

1. Before starting, the RNAClean XP beads should be warmed up to room temperature for at least 30 min prior to use. Vortex and shake the bottle well to fully resuspend the beads just before adding the beads to the RNA samples. Prepare fresh 80% ethanol for the cleanups.
2. Transfer the RNA samples to labeled microfuge tubes and add 90 μL (1.8×) resuspended beads to each sample. Mix well by pipetting up and down at least ten times.
3. Incubate for 15 min at room temperature.
4. Place the tubes onto a magnetic stand to separate the beads from the supernatant.
5. After the solution is clear, carefully remove and discard the supernatant without disturbing the beads.
6. Add 200 μL of freshly prepared 80% ethanol, and incubate for 30 s at room temperature. Use caution not to disturb the beads.
7. Remove the ethanol with a P200.
8. Repeat **steps 5** and **6** once for a total of two washes, and always use caution not to disturb the beads.
9. After the ethanol is removed from the second wash, briefly centrifuge the tubes in a Picofuge™ to pellet the beads and any remaining liquid. Place the microfuge tubes back on the magnet and remove any remaining liquid with a P20.
10. Air-dry the beads with the microcentrifuge tube lids open on the magnetic stand for 5 min until beads are dark brown but still glossy; do not over dry the beads.
11. Resuspend the beads in 7 μL of nuclease-free water. Pipette up and down ten times to resuspend the beads, close the microfuge tube lids, and incubate for 5 min in a microfuge rack.
12. Place the tubes back on the magnetic stand and wait until the solution clears (~5 min).
13. Transfer 5 μL of the purified RNA to a fresh 12-well strip tube. This is a good stopping point. Samples may be stored for up to 2 weeks in a −80 °C freezer or proceed to Subheading 3.4.

3.4 RNA Fragmentation and Illumina Library Preparation

1. Before starting, configure the following programs on a thermocycler with consideration for RIN values and target cDNA insert size (longer times will yield smaller cDNA inserts in final library amplicons), for setting the time at 94 °C in the *RNA_-Frag* program, and for setting the number of PCR cycles in the *NEB_PCR* program given starting amount of rRNA depleted RNA (> 50 ng = 7 cycles; > 10 ng = 10 cycles; > 1 ng = 13 cycles):

   A. ***RNA_Frag* (Fragments the RNA).**
   94 °C for 5 min; 4 °C hold; set time to 10 min for RIN > 7.

   B. ***NEB_First* (First Strand cDNA Synthesis).**
   25 °C for 10 min; 42 °C for 15 min; 70 °C for 15 min; 4 °C hold.

   C. ***NEB_Second* (Second Strand cDNA Synthesis).**
   16 °C for 60 min with the heated lid set to off.

   D. ***NEB_ATail* (End Prep Reaction).**
   20 °C for 30 min; 65 °C for 30 min; 4 °C hold.

   E. ***NEB_Lig1* (Adapter Ligation Reaction1).**
   20 °C for 15 min with the heated lid set to off.

   F. ***NEB_Lig2* (Adapter Ligation Reaction2).**
   37 °C for 15 min with the heated lid set to ≥45 °C.

   G. ***NEB_PCR* (PCR Reaction).**
   98 °C for 30 s; 7–13 cycles of: 98 °C for 10 s, 65 °C for 75 s; 65 °C for 5 min; 4 °C hold.

3.4.1 RNA Fragmentation

1. Thaw the treated RNA, if necessary, and centrifuge the strip tube at 1200 rpm for 1 min in a benchtop centrifuge. To each strip tube well containing 5 μL purified RNA, add the following reagents: 4 μL NEBNext First Strand Synthesis Reaction Buffer and 1 μL Random Primers. Place the strip tube into the thermocycler and run the ***RNA_Frag*** program (*see* **Note 7**).

3.4.2 First Strand cDNA Synthesis

1. Thaw the NEBNext Strand Specificity Reagent, and prepare a stock dilution of First Strand cDNA synthesis mix at a ratio of 2 μL NEBNext First Strand Synthesis Enzyme Mix for each 8 μL NEBNext Strand Specificity Reagent depending on the number of RNA samples being processed. Mix gently, but thoroughly by pipetting eight times and centrifuge briefly in a Picofuge™.
2. Add 10 μL of the First Strand cDNA synthesis mixture to each RNA sample, and gently pipette the entire volume ten times to mix thoroughly.
3. Seal the tubes with a fresh strip tube cap, centrifuge 1 min at 1200 rpm, place the strip tube in the thermocycler and run the ***NEB_First*** program.

3.4.3 Second Strand cDNA Synthesis

1. Retrieve the strip tube with the RNA from the thermocycler and centrifuge at 1200 rpm for 1 min in a benchtop centrifuge.
2. Prepare the second strand cDNA synthesis reaction on ice by adding the following components to each strip tube well containing the first strand reaction: 8 μL of NEBNext Second Strand Synthesis Reaction Buffer with dUTP Mix, 4 μL NEBNext Second Strand Synthesis Enzyme Mix, and 48 μL nuclease-free water. Gently pipette the entire volume six times to mix thoroughly.
3. Seal the tubes with a fresh strip tube cap, centrifuge 1 min at 1200 rpm, place the strip tube in the thermocycler and run the ***NEB_Second*** program.

3.4.4 AMPure XP Bead Clean-Up After Second Strand cDNA Synthesis

1. Before starting, the AMPure XP beads should be warmed up to room temperature for at least 30 min prior to use. Vortex and shake the bottle well to fully resuspend the beads just before adding the beads to the cDNA samples. Prepare fresh 80% ethanol.
2. Transfer the 80 μL samples into labeled microfuge tubes.
3. Add 144 μL of RNAClean XP beads to each sample and gently pipette up and down ten times to mix thoroughly.
4. Incubate the tubes in a microfuge rack for 5 min at room temperature.
5. Place the tubes on the magnetic stand for 5 min or until the solution is completely clear.
6. Remove and discard the supernatant from each sample. Leave the tubes on the magnet throughout the washing process.
7. Add 200 μL of freshly prepared 80% ethanol to each tube without disturbing the beads and incubate for 30 s.
8. Remove and discard the supernatant.
9. Repeat **steps** 7 and **8** for a total of two 80% ethanol washes.
10. Keep the tubes on the magnetic stand with the lids open for 5 min at room temperature to allow the beads to dry. Remove the tubes from the magnetic stand and place them in a microfuge rack.
11. Add 53 μL of resuspension buffer (0.1× TE Buffer) to each tube, and gently pipette the beads up and down ten times to mix thoroughly. Close the microfuge tube lids.
12. Incubate the tubes for 2 min at room temperature to elute the samples from the beads.
13. Place the tubes on the magnetic stand for 5 min or until the supernatant is clear.

14. Transfer 50 μl of supernatant (double-strand cDNA) from the tubes into a fresh 12-well strip tube and seal with a fresh strip cap. This is a safe stopping point. The cDNA samples may be stored in a −20 °C freezer for up to 1 week until ready to proceed to Subheading 3.4.5.

3.4.5 End Repair Prep of cDNA Library

1. Retrieve the strip tube with the RNA from the thermocycler (or the freezer if the samples were stored after Second Strand Synthesis), and centrifuge at 1200 rpm for 1 min in a benchtop centrifuge.
2. Prepare the End Repair Prep reaction on ice by adding the following components to each strip tube well containing 50 μL of the Second Strand Synthesis reaction: 7 μL of the NEBNext Ultra II End Prep Reaction Buffer and 3 μL of the NEBNext Ultra II End Prep Enzyme Mix, and gently pipette the entire volume 10 times to mix thoroughly.
3. Seal the tubes with a fresh strip tube cap, centrifuge 1 min at 1200 rpm, place the strip tube in the thermocycler, and run the ***NEB_ATail*** program.

3.4.6 Ligation Reaction

1. Retrieve the NEBNext Adapter for Illumina from the −20 °C freezer (*see* **Note 8**).
2. Make a diluted adapter stock using 0.5 μL of NEBNext Adapter for Illumina and 2.0 μL NEBNext® Adapter Dilution Buffer for each sample from the End Repair Prep reaction.
3. Prepare the ligation reaction on ice by sequentially adding the following components to each of the 60 μL End Repair Prep reactions: 2.5 μL of the diluted adapter, 1 μL of NEBNext Ligation Enhancer, and 30 μL of NEBNext Ultra II Ligation Master Mix. Gently pipette the entire volume up and down ten times to mix. Ensure the ligation reaction is thoroughly mixed, as inadequate mixing can significantly reduce ligation efficiency.
4. Seal the tubes with a fresh strip tube cap, centrifuge 1 min at 1200 rpm, place the strip tube in the thermocycler, and run the ***NEB_Lig1*** program.
5. Add 3 μL of USER Enzyme to the ligation mixture from the previous step, and gently pipette the entire volume up and down ten times to mix thoroughly. Seal the tubes with a fresh strip tube cap, centrifuge 1 min at 1200 rpm, place the strip tube in the thermocycler, and run the ***NEB_Lig2*** program.

3.4.7 AMPure XP Bead Clean-Up After Ligation Reaction

1. Before starting, remove the resuspension buffer and AMPure XP beads from 4 °C at least 30 min prior to use. Mix the beads vigorously by vortexing or shaking the bottle until the beads are well dispersed. Prepare fresh 80% ethanol.

2. Transfer the 96.5 μL samples into labeled microfuge tubes.
3. Add 87 μL of AMPure XP beads to each sample and gently pipette up and down ten times to mix thoroughly.
4. Incubate the tubes for 15 min at room temperature.
5. Place the tubes on the magnetic stand for 5 min or until the solution is completely clear.
6. Remove and discard the supernatant from each sample using a P200. Leave the tubes on the magnet throughout the washing process.
7. Add 200 μL of freshly prepared 80% ethanol to each tube without disturbing the beads and incubate for 30 s.
8. Remove and discard the supernatant using a P200.
9. Repeat **steps** 7 and **8** for a total of two 80% ethanol washes.
10. Keep the tubes on the magnetic stand with the lids open for 15 min at room temperature to allow the beads to dry. Remove the tubes from the magnetic stand and place them in a microfuge rack.
11. Add 17 μL of 0.1× TE to each tube, and gently pipette the beads up and down ten times to mix thoroughly. Close the microfuge tube lids.
12. Incubate the tubes for 2 min at room temperature to elute the samples from the beads.
13. Place the tubes on the magnetic stand for 5 min or until the supernatant is clear.
14. Transfer 15 μL of supernatant from each tube to a new labeled PCR strip tube. Seal the tubes with a fresh strip tube cap.
15. This is a safe stopping point and the samples may be frozen in a −20 °C freezer up to 1 month.

3.4.8 PCR Reaction

1. Before starting, remove the PCR Master Mix, the PCR Primer Cocktail and the samples (if they were frozen after the last step) from the −20 °C freezer.
2. Add 5 μL of an i5 index primer and 5 μL of an i7 index primer to each sample. Dual indexing requires that you use only one, unique i7 primer/index and one, unique i5 primer/index per sample (*see* **Note 8**).
3. Add 25 μL of NEBNext Ultra II Q5 Master Mix to each sample, and gently pipette the entire volume up and down ten times to mix thoroughly.
4. Seal the tubes with a fresh strip tube cap, centrifuge 1 min at 1200 rpm, and place the strip tube in the thermocycler and run the ***NEB_PCR*** program.

3.4.9 AMPure XP Bead Clean-Up After PCR

1. Before starting, the AMPure XP beads should be warmed up to room temperature for at least 30 min prior to use. Vortex and shake the bottle well to fully resuspend the beads just before adding the beads to the cDNA samples. Prepare fresh 80% ethanol.
2. Transfer the 50 μL samples into labeled microfuge tubes.
3. Add 50 μL of AMPure XP beads to each sample, and gently pipette up and down ten times to mix thoroughly.
4. Incubate the tubes for 15 min at room temperature.
5. Place the tubes on the magnetic stand for 5 min or until the solution is completely clear.
6. Remove and discard the supernatant from each sample. Leave the tubes on the magnet throughout the washing process.
7. Add 200 μL of freshly prepared 80% ethanol to each tube without disturbing the beads and incubate for 30 s.
8. Remove and discard the supernatant with a P200.
9. Repeat **steps** 7 and **8** for a total of two 80% ethanol washes.
10. Keep the tubes on the magnetic stand with the lids open for 15 min at room temperature to allow the beads to dry. Remove the tubes from the magnetic stand and place them in a microfuge rack.
11. Add 32.5 μL resuspension buffer to each tube, and gently pipette the beads up and down ten times to mix thoroughly. Close the microfuge tube lids.
12. Incubate the tubes for 2 min at room temperature to elute the samples from the beads.
13. Place the tubes on the magnetic stand for 5 min or until the supernatant is clear.
14. Transfer 30 μL of supernatant from each tube into a new labeled microcentrifuge tube. The library protocol is now complete. Store the libraries in a −20 °C freezer until ready to proceed with Subheading 3.4.10.

3.4.10 Library Validation and Sequencing

1. Each sample (1 μL) should be analyzed using an Agilent TapeStation or equivalent system. The resulting electropherogram should display a single peak centered around 300 bp, with a sharp onset at approximately 200 bp and a skewed tail extending toward larger amplicon sizes. An example of adaptered cDNA amplicons from a typical library is shown in Fig. 2.
2. Consult your Illumina sequencing provider for guidance on pooling samples before loading onto an Illumina flow cell. Aim to generate approximately 100 million paired-end, 150 nt sequencing reads per sample from the pooled libraries.

High Sensitivity D1000 ScreenTape

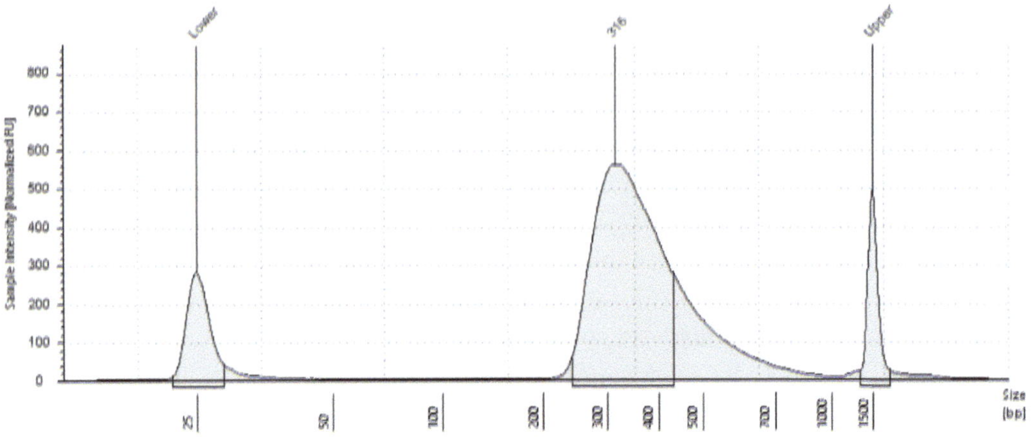

Sample Table

Well	Conc. [pg/µl]	Sample Description		
A1	1770	22585L1		

Peak Table

Size [bp]	Calibrated Conc. [pg/µl]	Assigned Conc. [pg/µl]	Peak Molarity [pmol/l]	% Integrated Area		
25	329	-	20200	-		Lower Marker
316	1770	-	8640	100.00		
1500	250	250	256	-		Upper Marker

Fig. 2 Library quality control to evaluate library size and molarity using the Agilent TapeStation system. A typical example of the final cDNA library product generated using the stranded total RNA workflow, and analyzed using the D1000 ScreenTape assay. The electropherogram of this sample shows that the maximum peak size typically occurs around 300 bp, as indicated in the displayed trace. The sample molarity is shown in the peak table as calculated from the maximum peak size seen by the TapeStation system

3.4.11 Bioinformatic Analysis of FASTQ Reads

Once the raw data in the form of FASTQ files is returned from the sequencing provider, it can be processed through a straightforward pipeline using established bioinformatics tools. These tasks are most efficiently performed on a Linux or MacOS computer with sufficient disk space to handle multi-gigabyte files and at least 32GB of RAM. Sequence quality assessment is crucial for evaluating the technical success of library generation. For instance, the FASTQC bioinformatics tool provides reports on per-base quality scores, per-sequence quality scores, and read duplication levels. An all-in-one tool for preprocessing FASTQ files is the *fastp* program [17], which offers functions for quality profiling, adapter trimming, read filtering, and merging of overlapped paired-end reads. For splice-aware alignment of the trimmed and merged FASTQ reads to the hg38 reference genome use the "Spliced Transcripts Alignment to a Reference" or *STAR* program [18]. STAR pre-compiled executables are available on GitHub: https://github.com/alexdobin/STAR/releases, with the static executables being the easiest to use. An annotation file in GTF format is required when generating genome indexes with *STAR*, and a good example is the GENCODE project comprehensive gene annotation GTF file for hg38, such as gencode.v47.annotation.gtf. When running a *STAR* mapping job, use the parameters specified in the STAR manual for the ENCODE standard options for long RNA-seq pipeline as a starting point. To maximize sensitivity for novel junction discovery, run STAR in 2-pass mode by including the parameter—twopassMode Basic.

Once BAM alignment files are generated, the read depth and splicing pattern across the *DMD* region can be displayed using the Integrative Genomics Viewer (IGV), including coloring alignments by read strand and examining "sashimi plots" with arcs representing reads spanning exon junctions and peaks representing exon coverage [19]. For generating coverage plot files, you can use *bedtools* to convert the BAM file to a bedGraph file, and to create a bigWig file, you can use *bedGraphToBigWig*, which is available from the UCSC genome browser tools download directory. The BAM file can be used as input for a more detailed examination of splice junction read counts and exon coverage using the R/Bioconductor *SGSeq* software package for analyzing splice events from RNA-seq data [20]. Read counts for gene, exon, and intron features can be extracted from the BAM file using the featureCounts program that is part of the Subread package and is available as a command-line program or through the R/Bioconductor *Rsubread* package [21]. To use *featureCounts*, you need the BAM file containing aligned reads and a list of genomic features in GTF, GFF, or SAF format. An SAF file for Dp427m exon and intron features can be created using coordinates from the gencode.v47.annotation.gtf file. The SAF file must include the following five columns for each exon or intron feature: GeneID, Chr, Start, End, and Strand. These read count files can be used to assess the relative abundance of exon and intron features, providing evidence for complex rearrangements or premature transcription termination due to mutations that lead to pseudo 3′-terminal exon usage [16].

4 Notes

1. 75% and 80% ethanol solutions are used throughout the protocols. These reagents need to be freshly prepared on the day they are used. The quantity that will be consumed for a protocol is determined by the number of samples being processed. The appropriate volumes should be prepared using the following ratios of absolute ethanol and RNase-free H_2O:

   75% ethanol: 7.5 mL absolute ethanol +2.5 mL RNase-free H_2O

80% ethanol: 8.0 mL absolute ethanol +2.0 mL RNase-free H_2O.

The diluted ethanol solutions should be discarded at the end of the day.

2. Use the smallest gauge needle as possible to break up the tissue. If the tissue fragments are not small enough, they can clog the needle. We often start with an 18 gauge needle to break up the tissue and then proceed to smaller-gauge needles.
3. A 3:1 ratio of TRIzol™ LS to aqueous phase should always be maintained.
4. The precipitation reaction can stably remain at −20 °C for several months.
5. Spectrophotometry results should indicate a 260 nM:280 nM absorbance ratio of >1.7 indicating a lack of protein contamination in the sample.
6. It is crucial to mix the RNA/probe hybridization reaction thoroughly prior to heating.
7. The fragmentation time can be adjusted based on the quality of the RNA. For RNA purified using the method described above, a 5 min fragmentation time has proven optimal for our samples. This duration yields the best cluster generation on NovaSeq 6000 flow cells. However, it does result in a moderate degree of overlap between 150 nt paired-end reads from the same amplicon.
8. When selecting a suitable set of multiplex adapters for the ligation and PCR steps, several factors should be considered. It is recommended to consult your Illumina sequencing facility for guidance on recovering 100 million paired-end reads per sample. Since samples will likely be pooled at the facility, they can advise on compatible adapter sets. We recommend using a dual-index adapter kit to ensure accurate de-multiplexing of your samples. If precise quantification is required, consider incorporating a unique molecular identifier (UMI) during the adapter ligation step. Both NEB and Illumina provide detailed documentation on the use of UMIs and i5/i7 index sequences. The NEBNext Multiplex Oligos for Illumina (Dual Index Primers Set 1) includes adapters and index primers optimized for multiplex sample preparation for next-generation sequencing on the Illumina platform. The dual-index primer strategy incorporates two eight-base indices, one in each primer. The i7 primers include indices located next to the P7 sequence, while the i5 primers feature indices next to the P5 sequence. Dual indexing is achieved by assigning a unique index to both ends of a sample being sequenced.

Acknowledgments

This work was supported in part by a grant from the National Institutes of Health (R01NS085238).

References

1. Koenig M, Monaco AP, Kunkel LM (1988) The complete sequence of dystrophin predicts a rod-shaped cytoskeletal protein. Cell 53:219–228
2. Roberts RG, Bobrow M, Bentley DR (1992) Point mutations in the dystrophin gene. Proc Natl Acad Sci USA 89:2331–2335
3. Bulman DE, Gangopadhyay SB, Bebchuck KG, Worton RG, Ray PN (1991) Point mutation in the human dystrophin gene: identification through western blot analysis. Genomics 10:457–460
4. Gardner RJ, Bobrow M, Roberts RG (1995) The identification of point mutations in duchenne muscular dystrophy patients by using reverse-transcription PCR and the protein truncation test. Am J Hum Genet 57:311–320
5. Prior TW, Bartolo C, Pearl DK, Papp AC, Snyder PJ, Sedra MS, Burghes AH, Mendell JR (1995) Spectrum of small mutations in the dystrophin coding region. Am J Hum Genet 57:22–33
6. Tuffery S, Bareil C, Démaillé J, Claustres M (1996) Four novel dystrophin point mutations: detection by protein truncation test and transcript analysis in lymphocytes from duchenne muscular dystrophy patients. Eur J Hum Genet 4:143–152
7. Matsuo M, Masumura T, Nishio H, Nakajima T, Kitoh Y, Takumi T, Koga J, Nakamura H (1991) Exon skipping during splicing of dystrophin mRNA precursor due to an intraexon deletion in the dystrophin gene of duchenne muscular dystrophy Kobe. J Clin Invest 87:2127–2131
8. Ikezawa M, Nishino I, Goto Y, Miike T, Nonaka I (1999) Newly recognized exons induced by a splicing abnormality from an intronic mutation of the dystrophin gene resulting in Duchenne muscular dystrophy. Hum Mutat 13:170
9. Tennyson CN, Shi Q, Worton RG (1996) Stability of the human dystrophin transcript in muscle. Nucleic Acids Res 24:3059–3064
10. Falzarano MS, Mietto M, Fortunato F, Farnè M, Martini F, Ala P, Selvatici R, Muntoni F, Ferlini A (2023) mRNA in situ hybridization exhibits unbalanced nuclear/cytoplasmic dystrophin transcript repartition in Duchenne myogenic cells and skeletal muscle biopsies. Sci Rep 13:15942
11. Hildyard JCW, Crawford AH, Rawson F, Riddell DO, Harron RCM, Piercy RJ (2020) Single-transcript multiplex in situ hybridisation reveals unique patterns of dystrophin isoform expression in the developing mammalian embryo. Wellcome Open Res 5:1–45
12. Flanigan KM, Dunn DM, Von Niederhausern A, Soltanzadeh P, Gappmaier E, Howard MT, Sampson JB, Mendell JR, Wall C, King WM, Pestronk A, Florence JM, Connolly AM, Mathews KD, Stephan CM, Laubenthal KS, Wong BL, Morehart PJ, Meyer A, Finkel RS, Bonnemann CG, Medne L, Day JW, Dalton JC, Margolis MK, Hinton VJ, Weiss RB, Bromberg M, Swoboda K, Kerr L, Gurvich O, Tuohy T, Taylor L, Zhao L, Hart K, Moural C, Hak K, Duval B, Hamil C, Mahmoud M, Aoyagi A, Viollet L, Gailey S, Lopate G, Golumbek P, Schierbecker J, Malkus B, Siener C, Baldwin K, Glanzman AM, Flickinger J, Naughton CE (2009) Mutational spectrum of DMD mutations in dystrophinopathy patients: application of modern diagnostic techniques to a large cohort. Hum Mutat 30:1657–1666
13. Tuffery-Giraud S, Béroud C, Leturcq F, Yaou RB, Hamroun D, Michel-Calemard L, Moizard MP, Bernard R, Cossée M, Boisseau P, Blayau M, Creveaux I, Guiochon-Mantel A, De Martinville B, Philippe C, Monnier N, Bieth E, Van Kien PK, Desmet FO, Humbertclaude V, Kaplan JC, Chelly J, Claustres M (2009) Genotype-phenotype analysis in 2,405 patients with a dystrophinopathy using the UMD-DMD database: a model of nationwide knowledgebase. Hum Mutat 30:934–945
14. Gonorazky H, Liang M, Cummings B, Lek M, Micallef J, Hawkins C, Basran R, Cohn R, Wilson MD, Macarthur D, Marshall CR, Ray PN, Dowling JJ (2016) RNAseq analysis for the diagnosis of muscular dystrophy. Ann Clin Transl Neurol 3:55–60
15. Gazzoli I, Pulyakhina I, Verwey NE, Ariyurek Y, Laros JFJ, 't Hoen PAC, Aartsma-Rus A (2016) Non-sequential and multi-step splicing of the dystrophin transcript. RNA Biol 13:290–305

16. Waldrop MA, Moore SA, Mathews KD, Darbro BW, Medne L, Finkel R, Connolly AM, Crawford TO, Drachman D, Wein N, Habib AA, Krzesniak-Swinarska MA, Zaidman CM, Collins JJ, Jokela M, Udd B, Day JW, Ortiz-Guerrero G, Statland J, Butterfield RJ, Dunn DM, Weiss RB, Flanigan KM (2022) Intron mutations and early transcription termination in Duchenne and Becker muscular dystrophy. Hum Mutat 43:511–528
17. Chen S, Zhou Y, Chen Y, Gu J (2018) Fastp: an ultra-fast all-in-one FASTQ preprocessor. Bioinformatics 34:i884–i890
18. Dobin A, Davis CA, Schlesinger F, Drenkow J, Zaleski C, Jha S, Batut P, Chaisson M, Gingeras TR (2013) STAR: Ultrafast universal RNA-seq aligner. Bioinformatics 29:15–21
19. Thorvaldsdóttir H, Robinson JT, Mesirov JP (2013) Integrative genomics viewer (IGV): high-performance genomics data visualization and exploration. Brief Bioinform 14:178–192
20. Goldstein LD, Cao Y, Pau G, Lawrence M, Wu TD, Seshagiri S, Gentleman R (2016) Prediction and quantification of splice events from RNA-seq data. PLoS One 11:e0156132
21. Liao Y, Smyth GK, Shi W (2014) FeatureCounts: an efficient general purpose program for assigning sequence reads to genomic features. Bioinformatics 30:923–930

Chapter 3

Preparation, Sectioning, and Histochemical Staining of Muscle Tissue

Emma C. Frair and Stefan Nicolau

Abstract

The collection and processing of frozen muscle tissue is essential for accurate downstream analysis of human and animal samples collected in Duchenne muscular dystrophy (DMD) research. Specific techniques are essential for preserving the tissue morphology and molecular markers that are vital for diagnostics, evaluation of disease progression, and preclinical in vivo research. Freezing muscle tissue in liquid nitrogen-cooled isopentane maintains tissue structure and preserves enzyme activity, which is crucial for enzyme histochemistry and other downstream protein analysis. In this chapter, we will discuss techniques for tissue handling, cryo-sectioning, and staining to minimize preparatory artifacts ensure optimal performance of downstream assays.

Key words Freezing, Sectioning, Staining, Cryostat, Histochemistry, Histology, Isopentane

1 Introduction

The collection and processing of frozen muscle tissue is essential for successful downstream analysis of human and animal samples collected in Duchenne Muscular Dystrophy (DMD) research [1]. In muscle disease research, unlike in many other fields, frozen muscle specimens are generally preferred over fixed tissue for a number of reasons: (1) freezing better preserves the architecture of muscle fibers and fascicles, which aids in the interpretation of histological changes and enables morphometric analyses. (2) Several histochemical stains routinely performed on muscle tissue rely on the activity of muscle enzymes, including nicotinamide adenine dinucleotide (NADH) tetrazolium reductase, succinate dehydrogenase (SDH), and myosin ATPase. Freezing is essential for preserving the activity of these enzymes. (3) Many of the of the antibodies and protocols for immunostaining of muscle proteins (in particular, dystrophin) were optimized on frozen tissue and may not perform as well on fixed specimens. (4) Lastly, freezing better preserves

Kevin Flanigan and Stefan Nicolau (eds.), *Duchenne Muscular Dystrophy: Methods and Protocols*, Methods in Molecular Biology, vol. 2975, https://doi.org/10.1007/978-1-0716-4811-7_3,

muscle RNA for transcriptomic analyses, which is of particular relevance for long transcripts such as dystrophin. The latter two techniques are discussed in greater detail in dedicated chapters (see *Manual Immunofluorescence Staining for Dystrophin and Key Dystrophin-Associated Proteins* and *Total RNA-seq as a Tool to Study DMD Splicing and Transcriptional Dynamics*). Although it is rarely the primary means of processing muscle samples, formalin fixation and paraffin embedding have the advantage of allowing larger samples to be processed, which allows detection of infrequent or patchy changes.

The most important aspect of muscle tissue processing is rapid freezing. If the sample freezes too slowly, ice crystals form in the muscle fibers, leading to artifactual vacuolation and distorting the architecture of the fibers. This chapter thus emphasizes a freezing technique using isopentane (2-methylbutane) pre-cooled in liquid nitrogen, which has been the preferred approach for decades. If supplies for this cannot be procured, alternative protocols exist using a slurry of dry ice and acetone or alcohol.

In order to allow cryo-sectioning, specimens need to be mounted to a chuck, typically made of wood, cork, or metal. This is best performed prior to freezing, in order to avoid accidental thawing of the specimen or introduction of artifacts at its outer edges. We favor mounting the specimen to the chuck using a small amount of tragacanth gum, rather than fully embedding it in a medium like OCT compound, as the latter interferes with some downstream molecular analyses.

This chapter provides detailed guidelines for muscle sample processing, including transfer, storage, cryo-sectioning, and staining, ensuring reliable downstream analysis and enhancing the accuracy of experimental outcomes. A protocol for hematoxylin and eosin (H&E) staining is included, which allows initial analyses of muscle morphology and identification of most artifacts introduced by freezing and sectioning. Protocols for other histochemical stains may be found in a number of printed sources [2, 3], as well as online on the Washington University Neuromuscular Disease Center website (https://neuromuscular.wustl.edu). Several myology textbooks contain more detailed descriptions of these techniques, as well as the range of pathological and artifactual changes that may be observed in muscle samples [2, 4–7].

2 Materials

2.1 Mounting and Freezing of Fresh Muscle Tissue

1. Normal saline, 0.9% sodium chloride solution.
2. Sterile petri dish.
3. Forceps: Sterile Adson forceps or equivalent.
4. Marking pen.

5. Sterile gauze.
6. Metric ruler.
7. Sterile dissection board.
8. Dissection microscope or magnifying monocle.
9. Dissection light or bright task lamp.
10. Saline-moistened gauze in a sterile petri dish: Dampen sterile gauze with physiological grade saline and wring it to remove excess liquid.
11. Chuck for mounting specimen: made of wood, metal, or cork, with a diameter of 10–20 mM, as required for the sample. Optionally, label the chuck with specimen identifiers using a permanent marker or a cryogenic self-laminating label.
12. 7% tragacanth gum solution: mix 70 g of tragacanth and 50 grains (3.24 g) of thymol in 1 L of milliQ water. Allow the mixture to incubate at room temperature, stirring every 30 min for up to 10 h. Aliquot the gum into ~50 mL conical tubes. For short-term storage, store at 4 °C, and for long-term storage, store at −20 °C.
13. Forceps: curved fine forceps, Adson forceps, or equivalent.
14. Dissection microscope or magnifying glass.
15. Isopentane (2-methylbutane).
16. Liquid nitrogen dewar.
17. Liquid nitrogen.
18. *250–500 mL modified plastic or metal beaker*: drill two holes at the top of the beaker to allow for strings to be tied with a suspended weight attachment.
19. Foerster Sponge Forceps (>10″).
20. Aluminum foil for specimen wrapping.
21. Dry ice in insulated cooler.
22. Plastic container for specimen storage.

2.2 Cryo-Sectioning and Slide Preparation of Frozen Muscle Specimens

1. Dry ice.
2. Styrofoam cooler or equivalent.
3. Aluminum foil: placed on top of the dry ice inside of the cooler to acclimate before coming to temperature.
4. Leica CM1950 cryostat or equivalent, capable of maintaining temperature between −30 °C and −15 °C.
5. Cut-resistant gloves.
6. Laboratory gloves and protective apparel.
7. Permanent or low-profile removable blades for the cryostat.
8. 70 mM glass roll plate.

9. Charged microscope slides (e.g., Superfrost™ Plus).
10. Small fine tipped paint brush.
11. Cleaning tools for the cryostat.
12. Glass cover slips (*see* **Note 1**).
13. Laundry antistatic spray.

2.3 Hematoxylin and Eosin Staining

1. Chemical filter paper.
2. Plastic chemical safe funnel.
3. Glass bottles.
4. Hematoxylin. Filter prior to use and replace weekly.
5. Diluted bluing reagent: 1:1 dilution of bluing reagent and distilled water.
6. Eosin-Y. Filter prior to use and replace weekly.
7. Ethanol alcohol gradient: 30% ethanol, 95% ethanol, and 100% ethanol.
8. Xylene.
9. Xylene-based mounting medium (e.g., Cytoseal).
10. Coplin jars, troughs, or other containers for staining. Fill these with reagents in the following sequence: hematoxylin, distilled water, diluted bluing reagent, distilled water, eosin-Y, 30% ethanol, 95% ethanol (3 jars), 100% ethanol (3 jars), and xylene (2 jars). Containers should be filled at least sufficiently to cover the sections.
11. Glass cover slips.
12. Wooden cotton-tipped applicator or equivalent.
13. Acetone-based nail polish (optional).
14. Brightfield microscope.

3 Methods

All steps are completed at room temperature in a laboratory with proper protective equipment, training, and compliant with institutional safety standards.

3.1 Mounting and Freezing of Fresh Muscle Tissue

1. Muscle samples should be collected on saline-moistened gauze in a petri dish (*see* **Note 2**).
2. Process and freeze tissue as soon as possible after collection, with a delay of no more than 1–2 h.
3. Trim the tissue to the desired size for mounting (*see* **Note 3**).

4. Fragments that are unsuitable for mounting may be wrapped in aluminum foil and snap-frozen for protein or nucleic acid extraction using the same procedure.
5. Orient the muscle tissue based on the direction of muscle fibers. It is standard practice for muscle to be mounted with the fibers perpendicular to the sectioning plane.
6. Place a small amount of tragacanth gum on top of the pre-labeled chuck, forming a dome approximately 5 mM in height (*see* **Note 4**).
7. The tissue to be frozen should be placed onto the mound of gum tragacanth. (The tissue should be inserted into the gum only enough to secure it but not entirely submerged).
8. Add isopentane to metal beaker inside a chemical fume hood. Lift beaker by the weighted strings, and gently submerge it into the liquid nitrogen. The weights should counterbalance the suspended beaker for stability while in the liquid nitrogen dewar. The beaker should be submerged sufficiently for the level of the surrounding liquid nitrogen to be above the isopentane in the beaker, but liquid nitrogen should not be allowed to spill into the isopentane.
9. To avoid introduction of freezing artifacts, allow the isopentane to cool to below −150 to −160 °C (*see* **Note 5**).
10. Long-handled forceps should be used to hold the chuck and submerge it face-down into the isopentane.
11. Hold the block submerged in isopentane for at least 30 s.
12. Muscle samples should be transferred to either liquid nitrogen or a − 80 °C freezer for long-term storage. Samples may be stored at temperatures up to −20 °C (e.g., in a cryostat or on dry ice) for short periods during transfers (*see* **Note 6**).

3.2 Cryo-Sectioning and Slide Preparation of Frozen Muscle Specimens

Protective equipment, including cut-resistant gloves, should be used when operating the cryostat. Users should familiarize themselves with the operation of their particular cryostat prior to use.

1. Adjust the cryostat's chamber temperature to the optimal range for cutting muscle, between −20 °C and − 25 °C (*see* **Note 7**).
2. Confirm that the glass anti-roll plate is clean and free of frost. Confirm that the edge of the anti-roll plate is not chipped. Brief warming can help eliminate frost before securely reinstalling.
3. Insert a fresh, sharp blade compatible with the cryostat, and configure the angle according to the instrument's specifications. The blade may need to be replaced during sectioning if it becomes dull.

4. Secure the chuck holding the muscle sample onto the head of the cryostat using an appropriate mechanism (*see* **Note 8**).
5. Allow the sample to adjust to the chamber's temperature for about 10–15 min.
6. Position the tissue block so the desired cutting plane is aligned with the blade.
7. Trim the sample until a flat, even surface is exposed. Sections cut while facing into the specimen may be collected for protein or nucleic acid extraction.
8. Muscle sections are generally cut at a thickness of 10 μm. Thinner sections (5–7 μm) may be preferred for specific downstream applications (*see* **Note 9**).
9. Sections will settle onto either the roll plate or cryostat stage. Sections may be lifted, turned, and flattened (if necessary) using fine paintbrushes.
10. Using fine-tipped brush, carefully lift the tissue section and lay it onto a pre-cooled microscope slide, making sure it remains flat and without folds (*see* **Note 10**). If adjustments are necessary, gently reposition the section with a fine brush or forceps. Ensure tissue sections are adequately adhered to prevent detachment during staining.
11. Alternatively, place the section on the cryostat stage, and lower a glass slide at room temperature onto the section to collect it. The section will immediately adhere to the slide.
12. Continue this process for additional sections, spacing them appropriately for downstream staining.
13. Slides may be stored at −20 °C or on dry ice for brief periods. Longer-term storage should be at −80 °C. Slides should be stained no later than a few months after sectioning, particularly for quantitative analyses.

3.3 Hematoxylin and Eosin Staining

The staining procedure is carried out under a chemical fume hood. All reagents should be collected and disposed of according to institutional guidelines.

1. Remove frozen slides with tissue sections from storage 15 min before starting the procedure and allow them to thaw completely. Ensure no condensation remains on the slides. Slides should be labeled in pencil or using a solvent-resistant marker.
2. Place slides into a slide holding basket that fits in the staining containers.
3. Submerge the basket in the hematoxylin solution for 6 min.
4. Transfer slides to the first distilled water jar. Rinse by filling the jar with clean distilled water until the solution runs clear.

5. Place the slide basket into the jar containing diluted bluing reagent. Dip slides up and down ten times.
6. Transfer the slide basket to a jar of clean distilled water and dip 15 times.
7. Submerge slides in eosin-Y and dip six times.
8. Process slides through the alcohol gradient, dipping 15 times in each jar: 30% ethanol, 95% ethanol (3 jars): 100% ethanol (3 jars).
9. Submerge slides in the first xylene jar and dip 15 times. Repeat with the second xylene jar.
10. After dipping in the second xylene jar, leave the slides submerged until mounting.
11. Apply a small dot of Cytoseal mounting medium to each tissue section. To avoid bubbles, seal the nozzle of the Cytoseal bottle when not in use, and store it upside down.
12. Place a glass coverslip over the slide. Use a clean wooden or cotton-tipped applicator to gently press down, avoiding direct pressure on tissue sections. Ensure no bubbles are present over the tissue sections (*see* **Note 11**).
13. (Optional) Apply a thin layer of clear acetone-based nail polish to the edges of the coverslip to secure it.
14. Place slides in a drying rack under the chemical hood until they are dry.
15. Once dried, store slides in a cool, dark location for long-term preservation.
16. Examine stained sections under a light microscope to evaluate muscle tissue morphology and identify any artifacts introduced by handling, freezing, sectioning, or staining (Fig. 1).

4 Notes

1. Most microscope objectives are optimized for #1.5 coverslips, which are 0.17 mM in thickness.
2. Manipulate the sample using forceps as gently as possible, in order to avoid introducing crushing artifacts.
3. Tissue blocks should be no larger than approximately 5 mM in diameter to allow them to freeze quickly enough. Multiple small fragments may be pressed together to form a larger block, or an excessively large sample may be divided into multiple blocks.
4. Dispensing the gum from a 5–10 mL syringe will help remove bubbles in the gum, which can cause cracking during freezing.

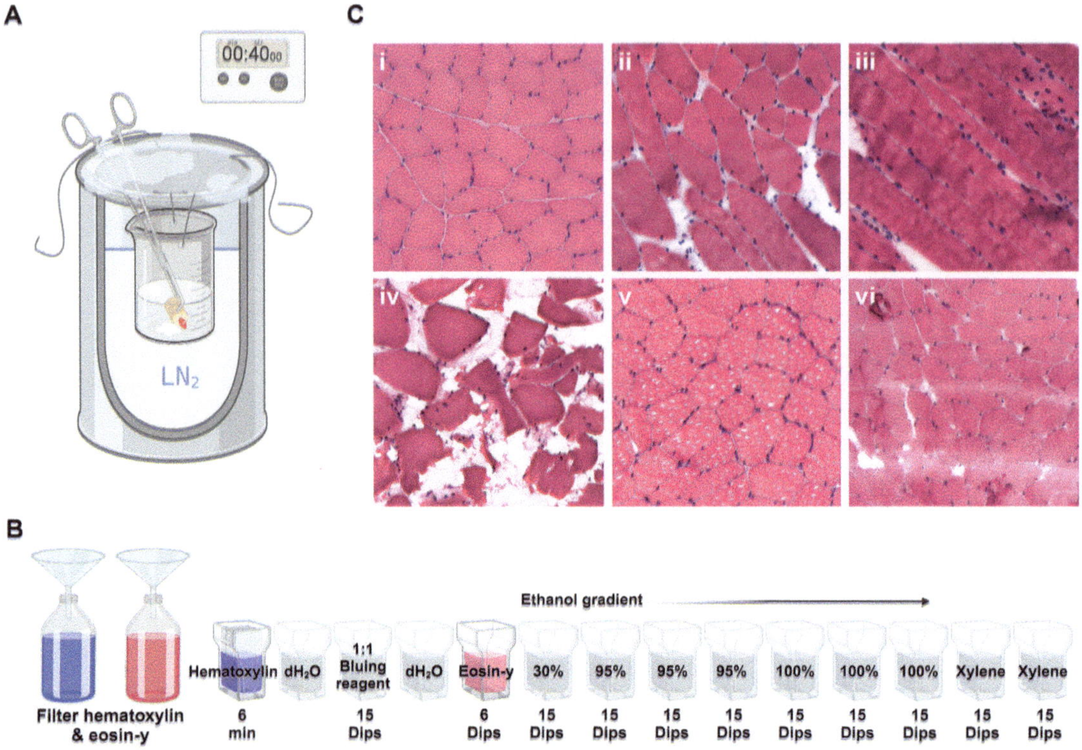

Fig. 1 Cryogenic preservation of muscle specimens and hematoxylin and eosin (H&E) staining. (**a**) Muscle specimens are frozen in chilled isopentane. A plastic or metal beaker approximately halfway filled with isopentane is suspended in a dewar filled with liquid nitrogen. The liquid nitrogen level in the dewar is above the level of the isopentane but below the holes or top rim of the beaker. The fresh muscle specimen mounted onto a wooden chuck with tragacanth gum is submerged into isopentane only once the isopentane is sufficiently chilled. (**b**) For H&E staining, slides are moved through a sequence of staining and wash reagents. (**c**) A series of common processing artifacts in muscle tissue are shown. (i) Muscle tissue without artifacts. (ii) Tissue in mixed orientation (cross-sectional and oblique). (iii) Tissue in longitudinal orientation. (iv) Tissue handled with excessive force, resulting in artifactual separation of muscle fibers. (v) Ice crystal artifact resulting from slow freezing (e.g., in insufficiently chilled isopentane). (vi) Sectioning artifact resulting from a dull or damaged blade. (Created in BioRender. Frair, E. (2025) https:/BioRender.com/ztuchaw)

5. Once the appropriate temperature is reached, isopentane thickens and white crystals form in the beaker, coating its perimeter.
6. When using dry ice to temporarily store or transport specimens, place a sheet of aluminum foil in the cooler on top of the dry ice, and ensure that all containers are pre-chilled before they come into contact with the specimens.
7. Lower temperatures may be preferred when sectioning samples with high fat content.
8. The mechanism for securing samples will depend on the cryostat's configuration. Examples include a vise clamp or using a drop of OCT compound to attach the sample to a metal chuck.

9. Due to the variable orientation of fibers, we suggest sectioning heart samples at 7 μm thickness in order to better visualize the sarcolemma of obliquely cut fibers.
10. In the protocols provided here, sections are collected and stained on slides, before a coverslip is mounted. Alternatively, sections may be collected and stained on coverslips and then mounted onto the slide. This allows sections from different samples or with different stains to be combined onto the same slide.
11. Slides may be cleaned of excess Cytoseal by agitating them in a jar of xylene.

References

1. Meng H, Janssen PM, Grange RW, Yang L, Beggs AH, Swanson LC, Cossette SA, Frase A, Childers MK, Granzier H, Gussoni E, Lawlor MW (2014) Tissue triage and freezing for models of skeletal muscle disease. J Vis Exp 89. https://doi.org/10.3791/51586
2. Dubowitz V, Oldfors A, Sewry CA (2021) Muscle biopsy: a practical approach, 5th edn. Elsevier, London
3. Sheehan DC, Hrapchak BB (1987) Theory and practice of histotechnology, 2nd edn. Battelle Press, Columbus
4. Carpenter S, Karpati G (2001) Pathology of skeletal muscle, 2nd edn. Oxford University Press, New York
5. Engel A, Franzini-Armstrong C (2004) Myology: basic and clinical, 3rd edn. New York, McGraw-Hill, Medical Pub. Division
6. Goebel HH, Sewry CA, Weller RO (2013) Muscle disease: pathology and genetics, 2nd edn. Wiley-Blackwell, Chichester
7. Karpati G, Hilton-Jones D, Bushby K, Griggs RC (2010) Disorders of voluntary muscle, 8th edn. Cambridge University Press, Cambridge/New York

Chapter 4

Manual Immunofluorescence Staining for Dystrophin and Key Dystrophin-Associated Proteins

Hui Meng, Emily M. Ott, Mariah J. Prom, Margaret J. Beatka, Tatyana A. Vetter, and Michael W. Lawlor

Abstract

The detection of dystrophin and the proteins of the dystrophin-associated protein complex (DAPC) by immunostaining allows for the detection of the presence and localization of these proteins in slide-mounted sections of skeletal muscle or heart tissue. This technique has been used historically for the diagnosis and characterization of muscle disorders including Duchenne muscular dystrophy and many of the limb girdle and congenital muscular dystrophies. The extensive usefulness of this technique has allowed the development of numerous antibodies that can be used in immunostaining assays. As an increasing number of research and translational science studies have recently focused on the restoration of dystrophin as a therapeutic strategy, high-quality immunostaining techniques and analysis remain sought-after scientific practices. This chapter describes useful procedures to perform the immunofluorescence staining for dystrophin and several other useful DAPC proteins, while also providing guidance based on our clinical diagnostic and translational science research experience from working in the muscular dystrophy field.

Key words Dystrophin, Immunostaining, Immunohistochemistry, DAPC, Pathology, Histology, Procedure

1 Introduction

Immunofluorescence and immunohistochemical staining (i.e., "immunostaining") has been performed to assess dystrophin positivity in the diagnostic and research settings for more than two decades, and yet it remains an extraordinarily useful tool in the study of Duchenne muscular dystrophy (DMD) [1–3]. Immunostaining involves the use of an antibody that recognizes a protein of interest to identify the presence of that protein on a slide-mounted tissue section. While immunostaining is often considered less quantitative than other methods to detect protein expression (such as mass spectrometry, western blot, or ELISA), it provides useful information about the localization of the target protein that cannot

Kevin Flanigan and Stefan Nicolau (eds.), *Duchenne Muscular Dystrophy: Methods and Protocols*, Methods in Molecular Biology, vol. 2975, https://doi.org/10.1007/978-1-0716-4811-7_4,

be obtained from those methods. In addition, immunostaining can be pursued using multiple primary antibodies, which allows the identification of multiple targets on the same slide using different fluorescence channels. This multiplex immunofluorescence approach is extremely useful for the study of disorders like DMD, as dystrophin deficiency produces a loss or inappropriate localization of other proteins that are related to disease pathogenesis. As a result, immunostaining has been an important diagnostic and research tool as the field has assessed the consequences of dystrophin deficiency (and dystrophin restoration) in muscle tissue.

Immunostaining panels related to the detection of dystrophin epitopes and other proteins of interest were previously a key assay in the diagnosis of DMD and other dystrophies, and a broad array of antibody tools have been developed. While the vast majority of DMD and other muscular dystrophy cases are now diagnosed through genetic testing without the need for a muscle biopsy, clinical immunostaining testing remains useful in a subset of patients with atypical presentations or where the impact of detected mutations is unclear. Additionally, as dystrophin-restoring therapies are increasingly under development or in use, immunostaining for dystrophin and other related proteins has become a key tool in establishing treatment efficacy in tissue.

Dystrophin is a key protein associated with membrane stabilization in skeletal muscle, and it performs this function in association with the other proteins of the so-called dystrophin-associated protein complex (DAPC). The DAPC acts as a connection between the cytoskeleton and the extracellular matrix to prevent shearing at the sarcolemmal membrane during muscle contraction. When dystrophin is present, dystrophin and DAPC proteins can be detected at the sarcolemmal surface. In the absence or marked deficiency of dystrophin, several other DAPC components are also decreased and absent [1]. In the context of treatment, the restoration of dystrophin can also lead to the correction of immunostaining findings in other DAPC components [4–9]. This has been used to argue for the functional relevance of therapies that produce shortened dystrophins (micro-dystrophin, mini-dystrophin, and exon skipped dystrophin), as the restoration of normal DAPC immunostaining suggests that these shortened dystrophins are capable of rebuilding the DAPC and preventing contraction-induced muscle injury. As the demonstration of functional efficacy in DMD animal models and patients can be challenging due to the slowly degenerative nature of the disease and inter-individual variation, immunostaining for dystrophin and other DAPC components can provide an extremely important surrogate functional endpoint to support therapeutic development in DMD. It should also be noted that immunostaining for DAPC components can be useful in diagnosing other forms of dystrophy, as mutations in the genes encoding DAPC components or proteins related to these components are known causes of some limb-girdle muscular dystrophies and congenital muscular dystrophies [10]. The use of these antibodies in other disorders is beyond the scope of this chapter, but it should be noted that they are only of diagnostic use in situations where dystrophin immunostaining is clearly appropriate.

This chapter will focus on techniques of immunofluorescence staining and visual interpretation for dystrophin and other DAPC components, with a specific emphasis on utilization in DMD. While it is also possible to use most of these antibodies to stain for these markers using immunohistochemical (IHC) staining, immunofluorescence staining is more broadly useful as it more easily supports multiplexing for colocalization studies and intensity-based automated quantification. The information provided here represents what we believe to be best practices in immunofluorescence staining, and the final section also provides our perspective on signal interpretation strategies based on our clinical diagnostic, basic science research, and translational science work.

2 Materials

1. Cryostat with anti-roll plate.
2. Slides rack and jars (e.g., plastic Coplin jar, slide stainer well with lid).
3. Humidified chambers capable of holding slides and protecting from light.
4. Fluorescence microscope system with microscope filters (minimum TRITC or Texas red, FITC/GFP, DAPI), objective lenses (minimum 10×, 20×) and a camera appropriate for fluorescent signal capture.
5. Optimal cutting temperature (OCT) compound.
6. Superfrost Plus slides; slides must be similarly charged or coated, or else sections will fall off during staining.
7. Chemically resistant marker, pencil, or slide printer.
8. Double-distilled water (ddH_2O) or laboratory tap deionized water.
9. 1× Tris-buffered saline (TBS). (e.g., BupH Dry-Blend TBS powder dissolved in ddH_2O).
10. 1× phosphate buffered saline (PBS) (e.g., BupH phosphate buffered saline packs).
11. Xylene.
12. 100% ethanol.
13. 95% ethanol.
14. Tween-20.
15. TBST: mix 1 L of TBS with 1 mL of Tween-20.
16. Methanol.

17. #1, 24 mM × 50 mM coverslips.
18. Fixation agent, if required (4% paraformaldehyde, 10% neutral buffered formalin, methanol, or acetone).
19. Blocking buffer (e.g., 2.5% goat serum in PBS).
20. Nucleic acid stain, if desired (e.g., DAPI).
21. Mounting medium: Fluoro-Gel or acceptable substitute.
22. Primary antibody solution: dilute the primary antibody or antibodies in TBST (see Table 1). Prepare 100 μL of solution for each slide to be stained.
23. Secondary antibody solution: dilute the secondary antibody or antibodies in TBST (see Table 2). Prepare 100 μL of solution for each slide to be stained.

Table 1
Information related to potentially useful primary antibodies

Target protein/ epitope	Antibody name	Species reactivity	Host species, Ig	Supplier	Catalog #	Dilution
Dystrophin/C-Terminus	*NCL-DYS2*	D, H, M, NHP	Mouse, IgG1	Leica	NCL-DYS2-CE	1:40
Dystrophin / Exon 44 (R17)	*MANEX44A*	D, H, M, NHP	Mouse, IgG2b	DSHB	MANEX44A (5B2)	1:10 (after 1:1 glycerol dilution)
Dystrophin / exons 10,11 (R1)	*MANEX1011b*	D, H, M, NHP	Mouse, IgG2a	DSHB	MANEX1011 B (1C7)	Determined during lot verification
Dystrophin/N-Terminus exon10,11	*DYSB*	D, M, H	IgG1κ	Leica	NCL-DYSB	Determined during lot verification
Dystrophin/Exon 43 (R16)	*MANDYS106*	H	Mouse, IgG2a	Millipore Sigma	MABT827	1:200
Dystrophin C-Terminus	*Dystrophin*	D, M, H	Rabbit	Abcam	ab15277	Determined during lot verification
Dystrophin/ Exon1, 1-10, aa1-68	*MANEX1A*	D, M, H	Mouse, IgG2b	DSHB	MANEX1A (4 C7)	Determined during lot verification
Dystrophin mid rod domain	*DYS1*	D, M, H	Mouse, IgG2a	Millipore Sigma	MAB1692	1:20
Dystrophin aa 3040-3320	*Dystrophin*	M, R, H	Rabbit	Abcam	ab275391	Determined during lot verification

(continued)

Table 1
(continued)

Target protein/ epitope	Antibody name	Species reactivity	Host species, Ig	Supplier	Catalog #	Dilution
β-Dystroglycan	*Beta-Dystroglycan*	D, M, H	Mouse, IgG2a	Leica	B-DG-CE	1:50
β-Sarcoglycan	*Beta-Sarcoglycan*	D, M, H, NHP	Mouse, IgG1	Leica	B-SARC-L-CE	1:50
Laminin alpha 2	*Laminin alpha 2*	M, H	Rat	Abcam	ab11576	1:400
Laminin	*Laminin*	D	Rabbit	Sigma	L9393	1:200
nNOS	*nNOS*	D, M, H, NHP	Rabbit	Sigma	N7280	1:1000
Spectrin	*Spectrin*	H	Mouse IgG2b	Leica	SPEC1-CE	1:100
Utrophin/N-Terminus	*NCL-DRP2*	D, M, H, NHP	Mouse, IgG1	Leica	DRP2-CE	1:10
γ-Sarcoglycan	*Gamma-sarcoglycan*	H	Mouse, IgG2b	Leica	NCL-g-SARC	1:100.
Dysferlin	*Dysferlin*	H	Mouse, IgG1	Leica	NCL-Hamlet-1	1:20
α-Dystroglycan	*Alpha-Dystroglycan clone IIH6*	H, M	Mouse, IgM	Sigma	05-593	1:50
α-Dystroglycan	*Alpha-Dystroglycan clone VIA4–1*	H, M, R, D	Mouse, IgG1	Sigma	05-298	1:30
α-Sarcoglycan	*Alpha-Sarcoglycan*	H,M, R,	Mouse, IgG1	Leica	NCL-L-a-SARC	1:200

Ig immunoglobulin, *D* dog, *H* human, *M* mouse, *NHP* nonhuman primate

Table 2
Information related to potentially useful secondary antibodies

Target species, Ig	Fluorescent dye conjugate	Host species	Supplier	Catalog #	Dilution
Mouse, IgG2b	Rhodamine Red-X	Goat	Jackson Immuno-Research	115-295-207	1:200
Mouse, IgG2b	AlexaFluor 594	Goat	ThermoFisher	A21145	1:400
Mouse, IgG2b	AlexaFluor 488	Goat	Thermo	A21141	1:400
Mouse, IgG1	AlexaFluor 594	Goat	Thermo	A21125	1:400

(continued)

Table 2
(continued)

Target species, Ig	Fluorescent dye conjugate	Host species	Supplier	Catalog #	Dilution
Mouse, IgG1	AlexaFluor 488	Goat	Thermo	A21121	1:400
Mouse, IgG2a	AlexaFluor 488	Goat	Jackson Immuno-Research	115-545-206	1:200
Mouse, IgG2a	Rhodamine Red-X	Goat	Jackson Immuno-Research	115-295-206	1:200
Mouse, IgG2a	AlexaFluor 594	Goat	Thermo	A21135	1:400
Rabbit, IgG	Rhodamine Red-X	Donkey	Jackson Immuno-Research	711-295-152	1:200
Rabbit, IgG	AlexaFluor 488	Goat	Thermo	A11008	1:400
Rabbit, IgG	AlexaFluor 594	Goat	Thermo	A11012	1:400
Rat, IgG	AlexaFluor 594	Goat	Thermo	A11007	1:200
Rat, IgG	AlexaFluor 488	Goat	Thermo	A11006	1:400
Mouse, IgG	AlexaFlour 488	Donkey	Jackson Immuno-Research	715-545-151	1:400

Ig immunoglobulin

3 Methods

3.1 Slide Preparation

1. Label slide(s) for tissue being sectioned with: Study Name/ Number, Animal/Subject ID, Tissue Type, and Date, and check with the study materials to confirm accuracy. Keep slides outside of the cryostat chamber to remain at room temperature (*see* **Note 1** for information on frozen tissue).
2. A slide should be prepared to test a no primary condition (to define secondary background staining).

3.2 Cryosectioning

1. Set the cryostat temperature to −23 °C and sectioning thickness to 8 μm. After the cryostat temperature is stable at −23 °C, transfer tissues on dry ice from a −80 °C freezer to the cryostat. (Cryostat settings can be found in *see* **Note 2**).
2. Specifics of cryostat operation will vary according to the cryostat used. It is advisable to use an anti-roll plate because the area of frozen muscle will not be surrounded by mounting medium and the muscle sections will tend to curl. Position the anti-roll plate to cut one flat section at a time from the tissue by slowly turning hand wheel. If necessary, use a small paint brush to smooth out folds in the section.

3. With the section on the cryostat platform: take a labeled slide, turn it upside down, and then place it directly on the sectioned tissue to pick the tissue off the platform. Repeat this step until you have the required number sections of tissue on slide(s). Slides can be used for staining or stored in a −80 °C freezer at this stage (*see* **Note 3**).

3.3 Fixation of Slides After Sectioning

1. While most dystrophin and DAPC antibodies do not require special fixation conditions to produce interpretable results, more optimal staining conditions may be obtained by performing different fixation conditions to the sections on the slides before staining is begun. The optimal staining condition will be different for different antibodies. Common fixation conditions include (1) air-dried slides, (2) 4% paraformaldehyde (PFA), 10% neutral buffered formalin (NBF) fixation, (3) 100% methanol fixation, or (4) 100% acetone.
   (a) For air dried slides: Place the slides on the lab bench for 30 min, and briefly rinse in TBST prior to starting stains.
   (b) For 4% PFA or 10% NBF fixed slides, incubate the slides in the fixative for 10–15 min and wash 3 × 5 min with TBST prior to starting stains.
   (c) For 100% methanol fixed slides, pre-chill 100% methanol in a −20 °C freezer, incubate the slides in the 100% methanol for 15 min in a −20 °C freezer, and wash 3 × 5 min with TBST prior to starting stains.
   (d) For 100% acetone fixed slides, pre-chill 100% acetone in a −20 °C freezer, add the slides, put the slides and acetone back in the freezer for 5–10 min, and wash 3 × 5 min with TBST prior to starting stains.

3.4 Staining Procedure

1. Air-dry the slides for 30 min.
2. Briefly rinse the slides in TBST in a Coplin jar, staining rack, or similar container.
3. Remove the slides from TBST, blot excess solution onto dry paper towels, and place the slides onto a rack in a humidified chamber.
4. Add four to five drops of blocking buffer to each slide, making sure to evenly cover each tissue section. Cover the humidified chamber and incubate at room temperature for 30 min.
5. Remove blocking buffer by tipping slide to the side and blotting solution off onto dry paper towels and return slide to humidified chamber. To prevent damage to sections, do not wipe away solution from the slides or over-dry the slides at any point.

6. Add 100 μL of primary antibody solution per slide, and carefully place a coverslip on top of the solution on the slide to allow the solution to cover all sections evenly. Make sure to limit bubbles on top of sections during pipetting or cover slipping (for primary antibody information, *see* **Note 4**, and for information on where to place coverslip, *see* **Note 5**). A discussion of additional markers that are often assessed in DMD is included in **Note 6**, as that may identify primary antibody options for staining in additional channels or other slides.
7. Perform a similar distribution of materials for any negative control slides (such as "no primary antibody" or isotype control slides). Repeat Subheading 3.4, **step 5** to add blocking buffer in place of the primary antibody Subheading 3.4, **step 6**.
8. Cover humidified chamber and incubate at room temperature for 2 h or at 4 °C overnight (16–18 h).
9. Remove the primary antibody solution by tipping the slide to the side, allowing the coverslip to slide off and blotting the solution onto dry paper towels. Wash the slides in TBST for 3 × 5–10 min in Coplin jars or a similar container. (Notes related to secondary antibody preparation can be found in **Note 7**).
10. Remove the slides from TBST, blot excess solution onto dry paper towels, and place the slides onto a rack in the humidified chamber. To prevent damage to the sections, do not wipe away liquid from the slides or over-dry the slides at any point.
11. Add 100 uL of secondary antibody solution per slide, and carefully place a coverslip on each slide to allow the solution to cover all sections evenly. Cover the humidified chamber and incubate at room temperature for 1 h. Limit the exposure of solutions and slides to light, as the secondary antibody solution contains light-sensitive elements. At the beginning of this incubation, remove Fluoro-Gel mounting medium from the refrigerator, and warm to room temperature on a lab bench prior to use.
12. Remove the secondary antibody solution by tipping the slide to the side, allowing the coverslip to slide off and blotting the solution onto dry paper towels.
13. Wash slides 3 × 10 min in TBST in Coplin jars.
14. Optional step: If nuclear counterstaining with DAPI is desired, replace Subheading 3.4, **step 13** with the following step:
    - Add 100 μL of 1:10,000 DAPI in TBST to each slide and incubate for 5 min. Then wash slides in TBST for 3 × 5–10 min in Coplin jars. Then proceed to Subheading 3.3, **step 14**.

15. After the final wash, rinse with ddH_2O in Coplin jar, and keep the slides in distilled water until ready to coverslip.
16. When cover-slipping, apply coverslips to the slides one at a time. Remove the slide from water, blot excess liquid onto dry paper towels, and place the slide on a flat surface. Add two to four drops of room temperature Fluoro-Gel mounting medium across each slide, and slowly place the coverslip onto the slide, allowing mounting medium to evenly spread across sections and making sure that no bubbles exist over the sections. Inspect the slide, and carefully remove any excess mounting medium by dabbing with a paper towel.
17. Slides can be incubated at room temperature in a closed slide folder for 30 min to help set the mounting medium. After this period, slides can be stored at 4 °C until evaluation. Allowing the mounting medium to continue setting overnight at 4 °C before evaluation can minimize the amount of mounting medium extrusion from the slides, but this is not strictly necessary if the timing of the experiment requires evaluation as soon as the cover slipping is complete.
18. Slides can be evaluated directly, or slide images can be evaluated using potentially less biased automated methods. (Details of different evaluation approaches are found in **Note 9**).

4 Notes

1. As noted below, some antibodies (including nearly all antibodies that detect dystrophin) only work well when using fresh frozen tissue (i.e., tissue that has been frozen without any sort of pre-fixation). To avoid freezing artifacts that would affect muscle morphology, however, it is necessary to freeze muscle tissue using isopentane, rather than freezing on dry ice, in a freezer, or by direct immersion in liquid nitrogen. Improper freezing will result in crystal formation within the myofiber cytoplasm and lead to extensive difficulties in interpreting most pathological findings. Areas around cell edges are less affected by freezing artifacts, so it is often possible to identify dystrophin positivity even in improperly frozen muscle. However, the detection of some markers (particularly neuronal nitric oxide synthase [nNOS]) will be impaired in improperly frozen muscle. A protocol and video that describes appropriate muscle freezing using isopentane are available at the following link: https://www.jove.com/t/51586/tissue-triage-and-freezing-for-models-of-skeletal-muscle-disease [11].

   Many of the commercially available antibodies to detect dystrophin can be used for the IF or IHC detection of dystrophin in fresh frozen tissue, but almost none of them work well in formalin-fixed paraffin-embedded (FFPE) tissue. Studies focusing on dystrophin expression should include frozen tissue collection strategies for any tissues in which dystrophin immunostaining is desired.

2. Skeletal muscle from different ages and disease states can contain different amounts of fat content. This can make sectioning difficult, but the ease of cutting may be improved by adjusting the cryostat temperature to lower than −23 °C (although this may vary with different models of cryostat). It may also be useful to change the section thickness to 10 μm or even thicker than 10 μm in specimens with high fat content, although one should consider the impact of thicker sections if tissue will be undergoing any quantitative assessment.
3. If frozen slides have been stored in −80 °C freezer for 1 month, particular attention should be paid to positive control conditions to ensure that tissue quality remains sufficient. Any concerns in this respect should lead to a recutting of tissue and repeat of a given staining procedure.
4. Recommended primary antibodies for some commonly used muscular dystrophy markers are described in Table 1. It is efficient to prepare primary antibody mixtures during the blocking incubation period (**step 4**). To prepare primary antibody mixtures, prepare 100 μL of primary antibody solution for each slide in TBST. An appropriate amount of additional primary antibody solution should be made up to account for potential volume loss. If multiple primary antibodies are being used, each should have either a different host species or same host species but different isotype/subclass so that they could be distinguished by separate secondary antibodies in later steps. Negative control conditions for primary antibodies can include "no primary antibody control" conditions (corresponding to using TBST in the solution instead of a primary antibody) or "isotype control" conditions (corresponding to the use of pooled, nonspecific antibodies of the same species and isotype as the primary antibody of interest, to ensure that the primary antibody signal is specific).

   For the detection of full-length dystrophin, the choice of primary antibody will likely relate most to the commercial availability, signal/noise characteristics, and the ability of a given antibody to be paired with antibodies to detect other markers of interest. For the detection of shortened dystrophins (including microdystrophin, mini-dystrophin, and exon-skipped dystrophin), the choice of antibody will additionally need to ensure that the target epitope is present in the shortened dystrophin protein [12]. Figure 1 shows a depiction of the target epitopes of some commonly used anti-dystrophin antibodies. Conversely, it is also possible to choose antibodies

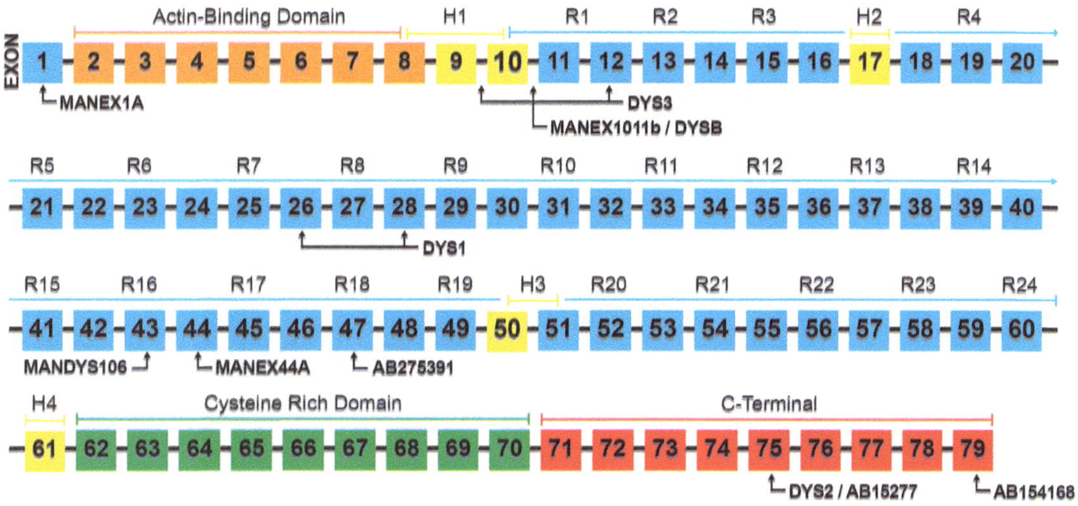

Fig. 1 Diagram of the dystrophin molecule and the epitope locations of some commonly-used antibodies. Boxes depict individual exons. Boxes colored orange denote the actin-binding domain. Boxes colored yellow denote hinge regions. Boxes colored green denote the cysteine rich domain, and boxes colored red denote the C-terminal domain. Blue boxes between exons 11-60 denote various rod domains. The binding sites of several commonly-used antibodies for the detection of dystrophin are identified with arrows

to detect epitopes that are not present in the shortened dystrophin of interest, and this would allow the recognition of dystrophin-positive fibers that are not related to the expression of the shortened dystrophin construct (essentially corresponding to revertant fibers) (Fig. 2) [13–15].

5. If slides will be viewed or imaged using an inverted microscope, the portions of the slide surface that will rest on the microscope stage should be left unobstructed. Coverslips and any label stickers should be placed in a manner that does not result in the slide resting on these raised surfaces when placed onto the stage to maintain a flat focal plane.

6. There are some special considerations in relation to the detection of other DAPC markers with immunostaining (Figs. 3 and 4). Beta-sarcoglycan:

   Beta-sarcoglycan is an excellent marker of DAPC expression in studies of DMD, particularly when evaluating dystrophin restoration. Factors that favor the use of beta sarcoglycan as a DAPC marker include (1) an easily predictable relationship of beta sarcoglycan brightness in areas that are either dystrophin-positive or dystrophin-negative, (2) the consistency of this phenotype across various dystrophin-deficient animal models and DMD patients, often using the same antibodies, and (3) the ability of beta sarcoglycan intensity to increase after dystrophin restoration. While beta sarcoglycan expression is still detectable at the sarcolemmal membrane in dystrophin

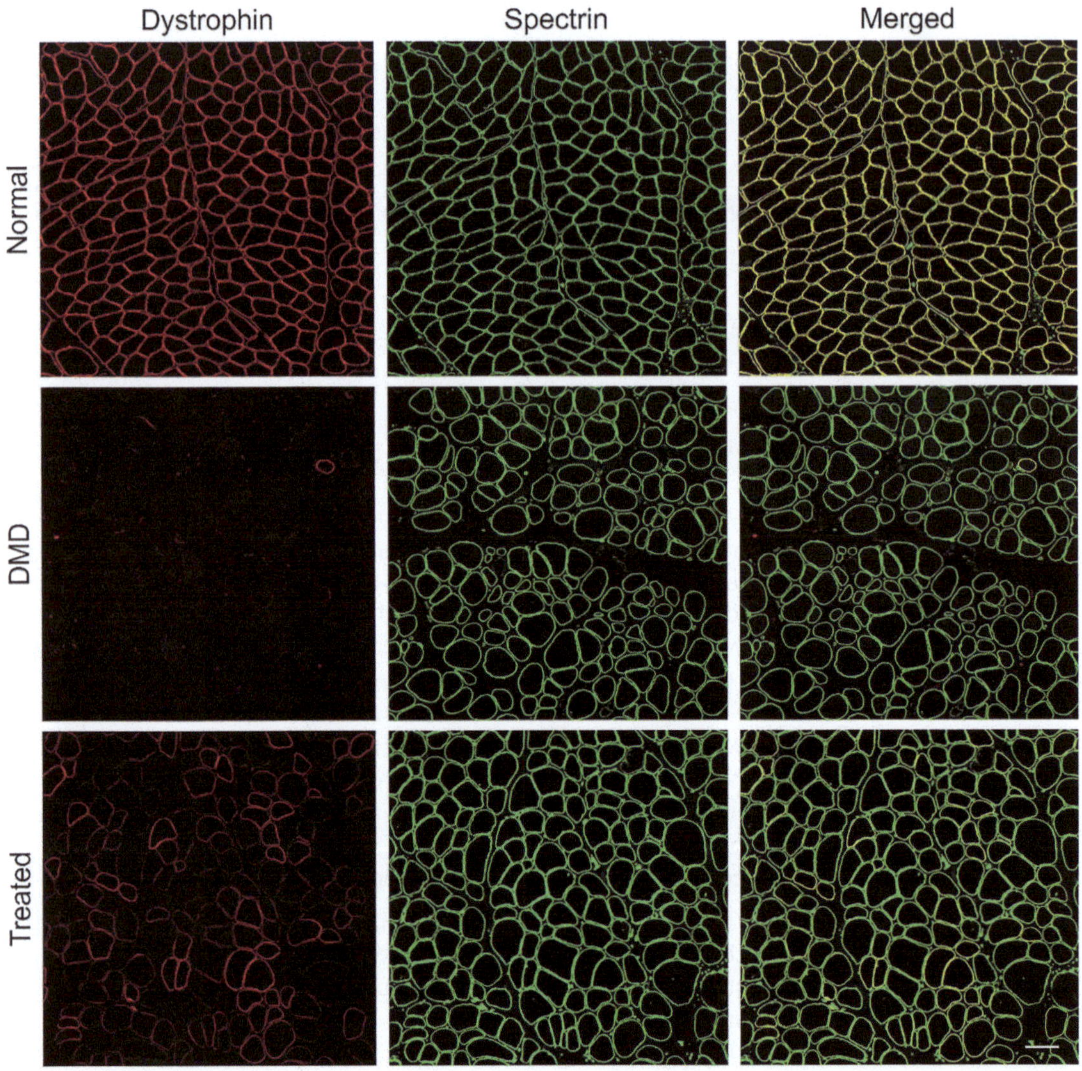

Fig. 2 Immunofluorescence staining for dystrophin and and a therapeutically restored (exon-skipped) shorter dystrophin protein. Dystrophin expression and localization was evaluated in non-diseased (normal) muscle and muscle from a Duchenne muscular dystrophy patient before (DMD) and after (Treated) treatment with a dystrophin-restoring gene therapy. The normal muscle displays dystrophin positivity in the sarcolemmal region of all myofibers, which is the appropriate localization. The pre-treatment biopsy image displays a single fiber that is positive for dystrophin, which corresponds to a revertant fiber. The treated biopsy image shows some level of positive sarcolemmal signal in most fibers, although the intensity of staining is highly variable. Immunofluorescence staining for spectrin was performed to identify the sarcolemmal regions of all fibers in the specimen, irrespective of dystrophin expression. The merged image shows the overlay of the dystrophin and spectrin channels. Bar = 100 μm

deficient muscle, the intensity of staining is markedly decreased (Fig. 3). This "dimming" of beta sarcoglycan intensity is easily recognizable in samples with dystrophin-positive (revertant or therapeutically positive) fibers, but the scoring of beta

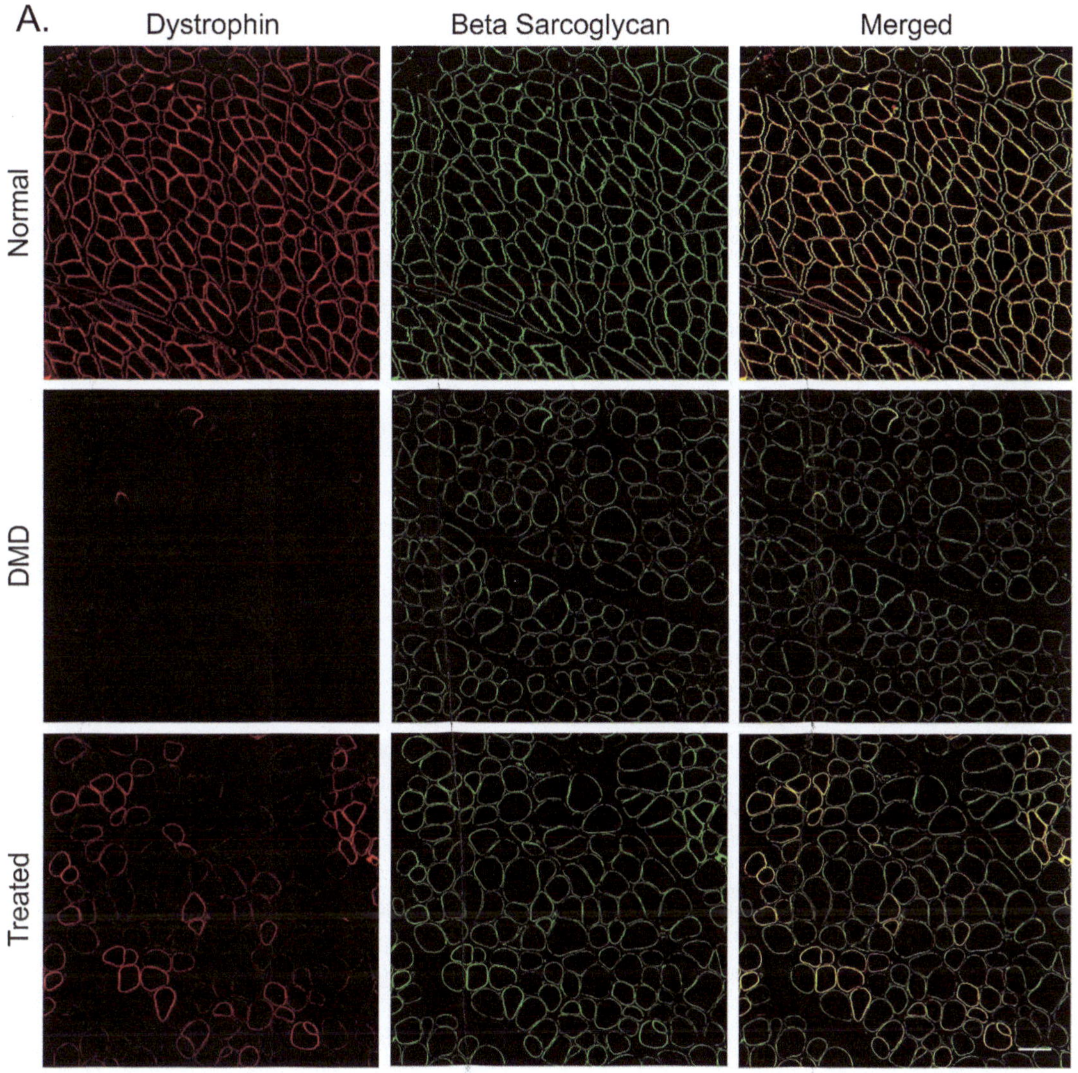

Fig. 3 Immunofluorescence staining for dystrophin and beta sarcoglycan. Dystrophin expression and localization was evaluated in non-diseased (normal) muscle and muscle from a Duchenne muscular dystrophy patient before (DMD) and after (Treated) treatment with a dystrophin-restoring gene therapy. Variable degrees of dystrophin positivity are observed in association with disease status and treatment. Immunostaining for beta sarcoglycan on the same slides illustrates a decrease in beta sarcoglycan fluorescence intensity in fibers that are deficient for dystrophin. This is particularly visible in the image from the treated patient's biopsy as the field contains myofibers across the entire range of dystrophin expression. The merged image illustrates the overlay between the green and red channels. Bar = 100 μm

sarcoglycan can be difficult in samples that are uniformly dim in intensity due to the complete lack of dystrophin-positive fibers. When scoring beta sarcoglycan positivity, positivity usually reflects brightly positive fibers and considers the dim fibers to be negative. Note that beta sarcoglycan intensity is not as

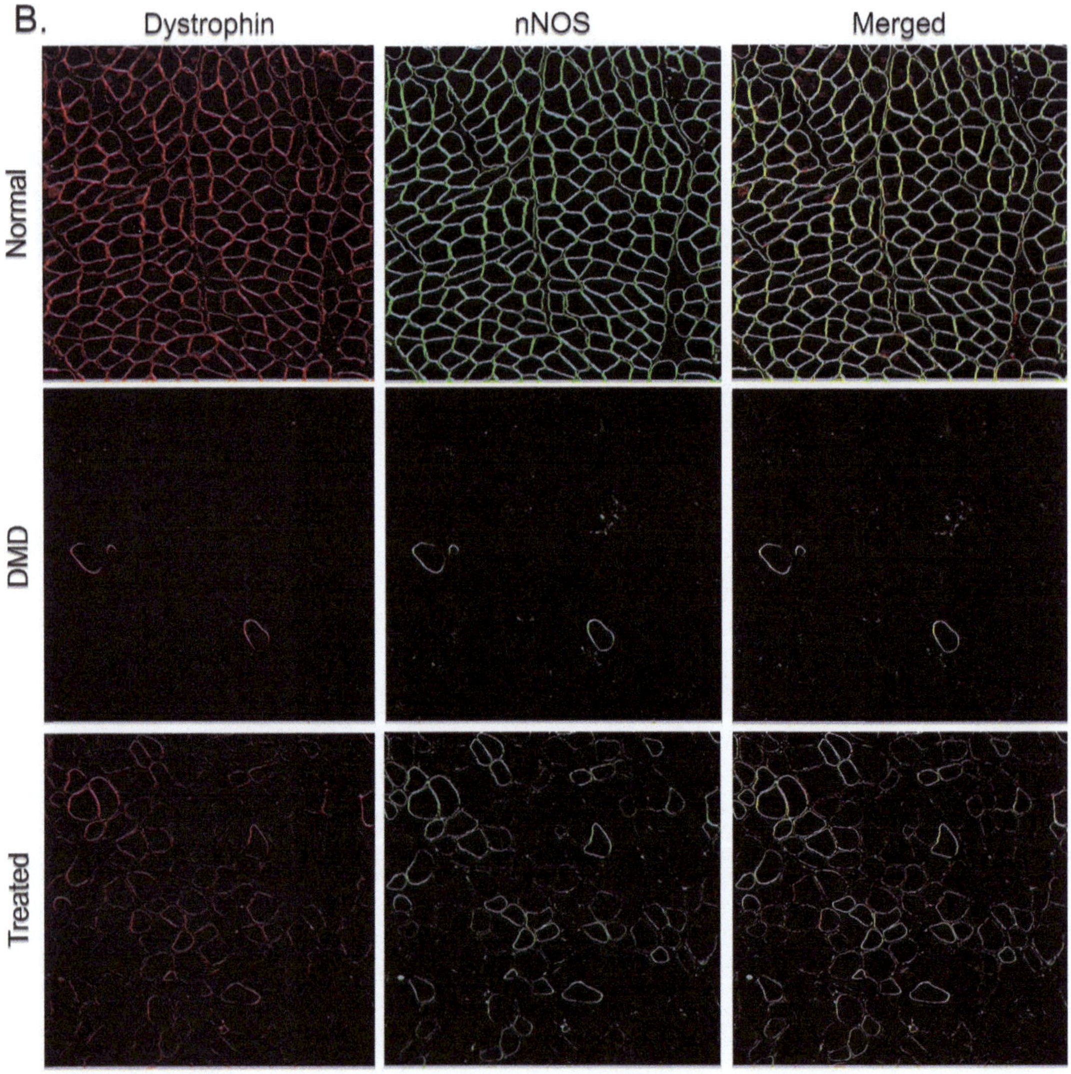

Fig. 4 Immunofluorescence staining for dystrophin and nNOS. Dystrophin expression and localization was evaluated in non-diseased (normal) muscle and muscle from a Duchenne muscular dystrophy patient before (DMD) and after (treated) treatment with a dystrophin-restoring gene therapy with an exon-skipping construct that includes the nNOS binding site. Variable degrees of dystrophin positivity are observed in association with disease status and treatment. Immunostaining for nNOS on the same slides illustrates a loss of nNOS positivity at the sarcolemma in fibers that are deficient for dystrophin. This is particularly visible in the image from the treated patient's biopsy as the field contains myofibers across the entire range of dystrophin expression. The merged image illustrates the overlay between the green and red channels. Bar = 100 μm

closely related to dystrophin expression in the heart, to the extent that it will not be uniformly bright in the presence of dystrophin or uniformly dim in dystrophin deficiency. As a result, it is not recommended for use as a DAPC marker in the heart.

nNOS: nNOS is a DAPC marker of interest in many studies due to its potential functional significance and the inclusion of nNOS binding domains in some therapeutic dystrophins (Fig. 4) [12]. nNOS can be evaluated by immunostaining to detect the presence and location of the protein, and the activity of nNOS can be demonstrated using a histochemical stain (4). However, it is worth noting that nNOS staining can be more difficult to evaluate than beta sarcoglycan because the intensity of staining is typically lower, and the ability to detect nNOS can be lost in areas of muscle that are improperly frozen. While it is possible to thaw and refreeze improperly frozen muscle [11], this does not restore nNOS positivity in muscle that was initially frozen improperly. As a result, the decision to evaluate nNOS should only be made in instances where there is confidence in appropriate tissue handling, and interpretation of nNOS results should include an identification of samples where greater than mild degrees of freezing artifact are present. Additionally, nNOS does not localize to the cell surface in cardiac myocytes, so it should not be used as a DMD-related histological marker when studying the heart.

Alpha dystroglycan: The evaluation of alpha dystroglycan positivity can be extremely challenging, as there is variability in immunostain intensity possible due to differences in glycosylation, in addition to the expected positivity differences related to the level of protein expression. Most studies and clinical laboratories using alpha dystroglycan staining use two different alpha dystroglycan antibodies, VIA-4 and IIH-6, which detect glycosylated alpha dystroglycan at different carbohydrate moieties [10, 16, 17]. The intensity of staining using each of these antibodies in normal muscle can be highly variable, and so in clinical practice, the only clearly diagnostically meaningful pattern is a marked decrease in the alpha dystroglycan staining using both antibodies (which would be consistent with alpha dystroglycanopathy). Alpha dystroglycan signals also decrease in the context of dystrophin deficiency, but the usefulness of using alpha dystroglycan as a DAPC marker is diminished by the variability in its positive signal in wild-type muscle. One would expect DMD muscle to have low alpha dystroglycan signals and for alpha dystroglycan positivity to increase with dystrophin restoration, but the interpretation of such signals is much more straightforward when using other DAPC markers to assess the impact of treatment. There is certainly usefulness to the use of alpha dystroglycan staining in the evaluation of alpha dystroglycanopathies, but this is beyond the scope of the current chapter.

Other DAPC targets (other sarcoglycans and dystroglycans, etc.): Additional DAPC immunostains are often pursued in the clinical or research setting and are potentially suitable for use in therapeutic studies of DMD [1, 2]. As noted above, in our experience these are less preferable than beta sarcoglycan or nNOS in studies that focus on only one to two DAPC components, either because the impact of dystrophin deficiency or restoration is less evident or because there are not good antibodies suitable for use across dystrophin deficient mice, canines, and humans.

7. Recommended secondary antibodies to use in immunostaining are summarized in Table 2. It is efficient to prepare secondary antibodies during the washes following primary antibody incubation (**step 8**). In preparing the secondary antibody mixture, prepare 100 μL of secondary antibody solution for each slide in TBST, plus some additional volume to account for potential volume loss. Secondary antibodies utilized should be targeted to the species and subclasses of the primary antibody (i.e., a mouse IgG anti-dystrophin primary antibody should be used with a Goat anti-mouse IgG secondary antibody). If multiple primary antibodies are being used, secondary antibodies utilized should be in separate color channels (i.e., AlexaFluor 488 and AlexaFluor 594). If multiplexing with a far-red secondary antibody (e.,g., AlexaFluor 647), AlexaFluor 568 should be substituted for AlexaFluor 594. The FITC/GFP range typically shows more autofluorescence in muscle than other channels, so using red AlexaFluor for dystrophin can provide more sensitivity to low-level positive signal. If negative control for primary antibody slide is used, secondary antibody should still be added, corresponding to the primary antibody with which other slides were stained. Note: since secondary antibodies contain light-sensitive fluorescent dye conjugates, make sure to limit the time solution is exposed to light and store in dark until use.
8. A nuclear counterstain like DAPI can be used to visualize the placement of nuclei in the section. However, while the use of DAPI can be straightforward, it is sometimes preferable to use the FITC/GFP channel for other markers or to avoid possible contributions of the DAPI stain or signal to the background observed on other channels.
9. The visual interpretation of dystrophin positivity can be approached in one of several ways. Overall positivity or negativity of the sample, without reference to the intensity or area of positivity involved. This is the approach commonly used in a diagnostic clinical setting, where dystrophin deficiency should manifest in negative staining across the entire sample, with the exception of revertant fibers that generally correspond to <5% of fibers (and usually <2% of fibers). The advantage here is that variation in tissue quality or freezing artifact is less likely to affect interpretation as long as a small portion of the tissue is

appropriate in quality. The disadvantage of this approach is that it is not sensitive to biologically relevant patchy positivity in dystrophin signal, which can be encountered in DMD carriers, some Becker MD patients, or patients with nonuniform therapeutic dystrophin restoration [2, 5]. Positivity related to the area of the sample that corresponds to positive fibers. This is the approach that we commonly use for visual estimation, and it involves estimating the area of the section that corresponds to dystrophin-positive fibers (for instance, by visualizing the section as a clock face and estimating the percentage of area that would be full if all positive areas were combined into a single area) [13–15]. This approach does not necessarily account for differences in positive intensity, as this can be difficult to assess visually. Positivity related to the number of positive fibers. This approach is straightforward when using automated image analysis methods but is difficult to apply by visual estimation because relating an area of positivity to a number of fibers can be difficult when fiber size is extremely variable (which is the case in DMD, as a result of chronic regenerative changes). It is feasible to manually count dystrophin-positive fibers when the number of fibers is very low, or by manual counting positive fibers in images. The total number of fibers in an area or sample can be identified by using another cell surface marker that remains positive in dystrophin deficiency, such as laminin or spectrin.

While manual counting is accurate, it is very laborious and may limit the area of the sample quantified. Large-scale quantification of the number and intensity of positive fibers is increasingly available using automated methods, as discussed in Chap. 5 of this book. Positivity in relation to the intensity of staining. Therapeutic increases in dystrophin positivity can relate to increases in (1) the number of fibers (or percentage of sample area) that is positive or (2) the intensity of positive staining within the positive fibers. In our experience, low levels of dystrophin restoration result in an expansion of the area of positivity first, with variable increases in intensity as the percentage of dystrophin restoration increases further.

It's important to note that visually interpretable differences in intensity are restricted to situations where dystrophin levels are still abnormally low. Despite considerable variation in dystrophin levels among non-DMD patients by western blot (corresponding to a variation of approximately 35% from the median sample, in an assessment of 29 non-DMD samples in our clinical laboratory), the intensity of immunostaining positivity did not appear to be visually different. This suggests that the intensity of immunostaining positivity for dystrophin plateaus at <65% in dystrophin content and so visual quantitative evaluations of dystrophin intensity are most likely to be useful in demonstrating expression differences at lower dystrophin levels (either due to low levels of dystrophin restoration, or early in the course of treatment).

It should also be noted that evaluating intensity quantitatively is very difficult when the difference in intensity is variable, as would be the case with varying degrees of dystrophin restoration in different fibers. While assessing a static difference in intensity can be straightforward (as is the case when evaluating "bright" vs. "dim" beta sarcoglycan areas), dystrophin restoration will result in a spectrum of dystrophin-positive intensities that will be difficult to visually score with accuracy. The most appropriate approach to quantitating intensity in this situation may involve the use of automated methods, as discussed in Chap. 5 of this book. Positivity while accounting for significant areas of non-muscle tissue in the sample. There is the potential for tissue samples to contain areas of non-muscle tissue where dystrophin could not be expressed, and the inability to recognize these areas may lead to an underestimation of the area of dystrophin positivity. This is especially a concern in studies where immunofluorescence staining is pursued without (1) other stains that might identify muscle vs. non-muscle areas or (2) a muscle membrane marker to distinguish the areas where muscle is present. In DMD, there is also the potential for degenerating muscle to be present, which may affect the ability to detect some muscle fiber markers. Additionally, there is the potential for fibrous or fatty tissue replacement of muscle, particularly in older canines or human patients. Overall, if the presence of non-myofiber tissue is a concern in a study, we would recommend the performance of a hematoxylin and eosin (H&E) stain or a skeletal muscle marker stain to allow for the assessment of sample quality and the presence of non-muscle tissue. Comparing the area of muscle tissue on H&E with the area of positivity on an immunostain can allow for correction factors to be applied for specific samples.

A straightforward way to assess positivity in muscle (either through the comparison of H&E and immunostained slides from adjacent areas, or through the inclusion of a skeletal muscle marker) is to perform slide scanning and to perform the comparisons using automated methods or through visual side-by-side comparisons on the computer. Note that we do not advocate using histological correction factors for non-muscle tissue to be applied to the correction of non-histological assays such as western blot, as this does not account for technical differences associated with those other assays. This is discussed more completely in Chap. 6 of this volume.

References

1. Ohlendieck K, Campbell KP (1991) Dystrophin-associated proteins are greatly reduced in skeletal muscle from mdx mice. J Cell Biol 115:1685–1694
2. Shahnoor N, Siebers EM, Brown KJ et al (2019) Pathological issues in dystrophinopathy in the age of genetic therapies. Annu Rev Pathol 14:105–126
3. Voit T, Stuettgen P, Cremer M et al (1991) Dystrophin as a diagnostic marker in Duchenne and Becker muscular dystrophy. Correlation of immunofluorescence and western blot. Neuropediatrics 22:152–162
4. Hakim CH, Burke MJ, Teixeira J et al (2023) Histological assessment of gene therapy in the canine DMD model. Methods Mol Biol 2587: 303–338
5. Lawlor MWS, Childers MK, Brown KJ (2019) Histological and biochemical evaluation of muscle gene therapy. In: Duan DMJR (ed) Muscle gene therapy, 2nd edn. Springer, pp 207–226
6. Mendell JR, Sahenk Z, Lehman K et al (2020) Assessment of systemic delivery of rAAVrh74. MHCK7. Micro-dystrophin in children with duchenne muscular dystrophy: a nonrandomized controlled trial. JAMA Neurol 77: 1122–1131
7. Sakamoto M, Yuasa K, Yoshimura M et al (2002) Micro-dystrophin cDNA ameliorates dystrophic phenotypes when introduced into mdx mice as a transgene. Biochem Biophys Res Commun 293:1265–1272
8. Shin JH, Pan X, Hakim CH et al (2013) Micro-dystrophin ameliorates muscular dystrophy in the canine model of duchenne muscular dystrophy. Mol Ther 21:750–757
9. Weisbart RH, Hansen JE, Nishimura RN et al (2005) An intracellular delivery vehicle for protein transduction of micro-dystrophin. J Drug Target 13:81–87
10. Dubowitz VDSCA (2013) Muscular dystrophies and allied disorders III: congenital muscular dystrophies and associated disorders. In: Muscle biopsy: a practical approach, 3rd edn. Elsevier, pp 302–330
11. Meng H, Janssen PM, Grange RW et al (2014) Tissue triage and freezing for models of skeletal muscle disease. J Vis Exp
12. Boehler JF, Brown KJ, Beatka M et al (2023) Clinical potential of microdystrophin as a surrogate endpoint. Neuromuscul Disord 33:40–49
13. Birch SM, Lawlor MW, Conlon TJ et al (2023) Assessment of systemic AAV-microdystrophin gene therapy in the GRMD model of Duchenne muscular dystrophy. Sci Transl Med 15: eabo1815
14. Hamm SE, Fathalikhani DD, Bukovec KE et al (2021) Voluntary wheel running complements microdystrophin gene therapy to improve muscle function in mdx mice. Mol Ther Methods Clin Dev 21:144–160
15. Hamm SE, Yuan C, McQueen LF et al (2023) Prolonged voluntary wheel running reveals unique adaptations in mdx mice treated with microdystrophin constructs +/− the nNOS-binding site. Front Physiol 14:1166206
16. Balci-Hayta B, Talim B, Kale G et al (2018) LARGE expression in different types of muscular dystrophies other than dystroglycanopathy. BMC Neurol 18:207
17. Jimenez-Mallebrera C, Torelli S, Feng L et al (2009) A comparative study of alpha-dystroglycan glycosylation in dystroglycanopathies suggests that the hypoglycosylation of alpha-dystroglycan does not consistently correlate with clinical severity. Brain Pathol 19:596–611

Chapter 5

Automated Quantification of Dystrophin Expression by Immunofluorescence in Humans and Animal Models

Tatyana A. Vetter and Michael W. Lawlor

Abstract

Imaging of dystrophin and related proteins using light microscopy is a key assay in neuromuscular pathology diagnostic and research laboratories. Immunofluorescence co-staining of dystrophin alongside markers like laminin or its glycoprotein binding partners allows for quantitative analysis of these markers in relation to each other, enabling the reporting of metrics like the percentage of dystrophin-positive fibers and dystrophin staining intensity at the sarcolemma. When designed carefully, automation of image analysis approaches allows for higher throughput, greater precision, and less bias than manual counting of muscle fibers. These automated approaches are especially valuable for rigorously and objectively testing the efficacy of potential therapeutic strategies in development for restoring dystrophin expression. This chapter presents a validated method, its modifications, and some alternatives for automated quantitative analysis of immunofluorescence images of dystrophin or other muscle markers co-stained with a sarcolemmal marker for muscle fiber detection.

Key words Image analysis, Microscopy, Quantification, Duchenne muscular dystrophy, DMD, Pathology

1 Introduction

Dystrophin and component proteins of the dystrophin-associated protein complex (DAPC) have been immunolabeled and imaged in muscle tissue since 1988 [1, 2]. In the decades since, standards have shifted from qualitative detection of dystrophin in single-field images to manual counting of dystrophin-positive fibers and ultimately to empirical quantification methods that can identify small differences in large images. Detection of dystrophin and its binding partners at the sarcolemma by microscopy is now a key element of most Duchenne muscular dystrophy (DMD) research projects, whether aiming to characterize or modify expression of these markers. Since functional benefits may be associated with differences in dystrophin expression as small as 1–5% of normal levels, unbiased

Kevin Flanigan and Stefan Nicolau (eds.), *Duchenne Muscular Dystrophy: Methods and Protocols*, Methods in Molecular Biology, vol. 2975, https://doi.org/10.1007/978-1-0716-4811-7_5,

quantitative approaches have become necessary to provide the sensitivity and precision required to answer contemporary research questions.

Although western blots have historically been viewed as the most quantitative technique for dystrophin expression analysis, their limitations include a lack of localization information, inconsistent sensitivity to very low protein expression, and the time and complexity associated with correctly extracting, separating, and transferring very large proteins. These limitations underscore the benefits of developing reliable and accessible quantification strategies for the more sensitive assay of localized dystrophin expression by immunofluorescence. Since immunofluorescence labels protein targets in situ, without relying on extensive sample handling and protein transfers, the process can be much simpler and more sensitive. Analysis of multiplexed fluorescence images can also quantify additional markers of dystrophic pathology, including central nucleation, fiber morphometrics, and areas of regeneration, painting a more complete picture of tissue characteristics.

A thoughtfully and elegantly developed quantitative approach to image analysis should have high throughput, minimize the role of operator judgment, and provide reproducible and sensitive results that accurately represent the whole sample. Ideally, any quantification approach should also leave a detailed audit trail of key outputs, allowing a qualified operator to verify the algorithm's accuracy or identify any shortfalls as a part of quality control measures before accepting the results as final. In recent years, multiple groups have published quantitative image analysis methods for dystrophin that fit many of these criteria, although some challenges related to objectivity, throughput, accessibility, or transparency remain common. The method presented in this chapter aims to address these challenges by quantifying the percentage of dystrophin-positive fibers and signal intensities at the sarcolemma of each fiber across whole tissue sections with minimal operator input. This quantification method has been used successfully for dystrophin and other sarcolemmal muscle markers in mouse muscles, clinical characterization studies, and gene therapy clinical trials with consistent and sensitive performance [3–9].

The platform used for developing and deploying the method described here, NIS-Elements, has a relatively shallow learning curve with a user-friendly graphical interface for creating and using algorithms, and can deliver a combination of visual and quantitative outputs that are highly useful for displaying and auditing the results. The three algorithms developed for performing this analysis process in NIS-Elements can be recreated by a proficient user of NIS-Elements following the details in this chapter. Alternatively, the algorithm files and instructions for importing and using them may also be requested from us. The algorithms are deployed in series and run fully autonomously, first to select the tissue to be analyzed, then to quantify dystrophin positivity within the fibers of the whole tissue area, and finally to quantify normalized dystrophin signal intensity within the same fibers. During the quantification process, the operator has the opportunity to exercise judgment only when removing certain tissue regions that are unsuitable for quantification due to artifact from the total tissue area that will be analyzed, ensuring objectivity and high throughput. Once quantification is complete, the same or different operator can perform a full audit of the algorithms' graphical and numerical outputs to ensure that they are reasonable and the data passes acceptance criteria. The chapter notes include modifications or alternatives depending on study goals or limitations due to tissue quality or control availability.

2 Materials

1. *Hardware*: A PC with substantial memory and a dedicated graphics card.
   (a) The minimum amount of RAM required for analysis is 8 times the file size of the largest image that may need to be analyzed (e.g., if the largest image to be analyzed is 4 GB, at least 32 GB of RAM should be available). Better specifications will increase analysis speed and throughput.
2. *Software*: NIS-Elements Advanced Research, v. 5.3 or newer, with the General Analysis 3 (GA3) module (Nikon Instruments).
3. *Algorithms*: (1) Accepted Region Selection; (2) Dystrophin Positivity Analysis; (3) Normalized Dystrophin Intensity Analysis.
   (a) The algorithms can be recreated based on the details in Tables 1, 2, and 3 by an operator generally familiar with image analysis and with NIS-Elements and the GA3 module. Alternatively, we may be contacted with requests for these algorithms.
4. *Images*: Multichannel fluorescence images with one channel for dystrophin and another for a sarcolemmal marker like laminin or spectrin. Spectrin shares the same subcellular localization with dystrophin, making it the theoretically better choice, but this distinction is not associated with a demonstrated difference in quantification results [8]. A greater variety of species options for primary laminin antibodies may result in laminin being the more practical choice for staining in mouse tissue or in combination with a third antibody.
   (a) Images must be in ND2, TIFF, or OME-TIFF format for compatibility with NIS-Elements.

Table 1
Accepted Region Algorithm

Operation	Purpose
1.1 *Smooth the sarcolemmal marker channel*	*Reduce impact of high-frequency noise and details*
1.2 Multilevel Otsu threshold on sarcolemmal marker, and keep 2 out of 3 layers	Automatically segment all sarcolemmal signal and create the "Accepted" binary layer
1.3 Clean, grow, smooth, and fill holes in the binary layer	Refine the "Accepted" binary layer and select the whole tissue
1.4 *Make inside and ring or use multiple erode iterations on tissue binary*	*Recede the "Accepted" layer from the tissue edge to avoid common edge artifacts*
1.5 Store the "Accepted" binary layer	Future use in Dystrophin Positivity Algorithm
Stored binary layers: "Accepted"	

(b) Images should be captured based on best practices for quantitative microscopy and saved and exported in ways that retain original bit depth and original distinct channels (*see* **Note 1** for details).

(c) The method has been deployed successfully on images with 0.65μm/pixel and higher image resolutions.

(d) Both confocal and widefield images are suitable for analysis. However, since confocal microscopes exclude out-of-focus planes, it is critical to ensure that the entire field/sample remains optimally in-focus during confocal acquisition to ensure that analysis results will be valid.

3 Methods for Building Algorithms

The tables in this section lay out the steps for reconstructing the GA3 algorithms that were utilized in our previous publications for dystrophin quantification [4–9].

3.1 General Recommendations

1. Depending on tissue and image characteristics (signal-to-noise ratio, magnification, artifacts, etc.), some steps from the original algorithms may be unnecessary, or additional steps may be required to optimize segmentation. The operator should strive to recognize these situations and adjust the algorithms accordingly using the large variety of analysis tools available in GA3. Steps that we use but don't consider critical to the algorithm are shown in italics—these steps may be omitted or replaced with others as needed.
2. The ability to rename nodes and their outputs can be helpful for staying organized while building GA3 algorithms. Tables 1, 2, and 3 include suggested names for binary layers and nodes in

Table 2
Dystrophin positivity analysis

	Operation	Purpose
2.1	*Rolling ball* correction on dystrophin (30 μm) and sarcolemmal marker (10 μm) channels	Remove background, but more conservatively for dystrophin channel
2.2	Roughly segment sarcolemmal and sarcoplasmic tissue regions as follows:	To enable the measurement of tissue background intensity only in sarcoplasmic regions
	2.2.1: Measure the 30th *quantile* intensity within "Accepted" layer for both channels 2.2.2: *Calculated Column*: [Quantile × 3] for both channels 2.2.3: *Simple Threshold*, using the calculated result as the dynamic parameter for the low threshold, on both channels 2.2.4: Take union ("or") of both channel binaries, *clean* and *dilate* the resulting sarcolemmal binary 2.2.5: *Invert* the sarcolemmal binary, making a new sarcoplasm binary layer, and *filter objects* in the resulting binary by size and shape (circularity and roughness) for likely muscle fibers	
2.3	Automatically threshold both channels based on tissue background intensities	Use an internally controlled, automated, unbiased threshold to segment sarcolemma and dystrophin
	2.3.1: Measure the 99th *quantile* and *standard deviation* of the intensity within the sarcoplasm binary 2.3.2: *Append* the measurements 2.3.3: *Calculate* a formula that uses both the quantile and standard deviation with certain multipliers to arrive at an appropriate threshold for each channel (e.g., [Quantile × 2+StDev × 1.5]) 2.3.4: *Simple Threshold*, using the calculated result as the dynamic parameter for the low threshold, on both channels to make "Dystrophin" and "Laminin"/"Spectrin" binaries	
2.4	*Clean "Dystrophin" and "Laminin"/"Spectrin"*	*Remove insignificant specks from binaries*
2.5	Process "Laminin"/"Spectrin" to refine selection of muscle fibers	Create a "Fibers" binary layer for measuring fiber size and sarcoplasmic signal intensities
	2.5.1: *Close "Laminin"/"Spectrin" binary and filter small objects* 2.5.2: *Invert* the "Laminin"/"Spectrin" layer, and make it into a new layer named "Fibers" 2.5.3: *Close holes and separate "Fibers" binary objects* 2.5.4: *Filter objects* in "Fibers" by size and shape	
2.6	Measure *Eq Diameter*, *min Feret's diameter*, and median intensity (*quantile*) of "Fibers" objects	Measure size (equivalent diameter and minimum Feret's diameters) and sarcoplasmic dystrophin channel intensity for each fiber
2.7	Make *inside and ring* from fiber binary objects	Make 4 μm "Ring" at the sarcolemma of each fiber
2.8	Take intersection ("*and*") of "Dystrophin" and "Laminin"/"Spectrin" binaries with the "Ring"	Attribute dystrophin and sarcolemmal marker signals to individual fibers
2.9	Use object parenting (*Aggregate Children*) to measure the combined lengths of "Dystrophin" and sarcolemmal binaries for each "Ring"	Measure the total sarcolemmal perimeter length and the dystrophin-positive perimeter length of each fiber

(continued)

Table 2
(continued)

Operation	Purpose
2.10 *Append columns* and *calculate* ratio (e.g., [SumDysLength/SumSarcolemmaLength × 100])	Determine the % dystrophin-positive perimeter of each fiber and complete "All Perimeters" table
2.11 Make new layer from "Ring" using Color by Value and recolor using % the dys-positive perimeter values	Create a "Color Ring" binary layer to display certain colors for any % positive perimeter >0
2.12 Measure *mean object intensity* of the dystrophin channel for each "Ring"	Measure mean sarcolemmal dystrophin signal for each fiber
2.13 *Append Columns* and *Calculate* the mean "Ring" dystrophin signal minus the median "Fibers" dystrophin signal	Subtract sarcoplasmic background fluorescence to determine the corrected dystrophin intensity per fiber
2.14 Append fiber intensities and size measurements	Make the "Fiber Size and Intensity" table
2.15 Make *Frequency Table* from "All Perimeters" and include a bin for each decile (e.g., 10–19.9%)	Create the "Frequency Table" to be used for PDPF calculations
2.16 *Count* the total fibers and *append* to "Frequency Table"	Includes the total fiber count for PDPF calculations
2.17 If desired, use *Export CSV* to export tables	Automatically export CSVs of tables to a directory
Stored binary layers: "Dystrophin," "Laminin"/"Spectrin," "Fibers," "Ring," "Color Ring"	
Stored tables: "All Perimeters," "Fiber Size and Intensity," "Frequency Table"	

specific instances, but operators should consider customizing node and binary layer names as needed beyond these suggestions.

3. Pay attention to ensuring that the original version of each channel, the necessary binary layers, and necessary data tables are stored or exported in each algorithm.

3.2 Accepted Region Algorithm

This algorithm creates a binary layer to automatically select the tissue section area for analysis. The purpose of first defining the analyzable area in this fashion is to allow the operator to exclude regions from this analyzable area without erasing those features from the image (Fig. 1). The algorithm should include steps shown in Table 1.

3.3 Dystrophin Positivity Algorithm

This algorithm creates several binary layers to annotate muscle fibers, their perimeters ("Ring"), and the dystrophin-positive and laminin/spectrin-positive segments around each perimeter (Fig. 2).

Table 3
Dystrophin and Spectrin/Laminin Intensity Algorithm

	Operation	Purpose
3.1	Rolling ball correction on dystrophin (30μm) and sarcolemmal marker (10μm) channels	Remove background, but more conservatively for dystrophin channel
3.2	From sources, pull in single binary nodes for "Fibers" and "Ring"	Reuse muscle fiber perimeter and sarcoplasm binary layers
3.3	Measure mean object intensity of the dystrophin channel for each "Ring"	Get mean sarcolemmal dystrophin signal for each fiber
3.4	Measure the median dystrophin intensity (quantile) of "Fibers"	Measure sarcoplasmic dystrophin intensity of each fiber
3.5	Append columns and calculate the mean "Ring" dystrophin signal minus the median "Fibers" dystrophin signal	Subtract sarcoplasmic background fluorescence to determine the corrected dystrophin intensity per fiber, and create the "Norm Intensity" table
3.6	Calculate the normalized dystrophin intensity by dividing the corrected dystrophin intensity by the mean of median normal control intensities (*100)	Create the "Norm_Dys" column for the "Norm Intensity" table
3.7	Rename the calculated node in Subheading 3, **step 6** to "normalization"	Clearly label the node where the normalization denominator will be entered during routine use
3.8	Use Color by Value on the "Ring" layer and recolor based on Norm_Dys values	Create a heatmap from the "Ring" binary layer to display certain colors for dystrophin intensities
3.9	If desired, use export CSV to export tables	Automatically export CSVs of tables to a directory
Stored binary layers: "Ring"		
Stored tables: "Norm Intensity"		

A colorful heatmap layer that assigns a custom color to each fiber perimeter based on its % dystrophin-positive perimeter is also created ("Color Ring"). Due to quirks in the functionality of the nodes that calculate % dystrophin-positive perimeter and assign the colors, "Color Ring" cannot capture fibers that have 0% dystrophin. Therefore, "Color Ring" must be viewed together with the "Ring" binary layer, and "Ring" should be colored appropriately corresponding to the color scale chosen for "Color Ring." This algorithm also provides the frequency table for calculating the % dystrophin-positive fibers (PDPF) and measures the mean perimeter dystrophin intensity corrected using the median sarcoplasmic dystrophin intensity for each fiber along with the fiber sizes. The five binary layers that are required to be stored from this algorithm are for auditing the results and for future use in the Normalized Dystrophin Intensity Algorithm. The algorithm should be comprised of the steps shown in Table 2.

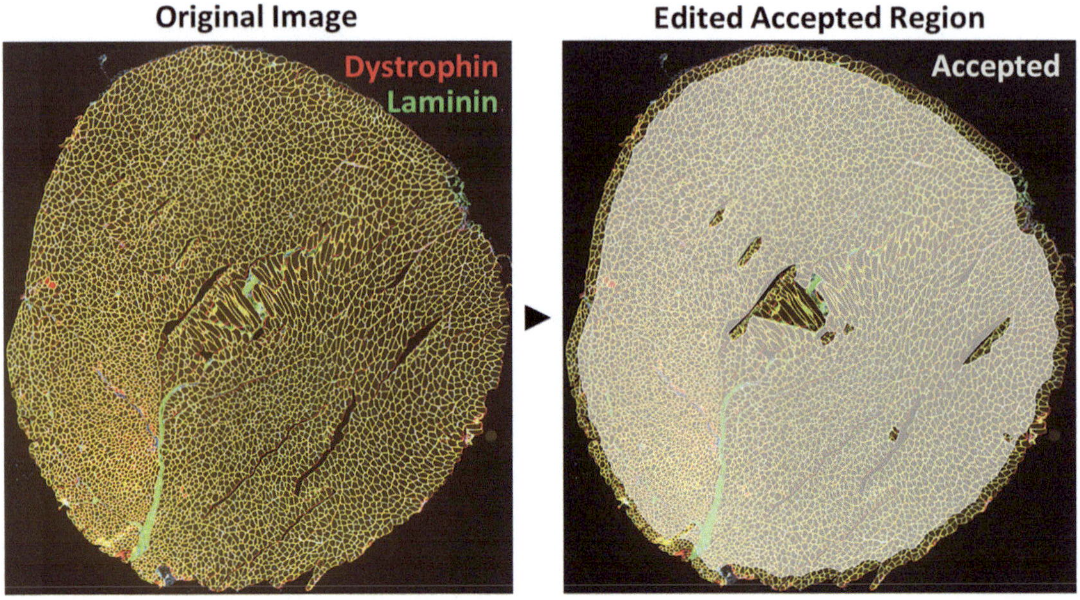

Fig. 1 Original image and corresponding accepted binary layer. *Left:* an example original input image for the Accepted Region algorithm. *Right:* corresponding example "Accepted" binary layer (white) after reasonable manual edits to exclude tissue artifacts from quantification

3.4 Normalized Dystrophin Intensity Algorithm

This algorithm pulls in the "Fibers" and "Ring" binary layers from the Dystrophin Positivity Algorithm, repeats the measurements and calculation for the corrected dystrophin intensity per fiber, and then normalized this corrected dystrophin intensity to that of normal controls within the same batch (Fig. 3). This results in the dystrophin intensity values being converted from fluorescence units to those reflecting the % of mean normal control intensities. The algorithm also recolors the "Ring" layer to create a new heatmap, this time depicting the normalized % dystrophin intensity for each fiber (Fig. 4). The algorithm should be comprised of the steps shown in Table 3. *See* **Note 2** for additional details related to controls.

4 Methods for Importing and Using the Algorithms

All steps take place in NIS-Elements unless otherwise specified and are described at the level of detail suitable for an operator with basic proficiency navigating the software. Basic proficiency can be developed through self-guided practice or through the use of official vendor support resources for NIS-Elements.

The three algorithms work in series, with the output from each algorithm serving as the input for the next one. Between

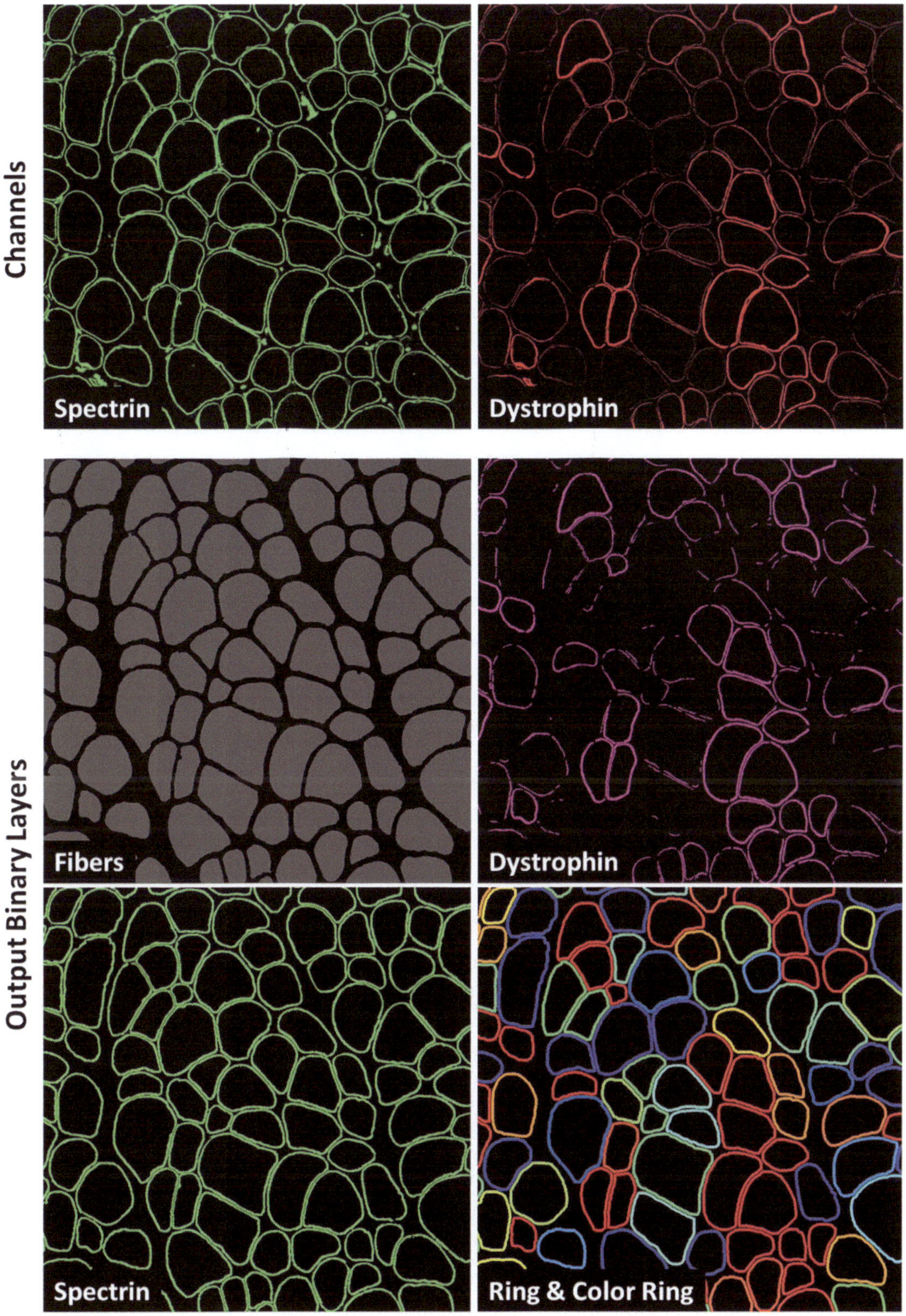

Fig. 2 Dystrophin positivity algorithm outputs. The original spectrin and dystrophin channels are shown along with the exported binary layers derived from these channels during the Dystrophin Positivity Analysis

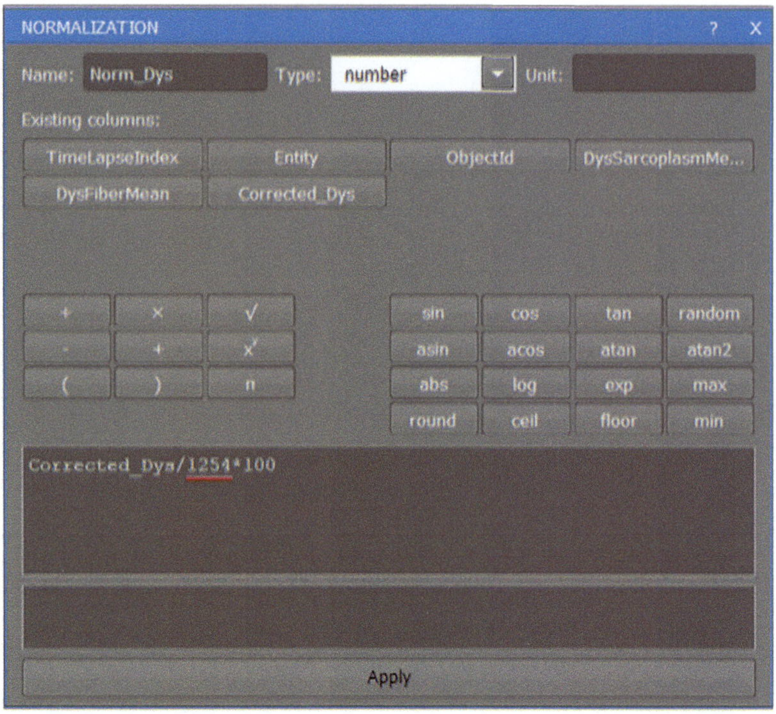

Fig. 3 Dystrophin intensity normalization node calculation. The intensity normalization calculation format should appear as shown. The red underline marks the value that should be calculated from the median normal control intensities in **step 7** of Subheading 4.4

algorithms, the operator has the opportunity to exclude artifacts from analysis and to calculate intensity normalization values based on a set of controls, but each algorithm runs fully autonomously once launched.

4.1 Importing Algorithms

1. Open Analysis Explorer (AE).
2. Create a new analysis by importing from a file.
3. Select all three algorithms for import.

4.2 Accepted Region Selection

1. Open an example image.
2. From AE, open the Accepted Region Analysis in Edit view.
3. Confirm that all algorithm steps appear downstream of the sarcolemmal marker channel. If the analysis steps fall downstream of the dystrophin channel or another marker, move the connector of the top node to the sarcolemmal marker channel instead, and save the change.
4. Open the Batch GA3 window, and queue up all images with the same channel combination to be processed using the Accepted Region Analysis.

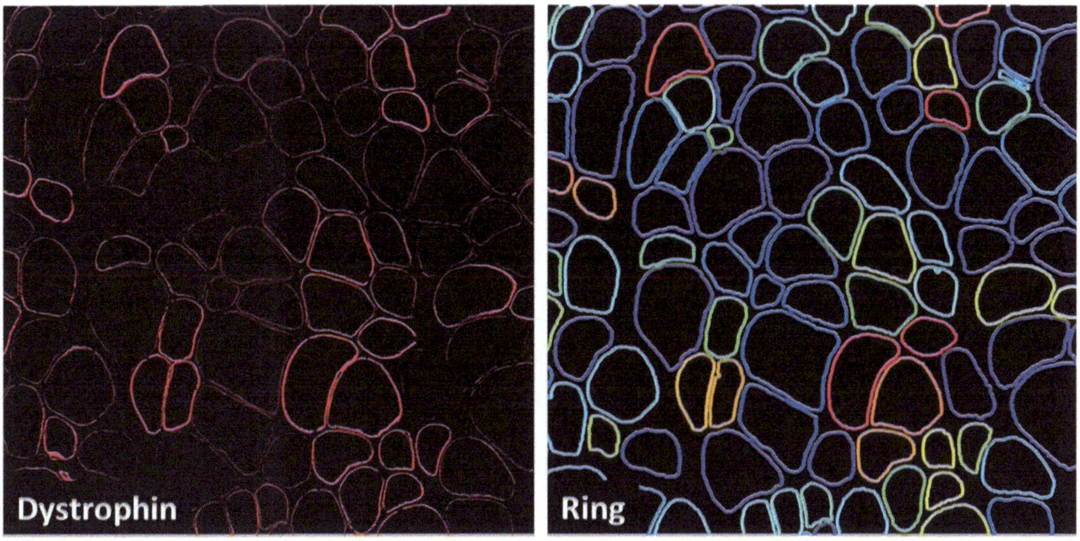

Fig. 4 Normalized dystrophin intensity algorithm outputs. The original dystrophin channel is shown along with the exported color-coded "Ring" binary layer derived from the dystrophin channel intensity and imported "Ring" layer

5. Ensure that original images will be kept, and launch the batch process. The process performs steps outlined in Table 1.
6. Once the batch process has finished, rename the folder containing the analyzed images, and ensure that all files appear in the folder. If the folder is left with its default name, its contents cannot be used as input in subsequent batch processes.
7. Open each processed image, and then launch the Binary Editor. Use the Binary Editor drawing tools to modify the Accepted binary layer to exclude artifacts that appear within the selected region. Artifacts include (but are not limited to): tissue folds and tears, mounting media bubbles, excessive handling or drying artifact, PAP pen marks, secondary antibody aggregates, stray clothing fibers, or any other extraneous features that may skew analysis (Fig. 1). Once the Accepted region is free of artifacts, leave the Binary Editor and save the changes to the image.

4.3 Dystrophin Positivity Analysis

1. Open an example image with the Accepted binary layer.
2. From AE, open the Dystrophin Positivity Analysis in Edit view.
3. Confirm that algorithm steps appear downstream of their correct respective channel based on labels in Edit view. If the analysis paths fall downstream of the wrong channels, switch all connectors from the wrong to the correct channels for each analysis path and save the change. It may be helpful to pull in a new *Channels* node to make the changes more cleanly.

4. Open the Batch GA3 window and queue up all images with the Accepted binary layer to be processed using the Dystrophin Positivity Analysis.
5. Ensure that original images will be kept, and launch the batch process. The process performs steps outlined in Table 2 and generates output images as shown in Fig. 2. If the process has failed, refer to **Note 3**.
6. Once the batch process has finished, rename the folder containing the analyzed images, and ensure that all files appear in the folder. If the folder is left with its default name, its contents cannot be used as input in subsequent batch processes.
7. The algorithm either stores tables in NIS-Elements or exports CSV files to a file directory. Confirm that each image is represented in NIS-Elements or in the file directory by these three tables:
   - "Frequency Table" contains fiber counts for each decile of dystrophin-positive perimeter (e.g., 30–39.9% positive, etc.) and the total fiber count. Allows for the calculation of fiber counts having dystrophin positivity above any decile threshold.
   - "All Perimeters" contains the source data for the frequency table. Enumerates every muscle fiber with its perimeter length based on the sarcolemmal marker, the sum of the length of its dystrophin-positive segments, and the ratio of dystrophin segment length to sarcolemmal length.
   - "Fiber Size and Intensity" enumerates every muscle fiber with its size (minimum Feret's diameter and equivalent diameter) and the dystrophin signal intensity at the sarcolemma and in the sarcoplasm.
8. To calculate the percentage of dystrophin-positive fibers (PDPF), determine how much dystrophin-positive perimeter should be required to consider a fiber positive overall. Typical positivity criteria have been 30% or higher for human biopsies and 50% or higher for mouse samples. Then add the total number of fibers having that perimeter positivity or higher in the Frequency Table, and divide that sum by the total number of fibers.

4.4 Normalized Dystrophin Intensity Analysis

1. Open the Fiber Size and Intensity table for each normal control.
2. For each normal control, calculate the *median* of the column showing the corrected dystrophin intensity (titled "Corrected_Dys" in our algorithms). This is the mean sarcolemmal dystrophin intensity minus the median sarcoplasmic dystrophin intensity in fluorescence units for each fiber.

3. Transfer the median corrected dystrophin intensity for each normal control into a separate table, and calculate the *mean* of these medians. This value will be used in **step** 7.
4. Open an example image with the stored binary layers from the Dystrophin Positivity Analysis.
5. From AE, open the Dystrophin Positivity Analysis in Edit view.
6. Confirm that algorithm steps appear downstream of their respective correct channel based on labels in Edit view. If the analysis paths fall downstream of the wrong channels, switch all connectors from the wrong to the correct channels for each analysis path and save the change. It may be helpful to pull in a new *Channels* node to make the changes more cleanly.
7. Open the parameters of the analysis node titled "Normalization," and enter the calculated mean of the normal control medians in the denominator position in the calculation (Fig. 3). Save the change.
8. Open the Batch GA3 window, and queue up all images with the Dystrophin Positivity Analysis binary layers to be processed using the Normalized Dystrophin Intensity Analysis.
9. Ensure that original images will be kept, and launch the batch process. The process performs steps outlined in Table 3 and generates output images as shown in Fig. 4.
10. Once the batch process has finished, ensure that all files appear in the folder containing the analysis output images, and rename the folder with an appropriate name. If the process has failed, refer to **Note 3**.
11. The algorithm exports CSV files to C:\Users\Public by default. Confirm that each image is represented in this location by an additional CSV file that ends with "Norm Intensity."

4.5 Depicting and Interpreting Analysis Results

1. The dystrophin positivity results should be shown as a single PDPF value per sample, and intensity results should be converted to a single mean value per sample in order to perform statistical comparisons between groups (Fig. 5c). Since each fiber is not an independent observation in the fiber-level data, only descriptive statistics (and not statistical hypothesis tests) can be performed on fiber-level data.
2. The perimeter positivities or normalized intensities for each muscle fiber in a sample or group can be presented as scatter plots, with the large number of fibers often creating the appearance of a violin plot (Fig. 5a, b). Cumulative or regular frequency histograms may also be used for displaying the fiber positivity or intensity distributions of each sample and can be created by converting the fiber-level data into histogram format. This conversion can be performed automatically in applications like GraphPad Prism.

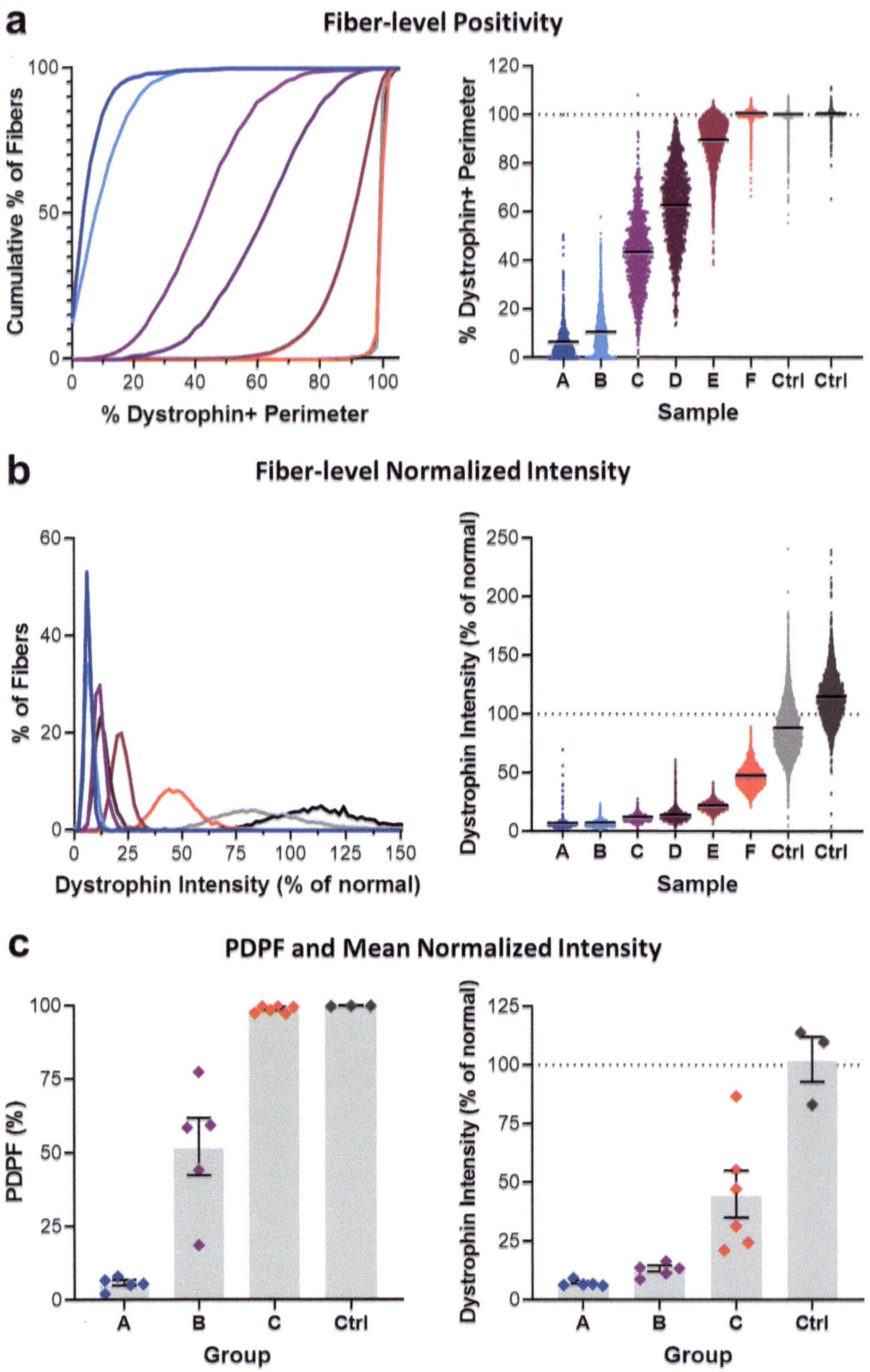

Fig. 5 Example graphs for displaying fiber-level and sample-level quantification results. Fiber-level positivity data **(a)** can be shown as cumulative histograms (*left*) or scatter dot plots (*right*) to display the fiber positivity on a continuous scale. Similarly, fiber-level intensity data **(b)** can also be shown as a histogram (*left*) or scatter dot plot (*right*) to show the range of varying intensities between fibers. For statistical comparisons and summary data, the dystrophin positivity should be reduced to a single PDPF value per sample (*left*), and the intensity data should be reduced to a single mean intensity value per sample (*right*)

3. The complete picture of dystrophin expression by immunofluorescence requires both the positivity and normalized intensity results. Although the key outcomes of studies seeking to restore dystrophin expression have been often reported on the basis of PDPF, fibers may be strongly positive or weakly positive, and this distinction is not reflected in PDPF data. For example, most Becker muscular dystrophy samples have almost all positive fibers, but often with around 50% or lower normalized intensity.

5 Method Modifications, Extensions, and Alternatives

5.1 Modifications

(a) Analysis in genetically homogeneous animal models can be performed as described, or it can be modified to utilize untreated disease control samples to calculate a higher dystrophin threshold based on nonzero basal dystrophin expression. For example, many mouse models display basal dystrophin expression at a very low level that can be detected easily by immunofluorescence in the majority of fibers, and it may be desirable to more selectively detect increases above this level for quantification. To perform this modification, an additional threshold calculation algorithm can be created (or requested from us) and deployed only on untreated dystrophin-deficient control samples to calculate a threshold that exceeds most of the low-level basal dystrophin expression in the samples. Once the threshold algorithm reports a suitable threshold for each negative control sample, the average can be calculated. An alternate simpler version of the Dystrophin Positivity Algorithm that allows for thresholds to be manually entered rather than calculated can be created (or requested from us) for this workflow, and the operator will enter the average threshold from the negative controls before saving the change and deploying the algorithm.

(b) Some antibodies provide much higher signal-to-background ratios that others, and different tissue qualities or staining methodologies may also cause the background to be higher or the specific staining less distinct. The same calculations for the dystrophin and sarcolemmal marker thresholds likely will not be appropriate for all studies and all antibodies. This threshold calculation can be relatively easily modified as needed. A control tissue of representative quality with fibers showing little to no dystrophin expression will be needed to make the calculation adjustment to the desirable level of sensitivity.

(c) Any time these algorithms are applied to a new magnification, antibody combination, or tissue type (e.g., fixed vs. fresh frozen, human vs. mouse, limb muscle vs. diaphragm, widefield vs. confocal), they will likely require minor adjustments to re-optimize fiber segmentation and marker detection.

(d) Throughout the process, some summary values are calculated as means and some as medians. These choices were made consciously based on which values should be sensitive to outliers and which ones should not. However, if preferable, only means could be calculated for all summary values to simplify the process and the reporting of methods and results.

(e) The third algorithm that provides normalized dystrophin intensity tables and heatmaps is optional if a normalized intensity heatmap isn't needed or if normal controls are not available. If normal controls are included but the heatmap is not desired, normalized dystrophin intensities can be calculated in Microsoft Excel or similar software instead of running the normalized intensity algorithm.

(f) This approach can be used successfully with images showing regions of interest (ROIs) instead of whole sections, although some algorithm modifications will be required. However, caution should be used when selecting ROIs to ensure that they do not misrepresent the dystrophin expression throughout the rest of the sample.

5.2 Extensions

(a) The same algorithms will be suitable with slight modifications for quantifying the expression of other markers localized at sarcolemma besides dystrophin, like the sarcoglycans, dystroglycans, neuronal nitric oxide synthase (nNOS), and others. For example, a similar approach has been used for quantifying Wisteria floribunda agglutinin (WFA)-FITC staining in the rAAVrh74.MCK.GALGT2 clinical trial and α-dystroglycan staining in ongoing preclinical mouse studies [3]. Various DAPC labels may also be added as a fourth marker and quantified with respect to dystrophin restoration, although this would require significant modification of the algorithms.

(b) This approach can also be used for quantifying dystrophin expression in cardiac sections. However, cardiomyocyte orientations change throughout the heart, and this myocyte segmentation strategy only works on areas where cardiomyocytes are cross-sectionally oriented. In mouse hearts, we sample multiple square ROIs from each heart section where all myocytes are cross-sectional, and pool analysis results from these ROIs to represent the whole sample [4, 5, 7]. Hearts often need to be imaged at higher magnifications for this process (we prefer 0.3μm/pixel), and myocyte segmentation characteristics should be adjusted to include shapes that are smaller and more irregular than skeletal muscle fibers. If the heart section in question does not have enough cross-sectional myocytes to be reliably analyzed, the alternative approach in Subheading 5, **step 5.1 (f)** should be used instead.

(c) If a DAPI stain is included for nuclear localization, the output images and binary layers from the second algorithm for dystrophin positivity may be used as input for an additional algorithm that quantifies internalized nucleation (also available on request from us) [4, 5].

5.3 Alternatives to the Described Process

(a) NIS-Elements was selected as the platform for this quantification approach due to having a remarkably powerful and flexible analysis module and user-friendly interface for a moderate price. However, a variety of other software platforms can also be used for analyzing dystrophin expression in multichannel muscle images. These platforms have a broad range of costs, learning curves, and capabilities. Some of these platforms are associated with publications that detail an analysis pathway designed for dystrophin quantification [10–14], and some are proprietary and available as a service [15]. Although these alternatives are outside the scope of this chapter, the more flexible software platforms may support the implementation of workflows similar to the process described here, since few (if any) of the analysis steps utilized in this method are unique to NIS-Elements. Similarly, many of the processes for dystrophin quantification published using other platforms can be recapitulated using NIS-Elements or some other alternative platforms.

(b) If the brightness of the sarcolemmal marker signal or the structure or quality of the tissue are not compatible with successful segmentation of individual muscle fibers, it's possible to instead quantify dystrophin positivity and intensity under the whole-tissue sarcolemmal mask. This process is derived from an earlier published method and could be readily performed on multiple available software platforms [8, 14]. It would entail (1) thresholding for and minimally processing the laminin- or spectrin-positive pixel mask, (2) thresholding for the dystrophin-positive pixels, and (3) measuring the mean dystrophin channel intensity under the sarcolemmal mask and the % overlap between the sarcolemmal and dystrophin masks. For this type of analysis, even normal tissue is unlikely to have 100% overlap, so the mean % overlap in normal tissue should be used to normalize the % overlap in the other groups.

6 Notes

1. *Suitable Image Characteristics:*
   (a) Multichannel image files should never be saved or exported for analysis in JPEG or any other RGB format. A single RGB image that shows signal from multiple channels lacks the original information that was distinct to each channel, and thus cannot be reliably quantified. Some proprietary microscopy software packages make exporting full bit-depth images with distinct original channels much less accessible than exporting RGBs. The software support team for the given software package should be asked for the correct workflow for exporting images without flattening all channels into RGB format or compressing bit depth.
   (b) Image bit depth should be as high as possible, since it determines the number of different intensity levels that can be present in the image. The algorithms were developed using 16-bit images that may have up to 65,536 different intensity values. An 8-bit image can only have up to 256 different intensity values per channel, and the algorithm behavior may not scale correctly for images that have much fewer and coarser intensity values. Images captured in 12-bit or 14-bit format should be suitable.
   (c) Whenever possible, images should be captured for analysis using a monochrome camera designed primarily for fluorescence imaging. Color cameras typically have multiple-fold lower sensitivity for fluorescence than quality monochrome cameras, and some color cameras only output 8-bit RGB images.
   (d) The camera exposure should be adjusted separately for each channel using a positive control sample to ensure that the camera dynamic range is used appropriately without saturating or underexposing pixels. Suitable settings should provide a large number of gray values for quantification (at least a few thousand, ideally more in 16-bit format) without significant risk of saturation even if a slightly brighter sample happens to be in the batch. Once selected at the beginning, the same settings should be used for every sample in the batch. It's not possible to quantitatively compensate for differences in imaging settings when comparing signal intensities.
   (e) Reasonable quality control measures, like confirming that whole fields of view are in focus and that adjacent fields are stitched together correctly without an offset or artifacts, are critical at the beginning and throughout the process. If

encountered, these issues should be resolved before continuing. Out-of-focus regions, misaligned fields, and severe windowpane artifacts will interfere with analysis and skew results.

2. *Appropriate Controls:*
   (a) Normal and negative (untreated or secondary-only) controls are not required for analysis of dystrophin positivity, since the automated thresholds are based on the sarcoplasmic signal intensity within each sample. Similarly, fiber dystrophin intensity can be calculated in unnormalized fluorescence units without using controls. However, normalized intensity quantification requires at least one suitable normal control as a reference for normalization.
   (b) Due to documented differences in intensities between normal control samples, including more normal controls will provide a more accurate estimation of normalized dystrophin intensities [8, 16].
   (c) Noticeable variations in staining should be expected between batches, and signal fading can skew results over time. Normal controls must be stained within the same batch as the samples of interest and imaged with identical settings and within the same timeframe. If these requirements are not met, there should be no expectation that the signal in the normal control images will serve as a suitable point of comparison for normalized intensity, since differences in staining or fading may confound the results.
   (d) Secondary-only staining and isotype controls are useful for documentation, but they should not be analyzed using these algorithms. The algorithms are developed to work on fully stained tissue with intact laminin or spectrin signal and will not perform accurately or predictably on images with only negative or nonspecific signal.
   (e) Markers like laminin and spectrin should not be used for normalization of dystrophin intensities without a sound rationale. The expression of these markers may change in the context of dystrophinopathies, and normalization of dystrophin against them may skew results inappropriately [8].
3. *Troubleshooting Failures:*
   (a) If attempts to perform a batch analysis return an error message or failed status, check the following:
      (i) Confirm that the algorithm does not show errors in red text in Edit view. Common algorithm errors include CSV file paths that are not accessible to the

current PC user, *Calculated* nodes where the calculation is missing a term, and backwards compatibility issues for algorithms created in versions of NIS-Elements that are newer than the version being used.

(ii) Confirm that the input and output file paths for the images and CSVs (including the image name and extension) are shorter than 256 characters. If these paths exceed 256 characters, programs may not be able to open or save the files successfully. If automatically appended analysis names cause the image file names to become too long, batch renaming can be performed using the "Rename-Item" command in Windows PowerShell to truncate or remove them.

(iii) Confirm that none of the CSV files that the algorithm may be trying to write to are currently open.

(b) If analysis results seem unreasonable or the tables only list blank or 0 values, check the following:

(i) Confirm that the algorithm Edit view shows the correct channels connected to the correct steps. Jumbled channels are the most common error by a large margin when using these algorithms.

(ii) Confirm that the batch process used images with the necessary binary layers as input, and that the binary inputs do not appear as "missing" in Edit view.

References

1. Bonilla E, Samitt CE, Miranda AF, Hays AP, Salviati G, DiMauro S, Kunkel LM, Hoffman EP, Rowland LP (1988) Duchenne muscular dystrophy: deficiency of dystrophin at the muscle cell surface. Cell 54(4):447–452. https://doi.org/10.1016/0092-8674(88)90065-7
2. Zubrzycka-Gaarn EE, Bulman DE, Karpati G, Burghes AH, Belfall B, Klamut HJ, Talbot J, Hodges RS, Ray PN, Worton RG (1988) The Duchenne muscular dystrophy gene product is localized in sarcolemma of human skeletal muscle. Nature 333(6172):466–469. https://doi.org/10.1038/333466a0
3. Flanigan KM, Vetter TA, Simmons TR, Iammarino M, Frair EC, Rinaldi F, Chicoine LG, Harris J, Cheatham JP, Cheatham SL, Boe B, Waldrop MA, Zygmunt DA, Packer D, Martin PT (2022) A first-in-human phase I/IIa gene transfer clinical trial for Duchenne muscular dystrophy using rAAVrh74.MCK.GALGT2. Mol Ther Methods Clin Dev 27:47–60. https://doi.org/10.1016/j.omtm.2022.08.009
4. Gushchina LV, Bradley AJ, Vetter TA, Lay JW, Rohan NL, Frair EC, Wein N, Flanigan KM (2023) Persistence of exon 2 skipping and dystrophin expression at 18 months after U7snRNA-mediated therapy in the Dup2 mouse model. Mol Ther Methods Clin Dev 31:101144. https://doi.org/10.1016/j.omtm.2023.101144
5. Gushchina LV, Vetter TA, Frair EC, Bradley AJ, Grounds KM, Lay JW, Huang N, Suhaiba A, Schnell FJ, Hanson G, Simmons TR, Wein N, Flanigan KM (2022) Systemic PPMO-mediated dystrophin expression in the Dup2 mouse model of Duchenne muscular dystrophy. Mol Ther Nucleic Acids 30:479–492. https://doi.org/10.1016/j.omtn.2022.10.025
6. Simmons TR, Vetter TA, Huang N, Vulin-Chaffiol A, Wein N, Flanigan KM (2021) Pre-clinical dose-escalation studies establish a

therapeutic range for U7snRNA-mediated DMD exon 2 skipping. Mol Ther Methods Clin Dev 21:325–340. https://doi.org/10.1016/j.omtm.2021.03.014

7. Stephenson AA, Nicolau S, Vetter TA, Dufresne GP, Frair EC, Sarff JE, Wheeler GL, Kelly BJ, White P, Flanigan KM (2023) CRISPR-Cas9 homology-independent targeted integration of exons 1-19 restores full-length dystrophin in mice. Mol Ther Methods Clin Dev 30:486–499. https://doi.org/10.1016/j.omtm.2023.08.009
8. Vetter TA, Nicolau S, Bradley AJ, Frair EC, Flanigan KM (2022) Automated immunofluorescence analysis for sensitive and precise dystrophin quantification in muscle biopsies. Neuropathol Appl Neurobiol 48(3):e12785. https://doi.org/10.1111/nan.12785
9. Wein N, Vetter TA, Vulin A, Simmons TR, Frair EC, Bradley AJ, Gushchina LV, Almeida CF, Huang N, Lesman D, Rajakumar D, Weiss RB, Flanigan KM (2022) Systemic delivery of an AAV9 exon-skipping vector significantly improves or prevents features of Duchenne muscular dystrophy in the Dup2 mouse. Mol Ther Methods Clin Dev 26:279–293. https://doi.org/10.1016/j.omtm.2022.07.005
10. Arechavala-Gomeza V, Kinali M, Feng L, Brown SC, Sewry C, Morgan JE, Muntoni F (2010) Immunohistological intensity measurements as a tool to assess sarcolemma-associated protein expression. Neuropathol Appl Neurobiol 36:265–274. https://doi.org/10.1111/j.1365-2990.2009.01056.x
11. Beekman C, Sipkens JA, Testerink J, Giannakopoulos S, Kreuger D, Van Deutekom JC, Campion GV, De Kimpe SJ, Lourbakos A (2014) A sensitive, reproducible and objective immunofluorescence analysis method of dystrophin in individual fibers in samples from patients with duchenne muscular dystrophy. PLoS One 9. https://doi.org/10.1371/journal.pone.0107494
12. Sardone V, Ellis M, Torelli S, Feng L, Chambers D, Eastwood D, Sewry C, Phadke R, Morgan JE, Muntoni F (2018) A novel high-throughput immunofluorescence analysis method for quantifying dystrophin intensity in entire transverse sections of Duchenne muscular dystrophy muscle biopsy samples. PLoS One 13:1–21. https://doi.org/10.1371/journal.pone.0194540
13. Scaglioni D, Ellis M, Catapano F, Torelli S, Chambers D, Feng L, Sewry C, Morgan J, Muntoni F, Phadke R (2020) A high-throughput digital script for multiplexed immunofluorescent analysis and quantification of sarcolemmal and sarcomeric proteins in muscular dystrophies. Acta Neuropathol Commun 8:1–16. https://doi.org/10.1186/s40478-020-00918-5
14. Taylor LE, Kaminoh YJ, Rodesch CK, Flanigan KM (2012) Quantification of dystrophin immunofluorescence in dystrophinopathy muscle specimens. Neuropathol Appl Neurobiol 38:591–601. https://doi.org/10.1111/j.1365-2990.2012.01250.x
15. Aeffner F, Faelan C, Moore SA, Moody A, Black JC, Charleston JS, Frank DE, Dworzak J, Kris Piper J, Ranjitkar M, Wilson K, Kanaly S, Rudmann DG, Lange H, David Young G, Milici AJ (2019) Validation of a muscle-specific tissue image analysis tool for quantitative assessment of dystrophin staining in frozen muscle biopsies. Arch Pathol Lab Med. https://doi.org/10.5858/arpa.2017-0536-OA
16. Aartsma-Rus A, Morgan J, Lonkar P, Neubert H, Owens J, Binks M, Montolio M, Phadke R, Datson N, Van Deutekom J, Morris GE, Rao VA, Hoffman EP, Muntoni F, Arechavala-Gomeza V (2019) Report of a TREAT-NMD/world duchenne organisation meeting on dystrophin quantification methodology. J Neuromuscul Dis 6:147–159. https://doi.org/10.3233/JND-180357

Chapter 6

Quantitative Evaluation of Dystrophin Expression Using SDS-PAGE Western Blot Methods

Margaret J. Beatka, Mariah J. Prom, Margaret Haberman, Jessica Sutton, J. Patrick Gonzalez, and Michael W. Lawlor

Abstract

SDS-PAGE Western blots allow for the identification of the presence and relative quantity of proteins of interest on the basis of their molecular weight and their recognition by specific antibodies. This technique has historically been useful in the diagnosis and study of Duchenne muscular dystrophy (DMD) and Becker muscular dystrophy (BMD), as it allows the recognition of clinically relevant abnormalities in the amount or size of dystrophin. While this technique is used infrequently for diagnosis at this point (as genetic testing for pathogenic variants of the *DMD* gene has become standard practice), it remains an important tool in basic science research and in the evaluation of therapies that restore or replace dystrophin expression. This chapter provides a technique that is suitable for the quantitative assessment of dystrophin expression in isolates of skeletal muscle tissue, based on our experience and guidance provided by the US Food and Drug Administration (FDA) for these types of assays. Topics associated with the design of quantitative Western blot techniques, options for loading control use, and the challenges associated with variable pathology within specimens are also discussed. Performing this technique can provide reproducible and reliable quantification of full-length or shortened dystrophin protein content in frozen muscle tissue.

Key words Dystrophinopathy, Microdystrophin, Molecular weight, Protein expression, Quantification, Protein levels, Duchenne muscular dystrophy, Becker muscular dystrophy, DMD, BMD, Electrophoresis, Antibodies

1 Introduction

Western blot using SDS-PAGE is a well-established method to assess protein content in tissue isolates, and it has been historically useful in the diagnosis of Duchenne muscular dystrophy (DMD), Becker muscular dystrophy (BMD), and carriers of dystrophin mutations [1, 2]. The technique involves the isolation of proteins from a tissue sample, followed by separation of denatured proteins according to molecular weight on a polyacrylamide gel. The proteins are then transferred to a membrane and exposed to antibodies to allow the detection of a protein of interest, and these antibodies

Kevin Flanigan and Stefan Nicolau (eds.), *Duchenne Muscular Dystrophy: Methods and Protocols*, Methods in Molecular Biology, vol. 2975, https://doi.org/10.1007/978-1-0716-4811-7_6,

can be visualized by one of several detection strategies. While the technique is technically complex, it is performed very broadly in the scientific community, and so SDS-PAGE Western blot data offers the advantage of being interpretable with a basic level of scientific expertise.

In the context of dystrophinopathy, SDS-PAGE Western blot has been used extensively in basic science research, translational research, and the diagnostic setting [3–7]. These studies often involve the direct detection of dystrophin protein, as most mutations in dystrophin result in a decrease in the amount of dystrophin protein. In addition, as a subset of dystrophin mutations result in shortened dystrophin proteins, SDS-PAGE Western blots may allow the identification of shortened dystrophin subtypes in some instances. The ability to differentiate full-length from shortened dystrophin has also been advantageous in studies that involve the development of shortened dystrophin (i.e., microdystrophin and mini-dystrophin, and exon-skipped dystrophin), although resolution may not be sufficient to differentiate dystrophin subtypes when the degree of shortening is minimal (as may be encountered with some exon skipping strategies). At present, SDS-PAGE is only minimally used in diagnostic work and is usually used to either understand whether potential dystrophinopathy patients with detectable dystrophin protein by immunostaining have decreased dystrophin levels or shortened dystrophin protein, or to establish the diagnosis of a dystrophinopathy in those patients with deep intronic mutations not detected by genomic DNA analysis [8]. Nevertheless, it remains an important research tool and has recently found new usefulness in the study and translation of therapeutic dystrophin restoration.

SDS-PAGE Western blot is currently one of several assays being used to provide quantification of dystrophin protein in research and therapeutic studies for DMD. Other techniques including capillary Western blot, mass spectrometry, and ELISA are also options in acquiring these data, but they are outside the scope of the current chapter. Some of these assays are considered more "quantitative" than SDS-PAGE Western blot, but the ability to use any of these techniques quantitatively depends on the experimental design and quality control parameters of the assay, rather than the assay itself. It is certainly true that these other techniques offer advantages in incorporating the necessary design and quality control parameters into their routine assay performance, and clinical trial studies in DMD are increasingly using mass spectrometry and capillary electrophoresis to assess dystrophin levels. However, an appropriately designed SDS-PAGE Western blot assay for dystrophin quantification would be equivalently quantitative if the assay showed appropriate performance on its quality control metrics. This chapter will describe a protocol that can be used for research and translational work, with additional information in the "notes" section to reflect our experience with performing quantitative dystrophin SDS-PAGE Western blots according to guidance provided by the Food and Drug Administration (FDA).

2 Materials

1. Protein extraction buffer (10 mL aliquot): Dilute 1 mL of 10× RIPA lysis buffer in 9 mL of ddH2O and add protease inhibitor tablets (1 tablet of Roche cOmplete Mini PI Cocktail and 1 tablet of Roche cOmplete ULTRA Mini PI Cocktail). Prepare the same day for tissue isolation and store at 4 °C.
2. 1× XT Tricine Running Buffer (1 L per running box): Dilute 50 mL of XT Tricine 20× running buffer in 950 mL of ddH2O. Store at room temperature.
3. 10% methanol 10× Tris/glycine transfer buffer (1 L per transfer box): Combine 100 mL of Tris/glycine 10×, 800 mL of ddH2O, and 100 mL of methanol. Cool to 4 °C prior to use and maintain at 4 °C for storage.
4. Tris buffered saline (TBS) (100 mL per membrane): Dilute 10 mL of 10× TBS in 90 mL ddH2O. Store at room temperature.
5. Blocking buffer (50 mL per membrane): Add 50 μL of Tween-20 to 49.95 mL of SuperBlock™ (Thermo Fisher, 37515) Blocking Buffer. Store at 4 °C.
6. Washing buffer (150 mL per membrane): Combine 10× TBS (15 mL), ddH2O (135.85 mL), Tween-20 (0.15 mL), Store at room temperature.
7. Antibody Solutions: See Table 1 for antibodies and instructions.
8. Enhanced Chemiluminescence Solution (ECL Prime) (2 mL per membrane), Kit (Cytivia): Reagent A (1 mL), Reagent B (1 mL). Store at room temperature and must be used within 10 min of preparation.
9. Lab Water: ddH2O.
10. Methanol.
11. BCA Protein Assay Kit.
12. BCA Standards.
13. XT sample buffer 4×.
14. XT reducing agent 20×.
15. Precision Plus Protein Dual Color Standards (Bio-Rad, 1610374).
16. Precision Plus Protein All Blue Standards (Bio-Rad, 1610373).

Table 1
List of primary and secondary antibodies for dystrophin Western blotting

Primary antibody	Host species	Supplier	Product number	Dilution	Secondary antibody	Host species	Supplier	Product number	Dilution
Dystrophin: Mandys106	Mouse, IgG2aκ	Millipore Sigma	MABT827	1:10,000 (not on website)	Peroxidase AffiniPure Donkey Anti-Mouse	Donkey, IgG	Jackson Immuno Research	715–035-150	1:10,000–1:200,000 (on website)
Dystrophin: Manex44a	Mouse, IgG1	DSHB	MANEX44A (5B2)	1:100	Peroxidase AffiniPure Donkey Anti-Mouse	Donkey, IgG	Jackson Immuno Research	715–035-150	1:10,000–1:200,000 (on website)
Loading control: B-tubulin HRP-conjugated	Rabbit	Abcam	Ab21058	1:1000 (on website)	N/A	N/A	N/A	N/A	N/A
Loading control: GAPDH HRP-conjugated	Mouse	Sigma	G9295	1:10,000	N/A	N/A	N/A	N/A	N/A

In our experience, Mandys106 works well for the detection of human dystrophin but does not detect mouse dystrophin. Manex44A detects mouse, canine, and human dystrophin well

Ig immunoglobulin, *N/A* not applicable

17. Tween-20.
18. ECL Prime Western Blotting Detection Reagent.
19. Cryostat.
20. Handheld homogenizer.
21. Disposable pestle mixers.
22. Centrifuge.
23. Microplate Absorbance Reader with a 560 nM filter.
24. Gel Loading Pipette Tips.
25. Criterion™ Vertical Midi-Format Electrophoresis Cell (Bio-Rad, 1656001).
26. BioRad 3–8% Criterion™ Tris-Acetate Gels, 18-well, 30 μL (Bio-Rad, 3450130).
27. PowerPac HC Power Supply (Bio-Rad, 1645052).
28. Criterion™ Blotter with Plate Electrodes (Bio-Rad, 1704070).
29. Immun-Blot® PVDF Membrane, 0.2 μm (Bio-Rad, 1620177).
30. Thick Blot Filter Paper, Precut, 9.5 × 15.2 cm.
31. Western Incubation Box, Clear, X-Large.
32. PerfectWestern 6 Sectional-Long, Clear Compartmented Container for Strips.
33. Analog rocking platform.
34. Magnetic stir bars.
35. Magnetic stirrer plate.
36. Flat tip forceps.
37. 96-well flat-bottom polystyrene plates.
38. Microplate absorbance reader with 560 nM filter.
39. Imaging system.
40. Imaging processing system (e.g., Bio-Rad Image Lab).
41. Plastic sleeves.

3 Methods

3.1 Sample Processing

1. Identify samples of interest for the experiment and prepare a map of the blots (*see* **Note 1** for blot design considerations as a bioanalytical assay and **Note 2** for general planning recommendations).
2. Set cryostat temperature to −23 °C and sectioning thickness to 8 μm.

3. Cut cryosections (refer to **Note 3** for tissue sampling options) into a precooled 1.5 mL microcentrifuge tube until sections reach approximately the 500 μL line. This can be approximately 100–200 sections from a 4 mM piece of tissue cut at 8 μm thick.
4. Store cryosections at −80 °C (*see* **Note 3** for suggestions on evaluating storage) or place on dry ice for subsequent protein extraction.

3.2 Protein Extraction

1. Pre-cool a temperature-controlled centrifuge to 4 °C.
2. Remove the microcentrifuge tubes containing sample cryosections from the freezer and keep them on dry ice until they are ready for use.
3. Transfer the microcentrifuge tubes containing sample cryosections from dry ice to wet ice and add ice cold protein extraction buffer (*see* **Note 4** for information about the protein extraction buffer and for volume recommendations for protein extraction).
4. Homogenize samples with a clean pestle for 30 s on wet ice varying the depth of the pestle (*see* **Note 5** for helpful notes on the homogenization process).
5. Centrifuge tubes at 18,000 × g for 10 min at 4 °C.
6. Maintain protein isolate tubes on wet ice for the remainder of this procedure.
7. Pipette the protein isolate layer (avoid the fatty top layer and bottom pellet, if applicable) into a diluted aliquot for protein determination and the remaining isolate volume into multiple undiluted aliquots (*see* **Note 6** for volume recommendations of the diluted aliquot and for information on freeze/thaw cycles of protein isolates).
8. Store the undiluted and diluted aliquots at −80 °C until needed (*see* **Note** 7 for notes on storage testing and recommendations).

3.3 Protein Determination

1. Remove diluted sample isolate tubes from storage and thaw on wet ice.
2. Vortex the tubes prior to loading the microplate.
3. Follow the manufacturer's instructions for the Microplate Procedure for the Pierce BCA Protein Assay Kit (*see* **Note 8** for guidance on proper protein quantification).

3.4 SDS-PAGE Sample Preparation

1. Prewarm a heating block to 95 °C.
2. Transfer undiluted protein isolate tubes from storage and thaw on wet ice.

3. Prepare SDS-PAGE samples by adding the components to a microcentrifuge tube in the following order: Protein extraction buffer, XT sample buffer 4× + XT reducing agent master mix 20×, and sample protein isolate. Ensure that each microcentrifuge tube will have enough to load 10ug of total protein per well (*see* **Note 9** regarding use of Western blot calculator for calculating volumes of each component).
4. Vortex each SDS-PAGE sample tube briefly, and then boil at 95 °C for 5 min.
5. Transfer (or discard) protein isolate tubes to storage based on remaining tube volume (*see* **Note 6** for notes on storing freeze/thawed protein isolates).
6. Briefly centrifuge tubes (5000 × g for 10 s) and keep tubes on wet ice prior to gel loading.

3.5 Sample Loading and Gel Electrophoresis

1. Prepare for gel electrophoresis by removing tape from the bottom of a 3–8% Tris-acetate pre-cast gel (*see* **Note 10** for selecting the appropriate gel type) and insert into an electrophoresis tank (*see* **Note 10** if running two gels in one tank).
2. Remove the comb that fills the loading wells.
3. Fill the inner chamber of each gel with 1× XT Tricine Running Buffer and then fill outer chamber to the indicated fill level on the tank.
4. Load the 20 uL of calibration standards, 20 uL of samples, and 7 uL of molecular weight ladders into wells (*see* **Note 11** for notes on sample loading order and on using molecular weight ladders) using a separate gel loading pipette tip for each well.
5. Place the lid on the tank and connect the power cables to the power supply.
6. Run the gel at a constant 80 volts for 60 min and then 120 volts for 35 min (*see* **Note 12** on gel electrophoresis).

3.6 Transfer Preparation and Cassette Assembly

1. Activate an appropriate number of 13 × 8.5 cm PVDF membrane(s) by soaking them in methanol for 15 min.
2. During this time, soak filter paper and foam pads (two pieces of filter paper and two foam pads per membrane) in cold transfer buffer.
3. After the 15-min activation of PVDF in methanol, soak the PVDF in transfer buffer for 5 min.
4. Fill a transfer tank half full with transfer buffer and place a magnetic stirring rod in the center of the tank.
5. Following gel electrophoresis, crack the gel cassette open, and cut off the lane dividers and the remaining blue loading dye artifact found horizontally at the bottom of the gel.

6. Assemble the transfer cassette in the following order (from bottom to top): black cathode (−), sponge, filter paper, SDS-PAGE gel, PVDF membrane, filter paper, sponge, red anode (+).
7. Use the roller horizontally to push out any bubbles in the cassette after placing the membrane onto the gel and following the final filter paper placement.
8. Close the cassette carefully and keep the components compressed against each other to prevent bubbles from entering.
9. Add transfer cassette(s) to a transfer tank apparatus, maintaining the gel orientation that was used in gel electrophoresis (*see* **Note 4** if running two gels in one tank).
10. Place a frozen ice block in the transfer tank, and add more transfer buffer in order to reach the indicated fill level on the tank.

3.7 Transfer Procedure

1. Place the transfer tank into a square foam ice bucket containing wet ice and onto a magnetic stirrer plate at 4 °C.
2. Place a stir bar into the transfer tank, set the stir plate to medium speed, and confirm that the stir bar is stirring.
3. Place the lid on the transfer tank, and connect the power cables to the power supply.
4. Run the transfer at a constant 80 volts for 2 h (*see* **Note 13** regarding transfer condition recommendations).
5. Record the amperage and wattage for the transfer at the start, after 15 min, and at the end (*see* **Note 14** for a note about voltage disruptions).

3.8 Transfer Cassette Disassembly

1. After transfer is complete, open the cassette, and use flat forceps to transfer the membrane(s) protein side up to a gel box containing transfer buffer.
2. Evaluate the transfer efficiency (*see* **Note 15** for notes on how to evaluate the transfer efficiency) by confirming the presence of the molecular weight ladder bands on the membrane and the absence of the molecular weight ladder bands on the gel.
3. Evaluate the membrane for physical abnormalities such as tears or bubbles (*see* **Note 16** for assessment).
4. Wash the membrane for 5 min in 1× TBS.
5. To evaluate full-length dystrophin and microdystrophin (>100 kDa) and other targets such as β-tubulin (<100 kDa), cut the membrane with a razor blade and straight edge between the 75 kDa and 100 kDa bands using the molecular weight ladders on each end of the blot as guides to maintain straight cuts.

3.9 Primary Antibody Incubation

1. For the dystrophin/microdystrophin membrane, incubate the membrane in 10 mL blocking buffer at RT for 1 h with constant agitation on a rocker (*see* **Note 17** for information about the use of a rocker).
2. For the β-tubulin membrane, incubate the membrane in 10 mL blocking buffer overnight at 4 °C with constant agitation on a rocker.
3. Pour off blocking buffer.
4. For the dystrophin/microdystrophin membrane, add the appropriate primary antibody solution (Table 1 and *see* **Note 18** for details on antibody selection), and incubate overnight in a multi-chamber box at 4 °C with constant agitation on the rocker.
5. For the β-tubulin membrane (*see* **Note 19** for details on loading control selection), add the appropriate primary antibody solution (Table 1), and incubate at RT for 1 h with constant agitation on a rocker.
6. Pour off and discard the primary antibody solution.
7. Wash the membrane three times for 10 min each with washing buffer.
8. Proceed to ECL incubation (**step 11**) for the β-tubulin membrane because the primary antibody is already conjugated and does not require secondary antibody incubation.
9. Pour off washing buffer, and add appropriate secondary antibody solution (Table 1) to the dystrophin/microdystrophin membrane, and incubate at RT for 1 h with constant agitation on a rocker.
10. Pour off the secondary antibody solution and wash membranes three times for 10 min each with washing buffer.
11. Remove the membrane from washing buffer and remove excess solution by dabbing corner of membrane on paper towel.
12. Place the membrane on a flat surface with the protein side up.
13. Incubate membranes in ECL solution for 5 min at RT (ensure the membrane is completely covered and there are no bubbles present in the solution).
14. Following incubation, remove excess solution by dabbing the membrane onto a paper towel. Do not allow the membrane to dry.
15. Place the membrane in a plastic sleeve, which prevents drying and allows the membrane to sit flat for imaging.
16. Remove bubbles between the plastic and the membrane.

3.10 Membrane Imaging

1. Place the membrane "protein side up" in the imaging area of a digital imager capable of capturing ECL-based signal, such as the Bio-Rad ChemiDoc.
2. Capture an autoexposure image and sequential images using the settings appropriate for chemiluminescent detection (*see* **Note 20** for image capture notes and recommendations).
3. Capture a colorimetric image to visualize the membrane and molecular weight markers (*see* **Note 21** for information on colorimetric image, Fig. 1).
4. Remove the membranes from the imager and transfer to 4 °C for short-term storage.

3.11 Quantification of Bands from Imaged Western Blots

1. Use image processing software (such as Bio-Rad Image Lab) to quantify and label bands of interest on the auto-exposed image for each membrane.
2. Define and label the lanes.
3. Identify the bands of interest within the lanes (Fig. 2). Full-length dystrophin is expected at 427 kDa, and microdystrophin around 150 kDa. β-tubulin is expected to be between 50 and 55 kDa.
4. Use a lane profile view to ensure that the band edges are appropriately defined to capture the whole protein band and

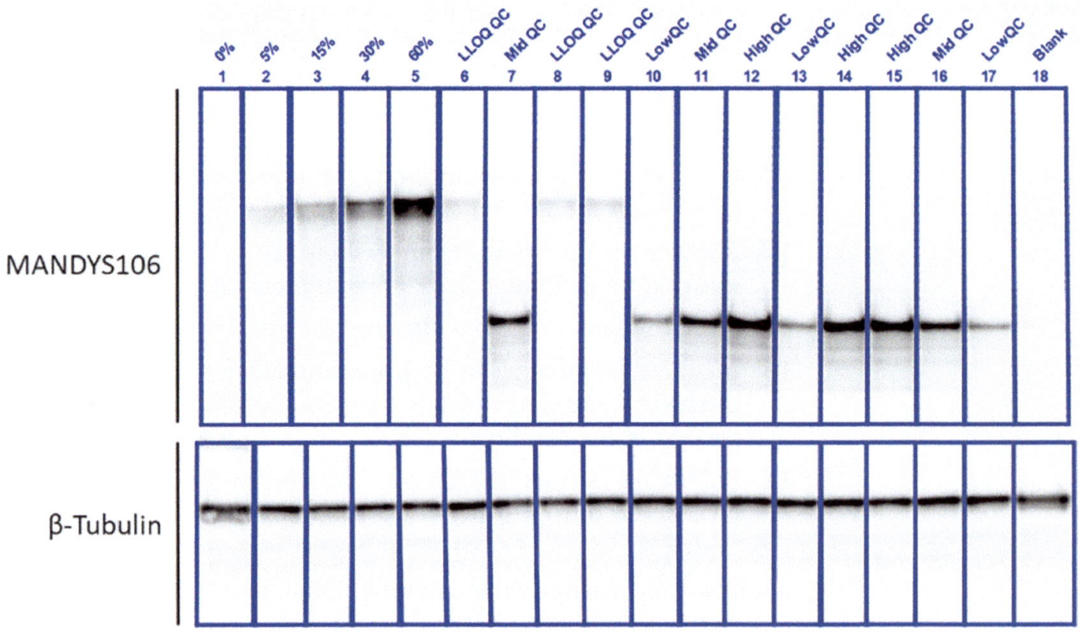

Fig. 1 Representative MANDYS106 and β-tubulin blots. This MANDYS106 blot image captures many of the necessary components of a quantitative bioanalytical assay design: a blank well (containing biological matrix of deficient tissue) in lane 1, a calibration curve containing the LLOQ (5%) and ULOQ (60%) in lanes 2–5, and replicates of a subset of QC levels (LLOQ, low, medium, and high) in lanes 6–17. The β-tubulin shows consistent loading across wells and proves to be an appropriate loading control marker

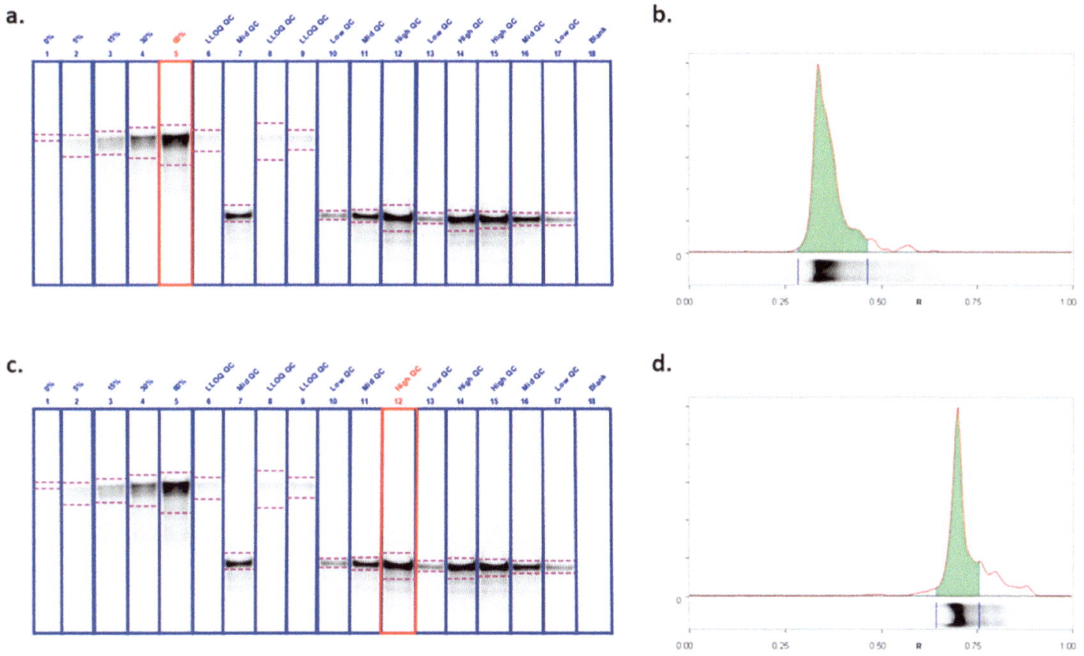

Fig. 2 Identification and integration of dystrophin and microdystrophin protein bands. Panels **a** and **c** demonstrate the identification of protein band edges (dotted magenta lines) that are within the established blot lanes (blue) or the highlighted (red) lane. Panels **b** and **d** are the corresponding lane profiles for the highlighted (red) lane in the corresponding panel **a** or **c**. The lane profile identifies the protein band (green area) between the blue band edges and under the red peak. The X axis captures the relative front where the distinction between protein size of dystrophin (panel **b**) and microdystrophin (panel **d**) can be seen

that the disk size for background subtraction is set appropriately (*see* **Note 22** for a background subtraction recommendation).

5. When applicable, use the software to apply the expected value (y) with the adjusted band volume (x) for the calibration curve points to produce a regression equation ($y = mx + b$) and R-squared value describing the linearity of the curve.
6. Export the adjusted volumes (total volume—background) of the protein bands to Excel for further evaluation.

3.12 Evaluating Western Blot Data

1. Use the linear regression equation from the calibration curve and the adjusted volume outputs for each sample to calculate the dystrophin content for the sample.
2. If the Western blot is being used as a bioanalytical assay, additional calculations (*see* **Note 23**) should be performed to evaluate the acceptability of the run (*see* **Note 24** for guidance on assay acceptance criteria).
3. If appropriate, correction of measured dystrophin levels for muscle tissue content may be considered.

4 Notes

1. Western blots are inherently limited by capacity restrictions due to a finite number of wells per gel, making prospective blot mapping critically important for the intended purpose of analysis. Based on the requirement for relative quantifications to be made within a blot, a number of wells need to be dedicated to calibration standards, quality control samples, blank(s), and molecular weight ladder(s), leaving few additional wells available for unknown samples (Fig. 3). As a result, considerations should include the number of calibration standards required, number of quality control levels and replicates, whether to run technical replicates of unknown samples within a single blot or across multiple blots, and acceptance criteria to incorporate to monitor assay performance.

   The use of a calibration curve with a range that covers the intended use of the assay helps to demonstrate accuracy, precision, and linearity within the analytical run. The following considerations are recommended for the calibration curve: be prepared in the same biological matrix as study samples, be composed of at least 4-6 nonzero points and a zero (blank) point, and be defined by the lower and upper limits of quantification (LLOQ and ULOQ).

   The use of quality control samples provides the ability to assess accuracy and precision of quantitative assessments made within a single run and across independent runs. The FDA's bioanalytical guidance on ligand binding assays, under which a Western blot falls, provides a framework for criteria to achieve

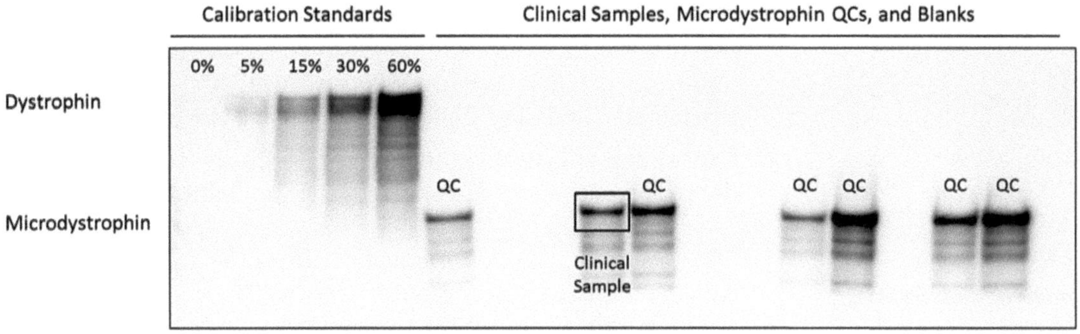

Fig. 3 Example microdystrophin clinical sample analysis blot. This is a representative blot for study sample analysis containing the necessary components to properly assess accuracy and precision of the run. Calibration standards are used to assess linearity, signal to noise, and linear regression is needed to quantify the microdystrophin content in the QCs and study samples. The QC levels have been run in duplicate in order to calculate %RE and %CV and evaluate the performance of the run. The non-calibration standard lanes have been randomized for the sample assessment. %RE = Percent relative error, %CV = % Percent coefficient of variation

when establishing a Western blot for quantitative purposes. However, due to differences between a Western blot and a more traditional quantitative ligand binding assay, such as an ELISA, especially regarding the number of total samples that can be analyzed in a single run, modifications to maintain assay rigor in the context are necessary. The following considerations are recommended for quality control samples: it should mimic study samples, a low QC level should be within three times of the LLOQ, a medium QC level should be around the geometric mean of the calibration curve range, and a high QC level should be at least 75% of the ULOQ.

Ligand binding assays tend to have a narrow assay range, it may be necessary to dilute a study sample for its concentration to fall within the range of the assay. When diluting samples, the operator should evaluate dilutional linearity and hook effect to confirm that the signal is not affected by dilution within the curve and that samples with signal absolute limit of quantification are not impacted by signal suppression.

2. The Western blot calculator (Fig. 4) can provide an efficient way of determining the volume of each component needed to load each well. In cases where different protein sources are being mixed to be loaded into the same well, this calculator provides loading information for up to three samples. Typically

Example of Western Blot Calculator										100% Protein in 25uL (ug)	Loading Buffer Conc.	Reducing Conc.	Protein to Load (ug)	Final Vol Per Rxn (uL)	Vol Per Well (uL)	
										12.5	4	20	10	25	20	
# of Wells	Well #	Sample #1	S1 Conc (ug/uL)	% of Sample	S1 Protein (ug)	Vol. of S1 (uL)	Sample #2	S2 Conc (ug/uL)	S2 Protein (ug)	Vol. of S2 (uL)	Sample #3	Vol. of S3 (uL)	Loading + Reducing	RIPA (uL)	Total Vol (uL)	Final Loaded Protein (ug)
1	1	Blank + Dual Color MW	1.0	0%	0	0	Dystrophin Negative Sample	2.0	12.5	6.3	Dual MW	7.6	7.5	3.7	25.0	10
1	2	Sample	1.0	100%	12.5	12.5							7.5	5.0	25.0	10
1	3	Sample	1.0	100%	12.5	12.5							7.5	5.0	25.0	10
1	4	Sample	1.0	100%	12.5	12.5							7.5	5.0	25.0	10
1	5	Sample	1.0	100%	12.5	12.5							7.5	5.0	25.0	10
1	6	Sample	1.0	100%	12.5	12.5							7.5	5.0	25.0	10
1	7	Sample	1.0	100%	12.5	12.5							7.5	5.0	25.0	10
1	8	Sample	1.0	100%	12.5	12.5							7.5	5.0	25.0	10
1	9	Sample	1.0	100%	12.5	12.5							7.5	5.0	25.0	10
1	10	Sample	1.0	100%	12.5	12.5							7.5	5.0	25.0	10
1	11	Sample	1.0	100%	12.5	12.5							7.5	5.0	25.0	10
1	12	Sample	1.0	100%	12.5	12.5							7.5	5.0	25.0	10
1	13	Sample	1.0	100%	12.5	12.5							7.5	5.0	25.0	10
1	14	Sample	1.0	100%	12.5	12.5							7.5	5.0	25.0	10
1	15	Sample	1.0	100%	12.5	12.5							7.5	5.0	25.0	10
1	16	Sample	1.0	100%	12.5	12.5							7.5	5.0	25.0	10
1	17	Sample	1.0	100%	12.5	12.5							7.5	5.0	25.0	10
1	18	Blank + All Blue MW	1.0	0%	0	0	Dystrophin Negative Sample	2.0	12.5	6.3	All Blue MW	7.6	7.5	3.7	25.0	10
												Extra Load + Reducing	100.0			
												Total: Load + Reducing	235.0			
												Loading Buffer	Reducing Agent			
												195.8	39.2			

Fig. 4 Western blot calculator example. This calculator displays example calculations for a single Western blot loaded with 10ug of protein per 20uL loaded in each well (makes 12.5ug in 25uL total). Ladders are shown in lanes 1 and 18. Lanes 2–17 show calculations for samples loaded at 100%

the protein sample of interest, a filler matrix if needed and possibly a ladder. Note: there is a minimum concentration required depending on the total protein amount loaded. If the calculator is showing a negative buffer volume, the concentration of one of the protein components is likely too low.

If a control curve is being used, it is best to include a 0% point (which could be in the lane containing ladder if needed) which contains 100% of the negative matrix sample to effectively analyze any background signal. If a control curve is not being used, a common quality control sample of known protein of interest content between all blots being compared is necessary. Comparing raw band volumes between blots is not advised due to the inconsistencies and general variability of Western blots.

3. Cryosections and crushed muscle can both be used to acquire skeletal muscle protein isolates. Starting material for protein extraction should be selected based on feasible tissue sampling options and the tools available for efficient homogenization. Muscle tissue typically requires strong mechanical homogenization techniques to break down muscle fibers to the extent that membrane proteins, such as dystrophin, can be released for analysis. This becomes an even greater challenge when processing muscle samples from Duchenne muscular dystrophy patients, where fibrofatty replacement of degenerated muscle fibers results in a physically harder sample. When processing larger tissue samples, bead-based and rotor-stator homogenization are commonly considered the most effective methodologies. Alternatively, generating cryosections of tissues for protein extraction upfront assists in the overall homogenization process, allowing for efficient extraction through less physically stringent methods such as handheld pestle-based homogenization. Cryosections typically result in high protein yields due to the increased surface area that comes in contact with the isolation buffer.

   The number of sections required will depend on the area of the sections. These tubes of sections can be stored at −80 °C for at least several weeks, until protein isolation is needed for the actual Western blot experiment. For longer-term storage, we recommend performing a study of sections from the same sample stored at various time lengths to characterize the stability of the protein of interest.

4. Monitoring sample stability is an important factor in quantitative Western blots that incorporate dystrophin and/or micro-dystrophin samples due to the high molecular weights of each protein and potential for degradation over time. While tissue stability, whether whole tissue or tissue cryosections, has proven to be a more reliable long-term format for storage to

maintain intact target proteins, protein extracts have shown a greater potential to demonstrate degradation over time. The operator should assess stability across the sample preparation and procedure to confirm that the processing and long-term storage do not impact the protein concentration/quality. The volume of protein extraction buffer to cryosection volume should be roughly 1:4, with a minimum 100 μL of protein extraction buffer being added.

5. Noting the color and presence of foam in the homogenization process helps indicate the quality of the sample being processed. Minimal foam and a milky color tend to indicate a sample that contains more fibrous and fatty tissue.
6. The protein determination (BCA) aliquot should contain a 1:5 ratio (e.g., 11 μL isolate to 44 μL ddH2O) of protein isolate to ddH2O. It is recommended to prepare enough volume to run replicate wells of this aliquot by BCA assay to provide increased confidence around the protein concentration of the sample. It is recommended that undiluted protein isolate be aliquoted in smaller (e.g., 25 μL) volumes in multiple tubes to avoid freeze/thaw cycles of the protein isolate that may cause degradation of the dystrophin protein. Freeze/thaw cycles can also be denoted on the tube when using that isolate tube.
7. Protease inhibitors are needed to maintain the stability of the isolated proteins in solution. A mix of two cocktails (Roche, cOmplete mini and Roche, cOmplete ultra mini) is suggested as it has been shown to provide the longest stability and preservation of the large dystrophin protein.
8. The following considerations are recommended for proper protein quantification: Technical duplicates of samples should be performed and the coefficient of variation (CV) between those duplicates should be below an absolute value of 20%, and the R-squared value of the BSA standard linear regression must be >0.9.
9. The SDS-PAGE samples should use the protein concentrations derived from the BCA assay to load 10 μg of total protein in 20 μL. It is recommended to use an excel calculator (see Fig. 4 for an example) to achieve this and determine necessary volumes of protein extraction buffer, XT sample buffer, and XT reducing agent, and sample protein lysate.
10. Evaluating different gel types: Based on the combination of proteins being analyzed on a single blot, different gel polyacrylamide concentrations should be considered. When analyzing very large proteins, such as dystrophin, low polyacrylamide concentrations of approximately 3–4% allow for efficient separation. However, when additional smaller proteins, such as microdystrophin or a loading control, are analyzed at the

same time, gradient gels incorporating multiple concentrations should be utilized. Gradients such as 3–8% allow for proper size separation across a wide molecular weight range and also prevent low molecular weight protein run-off. This can be useful in situations where low-molecular-weight proteins are being assessed on the same gels as full-length dystrophin and micro-dystrophin. When running two gels in one running or transfer box, it is recommended to differentiate between the two gels in the system by assigning gel 1 as the gel closest to the operator when the black contact is on the left and the red contact is on the right side of the tank. Gel 2 is assigned as the gel farthest from the operator when the black contact is on the left and the red contact is on the right side of the tank.

11. The following order is recommended for loading into the gel: load the calibration curve in the earlier wells, and then load the quality control samples and/or experimental samples in the following lanes. It is recommended to load two lanes of molecular weight ladders for each gel. For example, loading dual-color ladder in the first lane and all blue ladder in the last lane will allow the operator to orient the blot in downstream steps and help provide visual cues to cut the blot at the appropriate area (if necessary).

12. Running the gel at a lower voltage at the beginning of the electrophoresis procedure is a gentler treatment on the large dystrophin protein and more conducive to allowing the heavy protein to move through the gel. It is recommended to run the gel until the blue band of the sample buffer is approximately 1 cm above the bottom of the gel.

13. Protein size should be considered when evaluating both shortened dystrophins and full-length dystrophin, or when additional low-molecular-weight proteins are also quantified. Lower-molecular-weight proteins transfer more quickly than high-molecular-weight proteins. This makes the transfer of low-molecular-weight proteins less complicated. Transfer efficacy can be assessed by evaluating protein content in the gel (using stain-free gel technology or Coomassie Blue staining to image the blot after transfer) and on the transfer membrane (using the total protein visualization for stain free technology or Ponceau S staining) after transfer. As protein size is a factor, success should be judged based on the ability to transfer proteins of similar size or heavier in comparison to one's protein of interest. If a significant amount of protein is left in the gel after transfer, transfer performance can be optimized by changes in transfer time, transfer voltage, methanol concentration in the transfer buffer, and lowering the gel percentage.

    Increasing transfer time and increasing transfer voltage can transfer more protein onto the membrane from the gel. Common transfer issues like overheated buffer, melting gels, protein passing through the membrane, and gel sticking to the membrane can arise when transfer time and voltage are rapidly increased. These common transfer issues can be alleviated by understanding the trend of these two variables. Transfer time and transfer efficiency have an inverse relationship. As transfer voltage increases, the transfer run time should be decreased and vice versa.

While methanol increases protein binding to membranes, a higher methanol concentration in the transfer buffer can lead to protein precipitation and swelling in the gel. A decreased methanol concentration prevents protein precipitation and promotes migration of large proteins from the gel. Choosing a gel with a lower acrylamide content can also promote protein migration from the gel, but gels with lower acrylamide content are more fragile.

The operator needs to account for all these variables simultaneously since each plays a crucial role in incomplete transfer or over-transfer of proteins. If the operator is nervous about over transferring proteins, we recommend adding a second membrane to detect any protein that has precipitated past the first membrane. Conduct a Ponceau stain or imaging for stain free total protein visualization on the second membrane to detect proteins.

In general, high-molecular-weight proteins transfer best at a low voltage and a long transfer time, while low-molecular-weight proteins can be transferred at high voltage with a faster transfer time. "Turbo transfer" approaches can work well for proteins of low- to mid-molecular-weight but may not be useful for proteins of high-molecular-weight like dystrophin. Performance may be better with PVDF membranes rather than nitrocellulose due to the greater binding capabilities of PVDF.

The protein size differences between full-length and microdystrophin can become an issue. Different protein sizes tend to transfer unevenly and are not always consistent. When comparing samples to a control curve or quality control sample it is best to compare at the same molecular weight. While not always possible, note that if comparing at different molecular weights, it is recommended that transfer efficiency is examined.

14. It is recommended to record these power supply outputs to monitor disruptions in the transfer. Power supply disruptions are most likely to be seen in the first 15 minutes of the transfer procedure.

15. If a power failure occurs during gel running or protein transfer, the operator should assess whether the time of the failure can be established/understood. Once this is understood the operator should determine whether the experiment should be considered a failure or can continue. Presence of molecular weight ladders that have run to the appropriate location on the blot.

Alternatively, a Coomassie Blue stain may be used to visualize the protein in each lane.

Effective transfer of proteins can be analyzed by observing the transfer of the molecular weight ladders. Alternatively, it is possible to use stain-free gel technology (on gel types where this is available) or Ponceau protein stain to visually observe the proteins transferred to the membrane.

16. The operator should assess if there are membrane abnormalities that impact the region of interest and whether the membrane should continue to the immunoblotting procedure.
17. It is recommended to use a rocker instead of an orbital shaker because of the rocker's ability to completely disperse and mix the incubation solution.
18. Primary antibodies used for the evaluation of dystrophin expression should be selected based on the particular sequence of interest within the dystrophin protein, the specificity of the antibody in binding to the target epitope (and absence of cross-reactivity with coexisting structurally related molecules) in the Western blot format being used, and confirmation that the signal to noise ratio is at an appropriate level to maintain the desired sensitivity for low-level dystrophin expression. When utilizing Western blots for quantitative analysis of exon-skipped dystrophin or microdystrophin relative to a standard curve of full-length dystrophin, the primary antibody must recognize common epitopes that are present in both proteins to ensure accurate detection and quantification. Table 1 describes primary antibodies for the detection of dystrophin. Note that some antibodies (such as MANDYS106) will only detect human dystrophin and will not detect murine dystrophin.
19. Western blot assays may include loading controls or secondary markers. Loading control options include a house keeping protein or total protein staining. If a housekeeping protein (such as B-tubulin) is used, the protein size should be separated far enough away from protein of interest to allow for the blot to be cut and incubated in different primary antibodies. The loading control strip should also be imaged separately from the protein of interest blot strip. If total protein analysis is desired, a stain-free gel blotted with a low fluorescence PVDF is generally recommended over other total protein membrane staining options. The SDS-PAGE gel is activated prior to transfer with UV light, and the PVDF membrane is imaged post-transfer. Total protein should be used only when the entire sample is contained in the lane. If protein is run off the gel, or if the blot is cut, total protein should not be used. Total protein is the best option for normalizing Western blots as it allows detection of

proteins across a broad molecular weight range. When using housekeeping proteins, it is important to consider that disease, tissue type, or species type can impact the signal. Rather than normalizing, a loading control can be used to determine the variation in loaded protein between lanes. This allows the exclusion of some lanes if variation falls above a set level of variation. Typically, an absolute variation above 50% is considered a failure and that value should not be trusted.

20. The autoexposure setting estimates the optimal exposure time to ensure the best use of the dynamic range. It is recommended to capture a series of sequential 1 s exposures for a total of 30 seconds, in addition to the autoexposure image. Ideally the span of imaging should allow the visualization of all bands of interest in the gel regardless of their intensity relative to the control sample lanes. If an artifact outside the region of interest is driving the exposure for the blot, cover the artifact, and reimage to ensure the exposure is being driven by the protein bands.

21. The colorimetric image can later be merged with a chemiluminescent image to visualize the molecular weight ladders with the protein bands of interest.

22. It is recommended to use a disk size larger than your band width to avoid subtracting the true signal.

23. The FDA guidance recommends applying the following calculations to the calibration curve points and quality control samples to evaluate the accuracy and precision of the analytical run:

$$\text{Relative error} = \%\text{RE} = \frac{\text{mean calculated conc} - \text{nominal conc}}{\text{nominal concentration}} \times 100$$

$$\text{Coefficient of variation} = \%\text{CV} = \frac{\text{standard deviation } (n \text{ replicates})}{\text{calculated mean } (n \text{ replicates})} \times 100$$

Additionally, the following evaluations may be helpful in determining the performance of the assay:

$$\text{Signal-to-noise ratio} = \frac{\text{LLOQ Adj Volume}}{\text{Blank Standard Adj Volume}},$$

R-squared of the calibration curve linear regression.

24. The FDA's recommended assay acceptance criteria provide guidance for evaluation of an analytical run. The signal-to-noise ratio between the zero point and the LLOQ should be >5 to indicate the assay's ability to distinguish between signal and background. The R-squared of the regression equation should be >0.90 to indicate the linearity of the calibration curve.

Acceptance criteria suggestions for assessment of within-run accuracy and precision are as follows: the %RE of the calibration standards (besides the LLOQ and ULOQ) should be ±20%, the %RE of the LLOQ and ULOQ should be ±25%, and the %RE for at least 2/3 of the QCs (and at least 50% at each level) should be within ±20%.

To assess between-run accuracy and precision (if applicable), the overall mean %RE and %CV of all accepted run QCs should be within ±20%.

25. Correction of measured dystrophin levels for muscle content in a tissue sample has been used in some studies. Whether from a healthy donor or from a diagnostic or clinical trial biopsy, the proportion of muscle within a tissue sample may vary widely based on underlying disease pathology. In Duchenne muscular dystrophy, muscle fibers are replaced by fat and fibrosis as the disease progresses. This proportion of muscle within a given tissue specimen may need to be accounted for to accurately quantify proteins of interest. Various approaches may be taken to characterize muscle content within a tissue. By Western blot, other muscle proteins can also be analyzed in parallel to the target proteins and quantified relative to a standard. Additionally, orthogonal methods such as histological evaluations can be used to stain and quantify muscle and non-muscle areas within a tissue sample. However, it should be noted that proteins may not be extracted from non-muscle tissue to the same degree as muscle tissue, so using a correction factor based on histological area may not provide accurate correction of the degree of non-muscle protein in the protein isolate itself. If a correction for non-muscle tissue is desired, correction methods that assess non-muscle components after protein isolation are more likely to provide an accurate correction method. Regardless of the methodology used for this purpose, consistent application and description of results are critical to ensure accurate representation of reportable values.

References

1. Shahnoor N, Siebers EM, Brown KJ et al (2019) Pathological issues in dystrophinopathy in the age of genetic therapies. Annu Rev Pathol 14: 105–126
2. Voit T, Stuettgen P, Cremer M et al (1991) Dystrophin as a diagnostic marker in Duchenne and Becker muscular dystrophy. Correlation of immunofluorescence and western blot. Neuropediatrics 22:152–162
3. Lawlor MWS, Childers MK, Brown KJ (2019) Histological and biochemical evaluation of muscle gene therapy. In: Duan DMJR (ed) Muscle gene therapy, 2nd edn. Springer, pp 207–226
4. Mendell JR, Sahenk Z, Lehman K et al (2020) Assessment of systemic delivery of rAAVrh74. MHCK7. Micro-dystrophin in children with duchenne muscular dystrophy: a nonrandomized controlled trial. JAMA Neurol 77:1122–1131

5. Sakamoto M, Yuasa K, Yoshimura M et al (2002) Micro-dystrophin cDNA ameliorates dystrophic phenotypes when introduced into mdx mice as a transgene. Biochem Biophys Res Commun 293: 1265–1272
6. Shin JH, Pan X, Hakim CH et al (2013) Micro-dystrophin ameliorates muscular dystrophy in the canine model of duchenne muscular dystrophy. Mol Ther 21:750–757
7. Weisbart RH, Hansen JE, Nishimura RN et al (2005) An intracellular delivery vehicle for protein transduction of micro-dystrophin. J Drug Target 13:81–87
8. Waldrop MA, Moore SA, Mathews KD et al (2022) Intron mutations and early transcription termination in Duchenne and Becker muscular dystrophy. Hum Mutat 43:511–528

Chapter 7

Analysis of Dystrophin-Associated Glycoproteins: Focus on Expression and Glycosylation of Dystroglycan in Skeletal Muscle

Deborah Zygmunt, Patricia Lam, and Paul T. Martin

Abstract

The dystrophin-associated glycoprotein (DAG) complex includes a number of important transmembrane and membrane-associated proteins that contribute to the maintenance of muscle membrane stability and to signaling through the muscle cell membrane. DAG expression and post-translational modifications can be altered in skeletal muscle from patients with Duchenne muscular dystrophy (DMD) and other forms of muscular dystrophy, making it essential to have robust protocols to assess these features. While good antibody and DNA probes exist to assess expression of most DAGs, quantitative analysis of glycosylation and protein expression for α dystroglycan (αDG) has been problematic. Here we provide protocols to describe proper functional glycosylation and expression of αDG using immunostaining and immunoblotting, demonstrating how such protocols can identify glycosylation defects that give rise to muscular dystrophy.

Key words Immunostaining, Dystroglycan, Glycosylation, Dystroglycanopathy, Matriglycan

1 Introduction

The dystrophin-associated glycoprotein (DAG) complex includes a group of transmembrane and membrane-associated proteins that help convey both structural and signaling functions through the muscle membrane by binding to extracellular matrix (ECM) proteins, in particular laminin, and linking them through the muscle membrane to the F-actin cytoskeleton via interactions with dystrophin [1, 2]. These roles are paramount to dystrophin's function in the muscle sarcolemmal membrane; inactivating mutations in genes encoding DAGs or DAG modifiers give rise to forms of muscular dystrophy, much as mutations in dystrophin give rise to Duchenne muscular dystrophy [3–5]. Disease forms caused by DAG deficiencies may also display perturbations in neuronal development [6]. DAGs are often downregulated in DMD and DMD model

Kevin Flanigan and Stefan Nicolau (eds.), *Duchenne Muscular Dystrophy: Methods and Protocols*, Methods in Molecular Biology, vol. 2975, https://doi.org/10.1007/978-1-0716-4811-7_7,

skeletal muscle as the result of dystrophin deficiency [7, 8]. The extent of such DAG down-regulation for some components, in particular α dystroglycan (αDG), however, has been quite variable in a number of studies. DAGs include αDG, an extracellular membrane-associated protein; transmembrane proteins, including β dystroglycan (βDG), α, β, γ, and δ sarcoglycan, and sarcospan; and intracellular proteins, including α and β syntrophins and forms of α dystrobrevin. All DAGs are associated, either directly or indirectly, with dystrophin. In mild nonionic detergents such as digitonin, dystroglycan can be immunoprecipitated in a manner that includes its associations with all known DAGs and dystrophin [9, 10]. Such studies have also identified novel potential members of the DAG family and shown that altering cell surface glycosylation, for example, by overexpressing *GALGT2* (also called *B4GALNT2*), a glycosyltransferase known to modify αDG, can change DAG composition [11]. Use of stronger nonionic detergents (e.g., NP-40) allows for precipitation of slightly fewer DAGs but often in a way that still shows their binding in a protein complex (e.g., with wheat germ agglutinin (WGA)). Published protocols to demonstrate expression and membrane localization using immunostaining and immunoblotting for most DAGs are well established in mice and in humans and so will not be discussed here [1, 7, 12–20]. These published papers allow for careful assessment of protein and gene expression for dystrophin, utrophin, sarcoglycans, sarcospan, syntrophins, dystrobrevins, and neuronal nitric oxide synthase (nNOS).

The aspect we will focus on is how to describe proper functional glycosylation and expression of αDG, which has been a challenge for investigators for some time, including those studying DMD. Proper glycosylation of αDG is essential to the functioning of the DAG complex [15, 21]. αDG must be glycosylated with a phosphor-O-mannose-linked glycan, termed matriglycan, in order for ECM proteins, including laminins, agrins, and perlecans, to bind to it [22, 23]. Such interactions are essential for linking the ECM, through the DAG complex in the muscle membrane, to dystrophin and the F-actin cytoskeleton on the inside of the muscle cell. The essential component of matriglycan on αDG required for laminin binding is a repeating disaccharide of β1,3-linked xylose (Xyl) and α1,3-linked glucuronic acid (GlcA) synthesized by a dual function enzyme called LARGE [24, 25]. Almost two dozen genes are involved in the biosynthesis of matriglycan and its underlying glycan structures on αDG, and inactivating mutations in any of these genes can cause reduced matriglycan expression, which, in turn, reduces or eliminates ECM binding. Muscular dystrophy forms arising from such mutations are collectively termed dystroglycanopathies [6]. These include very severe congenital forms of muscular dystrophy where brain and eye development are affected in addition to skeletal muscles and less severe, often adult-onset, limb girdle forms of muscular dystrophy where eye and brain defects are absent and where skeletal, and sometimes also cardiac, muscle pathology remains [4, 6].

Assessment of functional αDG glycosylation is most often done using a combination of immunostaining and immunoblotting with the IIH6 monoclonal antibody. IIH6 binds to matriglycan [26, 27] and serves as a function-blocking antibody for laminin binding to αDG [21]. In dystroglycanopathies, IIH6 immunostaining and immunoblotting signals are reduced or absent [15]. Demonstrating absence of IIH6 signal is most often coupled with immunostaining and immunoblotting using an antibody to βDG, which is not glycosylated with matriglycan, and a polyclonal antibody to the αDG or the α/βDG polypeptide that recognizes the αDG protein sequence independent of glycosylation status [15]. Some monoclonal antibodies to αDG protein also can be used for this purpose [28]. Staining with antibodies that recognize the αDG polypeptide help determine if αDG expression is maintained even when functional glycosylation is eliminated. Western blots using an anti-αDG polypeptide antibody also demonstrate a reduced molecular weight for αDG protein when matriglycan is reduced or absent. In skeletal muscle, the molecular weight of αDG protein using a polypeptide antibody is reduced from a normal weight of 156 kDa. The αDG cDNA encodes a 72 kDa polypeptide, so protein blotted at this weight would indicate no glycosylation at all, but in most instances some glycosylation remains even without matriglycan expression, increasing the molecular weight somewhat. Blotting with antibodies to βDG remains unchanged at 43 kDa. Because IIH6 is an anti-glycan IgM antibody with relatively low binding affinity, investigators have often encountered problems with its use. Moreover, deficits in matriglycan expression are often not absolute, leaving a need to understand the extent of glycosylation loss and how it relates to disease severity. Here, we report on our successful protocol to use IIH6 and other reagents to assess αDG expression and glycosylation in mouse muscle, using the $FKRP_{P448L}$ knock-in model for limb girdle muscular dystrophy R9 (LGMDR9) [29] as a model for changes seen in dystroglycanopathies. In addition, we point out problems that often arise with the use of these reagents.

2 Materials

2.1 Immunofluorescence Staining

1. Cryostat capable of sectioning muscle in 10 μm sections (e.g., Leica CM1950).
2. ColorFrost Plus microscope slides.
3. Cover glass, #1.5.
4. Slide staining racks and dishes.

5. Low-salt Tris-buffered saline (LS-TBS): 50 mM Tris-HCl, 100 mM NaCl, pH 7.4. Prepare a 10× stock solution by combining 880 mL of 0.5 M Tris pH 6.8, 40 mL of 1.5 M Tris pH 8.8 and 116.88 g sodium chloride. Transfer to a graduated cylinder and bring the volume to 2 L using ultrapure water. Prepare a 1× working solution by diluting 200 mL of 10× stock solution to 2000 mL using ultrapure water. Store at room temperature.
6. Methanol: Stored at −20 °C.
7. Mouse on Mouse (M.O.M) blocking solution: 2 drops of Mouse on Mouse blocking reagent (Vector Labs) in 2.5 mL LS-TBS.
8. Blocking solution: 5% goat serum +3% bovine serum albumin (BSA) in LS-TBS. Add 5 mL goat serum and 3 g BSA in 80 mL LS-TBS. Mix to dissolve the BSA, and then use LS-TBS to bring the total volume to 100 mL. Keep the buffer on ice when staining and store the solution at 4 °C.
9. Antibody dilution solution: 1% BSA in LS-TBS. Dissolve 1 g BSA in 100 mL LS-TBS. Keep the buffer on ice when staining and store the solution at 4 °C.
10. Hydrophobic barrier PAP pen.
11. Anti-α-dystroglycan (αDG) antibody IIH6C4 (Sigma), mouse monoclonal IgM, store at −20 °C.
12. Anti-dystrophin antibody, rabbit monoclonal IgG (Abcam), store at −20 °C.
13. Alexa Fluor Plus 555 goat anti-mouse IgM secondary antibody, store at 4 °C.
14. Alexa Fluor Plus 647 goat anti-rabbit IgG secondary antibody, store at 4 °C.
15. ProLong Gold Antifade Mountant with DAPI, store at room temperature.
16. Widefield fluorescence microscope with an LED light source and 10× objective (e.g., Nikon Ti2-E widefield fluorescence microscope with a Lumencor SOLA LED engine and Hamamatsu ORCA Fusion camera and NIS-Elements AR software).

2.2 SDS Polyacrylamide Gel

1. Heat block.
2. Table-top centrifuge.
3. Electrophoresis power supply.
4. Western blot gel apparatus and tank.
5. Mini-PROTEAN 4–15% TGX stain-free precast acrylamide gels.
6. SDS-PAGE running buffer: 25 mM Tris, 192 mM glycine, 0.1% SDS, pH 8.3.

7. 4× Laemmli protein sample buffer: 277.8 mM Tris-HCL, pH 6.8, 44.4% (v/v) glycerol, 4.4% LDS, 0.02% bromophenol blue.
8. 2-Mercaptoethanol.
9. Protein standard that is visible with far red (IR680).
10. Stain-free imaging system.

2.3 Immunoblotting

1. Electrophoresis power supply.
2. Western blot transfer apparatus and tank.
3. Rocker.
4. Methanol.
5. Low-fluorescence PVDF membrane.
6. Western blotting filter paper.
7. Western blot transfer buffer: 25 mM Tris, 192 mM glycine, pH 8.3, 20% methanol.
8. Tris-buffered saline (TBS).
9. TBS with 0.1% Tween 20 (TBST).
10. 1% (v/w) sodium dodecyl sulfate (SDS) solution in water.
11. Blocking solution: Intercept TBS blocking buffer, or Every-Blot Blocking Buffer, or 5% milk in TBST.
12. Anti-α-dystroglycan antibody IIH6C4, mouse monoclonal IgM (Sigma 05-593-I).
13. Anti-β-dystroglycan antibody, mouse monoclonal IgG2a (Leica NCL-b-DG).
14. Anti-pan-dystroglycan (α and β, not glycosylation-dependent), sheep polyclonal (R&D AF6868).
15. Anti-GAPDH antibody, rabbit polyclonal (Sigma G9545), or hFAB rhodamine anti-GAPDH IgG (Bio-Rad 12004168).
16. IRDye 800CW goat anti-mouse IgM (μ chain-specific) secondary antibody.
17. IRDye 680RD goat anti-mouse IgM (μ chain-specific) secondary antibody.
18. IRDye 800CW goat anti-mouse IgG_{2a}-specific secondary antibody.
19. Dylight 800 donkey anti-sheep IgG (H + L) secondary antibody.
20. IRDye 680RD goat anti-rabbit IgG (H + L) secondary antibody.
21. Western blot imager capable of imaging infrared fluorescent dyes (e.g., ChemiDoc MP or Odyssey Imagers).

3 Methods

3.1 Immunostaining (IIH6 and Dystrophin)

Immunostaining with IIH6 is used to define the extent of expression of matriglycan, while staining with dystrophin is used to mark all myofiber membranes. In instances where dystrophin is absent or perturbed, as in DMD muscle, β spectrin can be used as a muscle membrane marker instead. 4′,6-diamidino-2-phenylindole (DAPI) is added to stain nuclei. Carry out all steps at room temperature unless otherwise noted.

1. Cut 10 μm sections of muscle using a cryostat and place three sections on a slide. Circle sections using a hydrophobic barrier PAP pen.
2. Wet the sections by dipping in LS-TBS.
3. Fix the sections in ice cold methanol for 10 min (*see* **Note 1**). Wash the slides 3 times for 5 min each in LS-TBS (*see* **Note 2**).
4. Block the sections using Mouse on Mouse blocking solution. Incubate the slides for 1 h at room temperature (*see* **Note 3**). After incubation, wash the slides with LS-TBS as previously. Block again with 5% goat serum/3% BSA in LS-TBS. Incubate for 1 h at room temperature. Do not wash slides after this blocking step.
5. Dilute primary antibodies in 1% BSA in LS-TBS. Dilute IIH6 at 1:100, and anti-dystrophin at 1:200. Incubate one section with both IIH6 (*see* **Note 4**) and anti-dystrophin, one section with only anti-dystrophin, and one section with no primary antibodies (*see* **Note 5**). Incubate the primary antibodies at room temperature for 2 h or at 4 °C overnight. Wash as before.
6. Dilute secondary antibodies in 1% BSA in LS-TBS. Dilute anti-mouse IgM at 1:1000 and anti-rabbit IgG at 1:500. Incubate all sections with the same antibody mixture at room temperature for 1 h. Wash as before (*see* **Note 6**).
7. Coverslip and mount the slides using ProLong Gold antifade mountant with DAPI. Let the slides sit at room temperature in the dark overnight to cure before imaging.
8. Image the slides using an appropriate widefield fluorescent microscope with an LED light source under 10× magnification. Determine proper exposure times on control slides, and image all sections using the same exposure times (*see* **Note 7**). An example of staining in a mouse model of LGMDR9 (before and after treatment) is shown in Fig. 1.

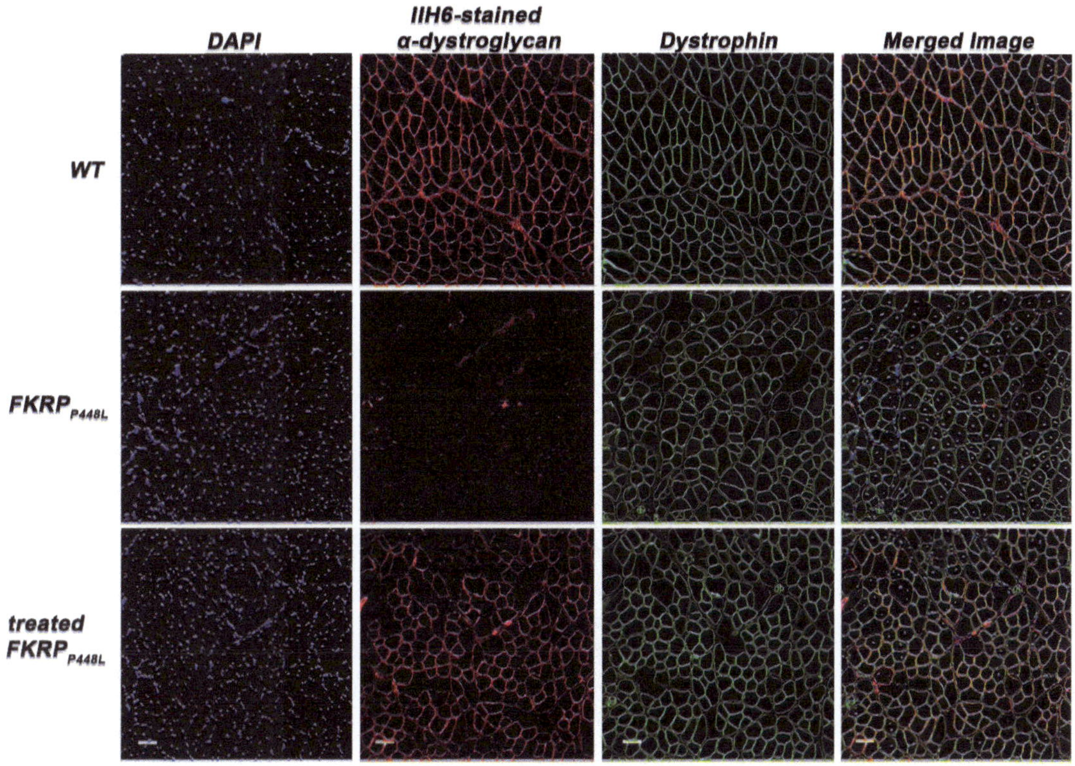

Fig. 1 Immunostaining for IIH6 and dystrophin. Skeletal muscle sections from wild-type (WT) mice, $FKRP_{P448L}$ mice (a model for LGMDR9), and treated $FKRP_{P448L}$ mice given human *FKRP* gene therapy are shown. Single-channel staining at constant exposure times for dystrophin, a marker for the muscle sarcolemmal membrane (green), IIH6, which stains matriglycan on α dystroglycan (red), and DAPI, a marker for nuclei (blue), are shown, along with a merged image on the right. Bar is 100 μm

3.2 Immunoblotting (IIH6, αDG Protein, βDG Protein)

1. Dilute 40 μg of muscle protein (measured using Bradford or BCA assay) with 4× Laemmli sample buffer with 10% 2-mercaptoethanol. Heat the protein at 98°C for 10 min, and then centrifuge at 16,000 × g for 5 min at 4°C.
2. Prepare TGX-stain-free gel (*see* **Note 8**): carefully remove the comb by pulling directly up, and remove the tape from the bottom on the gel. Place the gel in the electrophoresis tank and fill with SDS-PAGE running buffer. Rinse out the wells with running buffer.
3. Load protein samples along with an appropriate protein standard. Perform electrophoresis at 80 volts for 15 min, and then increase to 140 volts for 45 min or until the dye front reaches the bottom.

4. Carefully pry the gel plates apart and remove the gel (*see* **Note 9**). Place in a tray containing running buffer. Image the gel on an imager capable of imaging stain-free gels (e.g., ChemiDoc MP). Verify that proteins are appropriately separated before proceeding to immunoblotting.
5. Activate the LF-PVDF membrane by soaking in methanol for 15–30 s (*see* **Note 10**). Use clean forceps to handle the membrane. Soak gel, activated membrane, filter paper, and transfer sponges in transfer buffer for 15–20 min at room temperature.
6. Assemble the transfer stack and place in wet transfer apparatus. Run at 340 mA for 1–1.5 h at 4°C.
7. Disassemble the blotting sandwich, and place the gel and membrane in a container containing distilled water (*see* **Note 11**). Image the membrane and gel (optional) on an imager capable of imaging stain-free gels and blots to confirm transfer of proteins to membrane.
8. Block the membrane in blocking buffer with gentle shaking for 1 h at room temperature.
9. Dilute primary antibodies in blocking buffer with 0.2% (v/v) Tween 20. Incubate overnight at 4°C with gentle shaking. Dilution factors are as follows: IIH6: 1:1000; β-DG: 1:500; pan(α/β)-DG: 1:500; rabbit GAPDH: 1:50,000 (*see* **Note 12**).
10. Wash the membrane with TBST six times, 5 min each wash (*see* **Note 13**).
11. Incubate in secondary antibody for 1 h at room temperature with gentle shaking: dilute the appropriate IRDye secondary 1: 10,000–1:15,000 in blocking buffer with 0.2% Tween 20 and 0.01% SDS (IIH6: goat anti-mouse IgM; β-DG, goat anti-mouse IgG2a; panDG, donkey anti-sheep IgG; GAPDH, goat anti-rabbit IgG). If using hFAB rhodamine Anti-GAPDH for multiplexing, dilute 1:2500–1:5000.
12. Wash the membrane with TBST six times, 5 min each wash.
13. Rinse the membrane with TBS to remove residual Tween 20.
14. Image the membrane on an appropriate imager, with multiplex channels for the fluorescent secondary antibodies and the stain-free blot (*see* **Note 14**). An example of a western blot in a mouse model of LGMDR9 (before and after treatment) is shown in Fig. 2.
15. Total protein normalization is the preferred method over normalization to a house-keeping protein for Western blot quantification.

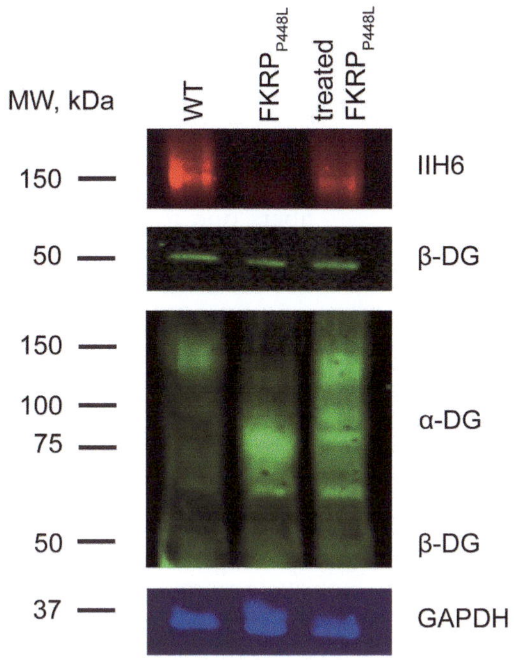

Fig. 2 Western blots for IIH6, βDG, pan-DG, and GAPDH. Protein lysates were made from mouse skeletal muscles from wild type (WT), $FKRP_{P448L}$ mice, and treated $FKRP_{P448L}$ mice given human *FKRP* gene therapy. IIH6 demonstrates changed matriglycan expression (functional glycosylation), βDG shows no change, and pan-DG demonstrates αDG protein expression. For pan-DG, natively glycosylated αDG (top band) runs at about 156 kDa, while lower-molecular-weight bands reflect underglycosylated protein. $FKRP_{P448L}$ muscle shows a clear band-shift to lower-molecular-weight αDG protein with the pan-DG antibody, consistent with reduced functional glycosylation. This is also demonstrated by the reduced IIH6 expression. GAPDH is shown as a control for protein loading and transfer. Note that treatment with *FKRP* gene therapy produces only partial recovery of the wild-type pattern. *MW* molecular weight, *kDa* kilodaltons, *WT* wild type, *β-DG* β dystroglycan, *α-DG* α dystroglycan, *GAPDH* glyceraldehyde-3-phosphate dehydrogenase

4 Notes

1. Fixing with 2% paraformaldehyde in PBS causes an increase in nonspecific background staining. If the tissues need to be fixed with paraformaldehyde, a second fixation with ice-cold methanol is recommended. There will still be higher nonspecific background staining, but it will be reduced by the methanol fixation.
2. Washing with PBS or TBS with a salt concentration of 150 mM has been tested. The membrane staining is not as crisp as when washing with the low salt conditions.

3. Mouse on Mouse blocking reagent is necessary to prevent background staining of mouse Fc receptors with the secondary anti-mouse antibody. Fc receptors are often upregulated in dystrophic skeletal muscle. Adjustments should be made if needed to prevent background staining with secondary antibody alone. Intracellular background staining in skeletal myofibers due to mouse Ig uptake into damaged myofibers also presents a background issue if secondary antibody binding is not blocked appropriately.
4. IIH6 is an anti-mouse IgM antibody that binds to matriglycan, the β1,3Xylα1,3GlcA motif on αDG that allows for laminin binding. This antibody is only stable at 4 °C for several weeks. Some lots of commercial antibody vary in quality, showing variable matriglycan binding (and sometimes showing no binding at all). It is highly recommended that all lots of IIH6 antibody be checked regularly for ability to stain matriglycan by comparing staining of Chinese Hamster Ovary (CHO) cells, where staining is minimal, to CHO cells stably transfected with LARGE, the gene that encodes the tandem enzyme that creates matriglycan, where staining is robust [30]. This is an excellent way to ensure that the antibody is capable of binding the IIH6 epitope using a very strong positive control.
5. The purpose of staining one section with dystrophin alone is to be able to quantitate the level of background IgM staining. If quantitation of IIH6 intensity is desired, the background level of IgM staining can be quantitated in the dystrophin only section and subtracted from the IIH6 intensity level in the immunostained section.
6. If anti-mouse IgM secondary antibody is not available, it is possible to visualize IIH6 staining using an anti-mouse IgG (H + L) antibody, but the signal is not as strong as using an IgM-specific secondary antibody.
7. If the researcher desires to quantitate the intensity of IIH6 staining, we recommend imaging using the Nikon Ti2-E widefield fluorescence microscope with a Lumencor SOLA LED engine and Hamamatsu ORCA Fusion camera and a Nikon Plan Apochromat Lambda 10x objective. These images can then be analyzed using the NIS-Elements AR software for data including but not limited to IIH6 intensity, myofiber diameter, and central nuclei analysis. Methanol fixation can sometimes lead to a more punctate staining pattern with the IIH6 antibody than is seen using a dystrophin antibody. Because of this, it is recommended that quantification of staining for IIH6-positive fibers be set at 50% or more staining of dystrophin-positive fibers. This will ensure that minor

interruptions in membrane staining do to not fail to count positively stained myofibers.

8. TGX stain-free gels allow for normalization to total protein and verification of successful protein transfer to PVDF membrane without additional staining. If not available, any commercially available 4–15% polyacrylamide gel is acceptable, but a total protein stain would be required to normalize expression.
9. It is helpful to remove the bottom gel ridge and excise off the wells to allow for easier handling. We also make at notch at the top right corner so that the orientation of the gel is always known. Placing the gel under a gentle stream of distilled water also helps with removing the gel from the plate.
10. Use of low-fluorescence PVDF membrane is essential to reduce background autofluorescence when using fluorescent secondary antibodies.
11. Clip the top right corner of the membrane so that you can maintain correct orientation of the membrane.
12. Multiplexing of various antibodies is achievable as long as appropriate secondary antibodies that fluoresce at different wavelengths are used. We have successfully imaged IIH6 and β-DG, IIH6 and GAPDH, β-DG and GAPDH, pan-DG and GAPDH, and IIH6 and β-DG and GAPDH. When multiplexing with IRDye 680 and IRDye 800, the IRDye 800 secondary antibody should be used to detect DG (the lower-abundance protein target), and IRDye 680 should be used to detect GAPDH (the more abundant protein).
13. Unlike immunostaining, we do not see a difference in IIH6 for immunoblotting when comparing LS-TBS and TBS, so TBS is used here.
14. When using αDG or α/βDG anti-peptide antisera to blot skeletal muscle protein lysates, quantify natively glycosylated αDG as the band centered on 156 kDa. This band is quite thick due to its high level of glycosylation and so may present as a band ranging from about 140–180 kDa. Quantify underglycosylated αDG protein as all bands between 72 kDa and 140 kDa. Always run a wild-type muscle control so that you can compare underglycosylated pattern to a native glycosylation pattern. Underglycosylated αDG typically presents as thick protein band centered at 100–120 kDa, but it can include lower molecular weight bands down to 72 kDa, which is the molecular weight of αDG when no glycosylation is present. If using an antiserum that recognizes βDG as well αDG, βDG will present as a 43 kDa band. If using an antiserum only to αDG polypeptide, and then quantify βDG using a monoclonal antibody specific to the βDG protein sequence. Keep in mind, however, that the C-terminus of βDG protein, where

antibodies are often made, can contain posttranslational modifications (e.g., tyrosine phosphorylation) that may alter antibody affinity [31]. Also note that in certain situations (e.g., cancer cachexia), βDG glycosylation may be increased [32]. This can also be seen if βDG is expressed in the absence of αDG [33].

References

1. Ervasti JM, Campbell KP (1991) Membrane organization of the dystrophin-glycoprotein complex. Cell 66(6):1121–1131
2. Michele DE, Campbell KP (2003) Dystrophin-glycoprotein complex: post-translational processing and dystroglycan function. J Biol Chem 278(18):15457–15460
3. Wein N, Alfano L, Flanigan KM (2015) Genetics and emerging treatments for Duchenne and Becker muscular dystrophy. Pediatr Clin N Am 62(3):723–742. https://doi.org/10.1016/j.pcl.2015.03.008
4. Godfrey C, Foley AR, Clement E, Muntoni F (2011) Dystroglycanopathies: coming into focus. Curr Opin Genet Dev 21(3):278–285. https://doi.org/10.1016/j.gde.2011.02.001
5. Liu W, Pajusalu S, Lake NJ, Zhou G, Ioannidis N, Mittal P, Johnson NE, Weihl CC, Williams BA, Albrecht DE, Rufibach LE, Lek M (2019) Estimating prevalence for limb-girdle muscular dystrophy based on public sequencing databases. Genet Med 21(11): 2512–2520. https://doi.org/10.1038/s41436-019-0544-8
6. Kanagawa M (2021) Dystroglycanopathy: from elucidation of molecular and pathological mechanisms to development of treatment methods. Int J Mol Sci 22(23). https://doi.org/10.3390/ijms222313162
7. Matsumura K, Ervasti JM, Ohlendieck K, Kahl SD, Campbell KP (1992) Association of dystrophin-related protein with dystrophin-associated proteins in mdx mouse muscle. Nature 360(6404):588–591
8. Ervasti JM, Ohlendieck K, Kahl SD, Gaver MG, Campbell KP (1990) Deficiency of a glycoprotein component of the dystrophin complex in dystrophic muscle. Nature 345(6273): 315–319. https://doi.org/10.1038/345315a0
9. Johnson EK, Zhang L, Adams ME, Phillips A, Freitas MA, Froehner SC, Green-Church KB, Montanaro F (2012) Proteomic analysis reveals new cardiac-specific dystrophin-associated proteins. PLoS One 7(8):e43515. https://doi.org/10.1371/journal.pone.0043515
10. Johnson EK, Li B, Yoon JH, Flanigan KM, Martin PT, Ervasti J, Montanaro F (2013) Identification of new dystroglycan complexes in skeletal muscle. PLoS One 8(8):e73224. https://doi.org/10.1371/journal.pone.0073224
11. Yoon JH, Johnson E, Xu R, Martin LT, Martin PT, Montanaro F (2012) Comparative proteomic profiling of dystroglycan-associated proteins in wild type, mdx, and Galgt2 transgenic mouse skeletal muscle. J Proteome Res 11(9): 4413–4424. https://doi.org/10.1021/pr300328r
12. Flanigan KM, Vetter TA, Simmons TR, Iammarino M, Frair EC, Rinaldi F, Chicoine LG, Harris J, Cheatham JP, Cheatham SL, Boe B, Waldrop MA, Zygmunt DA, Packer D, Martin PT (2022) A first-in-human phase I/IIa gene transfer clinical trial for Duchenne muscular dystrophy using rAAVrh74.MCK.GALGT2. Mol Ther Methods Clin Dev 27: 47–60. https://doi.org/10.1016/j.omtm.2022.08.009
13. Vetter TA, Nicolau S, Bradley AJ, Frair EC, Flanigan KM (2022) Automated immunofluorescence analysis for sensitive and precise dystrophin quantification in muscle biopsies. Neuropathol Appl Neurobiol 48(3):e12785. https://doi.org/10.1111/nan.12785
14. Vainzof M, Passos-Bueno MR, Canovas M, Moreira ES, Pavanello RC, Marie SK, Anderson LV, Bonnemann CG, McNally EM, Nigro V, Kunkel LM, Zatz M (1996) The sarcoglycan complex in the six autosomal recessive limb-girdle muscular dystrophies. Hum Mol Genet 5(12):1963–1969
15. Michele DE, Barresi R, Kanagawa M, Saito F, Cohn RD, Satz JS, Dollar J, Nishino I, Kelley RI, Somer H, Straub V, Mathews KD, Moore SA, Campbell KP (2002) Post-translational disruption of dystroglycan-ligand interactions in congenital muscular dystrophies. Nature 418(6896):417–422
16. Crosbie RH, Straub V, Yun HY, Lee JC, Rafael JA, Chamberlain JS, Dawson VL, Dawson TM, Campbell KP (1998) mdx muscle pathology is

independent of nNOS perturbation. Hum Mol Genet 7(5):823–829
17. Crosbie RH, Lebakken CS, Holt KH, Venzke DP, Straub V, Lee JC, Grady RM, Chamberlain JS, Sanes JR, Campbell KP (1999) Membrane targeting and stabilization of sarcospan is mediated by the sarcoglycan subcomplex. J Cell Biol 145(1):153–165
18. Peters MF, Adams ME, Froehner SC (1997) Differential association of syntrophin pairs with the dystrophin complex. J Cell Biol 138(1): 81–93
19. Peters MF, Sadoulet-Puccio HM, Grady MR, Kramarcy NR, Kunkel LM, Sanes JR, Sealock R, Froehner SC (1998) Differential membrane localization and intermolecular associations of alpha-dystrobrevin isoforms in skeletal muscle. J Cell Biol 142(5):1269–1278
20. Xu R, DeVries S, Camboni M, Martin PT (2009) Overexpression of Galgt2 reduces dystrophic pathology in the skeletal muscles of alpha sarcoglycan-deficient mice. Am J Pathol 175(1):235–247. https://doi.org/10.2353/ajpath.2009.080967
21. Ervasti JM, Campbell KP (1993) A role for the dystrophin-glycoprotein complex as a transmembrane linker between laminin and actin. J Cell Biol 122(4):809–823
22. Yoshida-Moriguchi T, Campbell KP (2015) Matriglycan: a novel polysaccharide that links dystroglycan to the basement membrane. Glycobiology 25(7):702–713. https://doi.org/10.1093/glycob/cwv021
23. Yoshida-Moriguchi T, Yu L, Stalnaker SH, Davis S, Kunz S, Madson M, Oldstone MB, Schachter H, Wells L, Campbell KP (2010) O-mannosyl phosphorylation of alpha-dystroglycan is required for laminin binding. Science 327(5961):88–92. https://doi.org/10.1126/science.1180512
24. Inamori K, Yoshida-Moriguchi T, Hara Y, Anderson ME, Yu L, Campbell KP (2012) Dystroglycan function requires xylosyl- and glucuronyltransferase activities of LARGE. Science 335(6064):93–96. https://doi.org/10.1126/science.1214115
25. Goddeeris MM, Wu B, Venzke D, Yoshida-Moriguchi T, Saito F, Matsumura K, Moore SA, Campbell KP (2013) LARGE glycans on dystroglycan function as a tunable matrix scaffold to prevent dystrophy. Nature 503(7474): 136–140. https://doi.org/10.1038/nature12605
26. Sheikh MO, Capicciotti CJ, Liu L, Praissman J, Ding D, Mead DG, Brindley MA, Willer T, Campbell KP, Moremen KW, Wells L, Boons GJ (2022) Cell surface glycan engineering reveals that matriglycan alone can recapitulate dystroglycan binding and function. Nat Commun 13(1):3617. https://doi.org/10.1038/s41467-022-31205-7
27. Chandel I, Campbell KP (2023) Identification of matriglycan by dual exoglycosidase digestion of alpha-dystroglycan. Bio Protoc 13(18): e4827. https://doi.org/10.21769/BioProtoc.4827
28. Humphrey EL, Lacey E, Le LT, Feng L, Sciandra F, Morris CR, Hewitt JE, Holt I, Brancaccio A, Barresi R, Sewry CA, Brown SC, Morris GE (2015) A new monoclonal antibody DAG-6F4 against human alpha-dystroglycan reveals reduced core protein in some, but not all, dystroglycanopathy patients. Neuromuscul Disord 25(1):32–42. https://doi.org/10.1016/j.nmd.2014.09.005
29. Blaeser A, Keramaris E, Chan YM, Sparks S, Cowley D, Xiao X, Lu QL (2013) Mouse models of fukutin-related protein mutations show a wide range of disease phenotypes. Hum Genet 132(8):923–934. https://doi.org/10.1007/s00439-013-1302-7
30. Yoon JH, Xu R, Martin P (2013) A method to produce and purify full-length recombinant alpha dystroglycan: analysis of N- and O-linked monosaccharide composition in CHO cells with or without LARGE overexpression. PLoS Curr 5. https://doi.org/10.1371/currents.md.3756b4a389974dff21c0cf13508d3f7b
31. James M, Nuttall A, Ilsley JL, Ottersbach K, Tinsley JM, Sudol M, Winder SJ (2000) Adhesion-dependent tyrosine phosphorylation of (beta)-dystroglycan regulates its interaction with utrophin. J Cell Sci 113(Pt 10): 1717–1726
32. Acharyya S, Butchbach ME, Sahenk Z, Wang H, Saji M, Carathers M, Ringel MD, Skipworth RJ, Fearon KC, Hollingsworth MA, Muscarella P, Burghes AH, Rafael-Fortney JA, Guttridge DC (2005) Dystrophin glycoprotein complex dysfunction: a regulatory link between muscular dystrophy and cancer cachexia. Cancer Cell 8(5):421–432. https://doi.org/10.1016/j.ccr.2005.10.004
33. Jayasinha V, Nguyen HH, Xia B, Kammesheidt A, Hoyte K, Martin PT (2003) Inhibition of dystroglycan cleavage causes muscular dystrophy in transgenic mice. Neuromuscul Disord 13(5):365–375

Chapter 8

Quantification of Dystrophin Expression by Capillary Immunoassay

Stefan Nicolau

Abstract

Dystrophin quantification by western blot has been a mainstay of dystrophinopathy research for decades. More recently, the capillary western immunoassay has emerged as an alternative to traditional western blots. With this technique, electrophoresis and detection steps all take place in glass capillaries instead of gels and membranes. Advantages include the ability to perform quantification on very small amounts of protein lysate, as well as high-throughput and reproducibility. We provide here an optimized protocol for capillary western quantification of dystrophin expression, as well as validation data characterizing the performance of the assay.

Key words Protein, Antibody, Immunoprobing, Cell culture, Muscle

1 Introduction

Western blotting has been one of the mainstays of dystrophin protein quantification for decades, enabling the discovery of key disease mechanisms in Duchenne muscular dystrophy and the evaluation of many molecular therapeutics. Indeed, dystrophin expression has been a common endpoint in clinical trials and served as the basis for regulatory approval of several exon skipping drugs and gene therapy [1–5]. Owing to the large size of the dystrophin protein (427 kDa for the muscle isoform), however, this is a time-consuming and technically challenging process. Indeed, questions regarding dystrophin quantification have featured prominently in discussions surrounding the approval of dystrophin-restoring therapies [6–8]. More recently, the capillary western immunoassay has emerged as an alternative to traditional western blots and has been employed in clinical and preclinical studies of dystrophin expression [9–12]. During a capillary western immunoassay, an electrical gradient induces migration of proteins inside a glass capillary filled with matrix, instead of a gel. Proteins are then

Kevin Flanigan and Stefan Nicolau (eds.), *Duchenne Muscular Dystrophy: Methods and Protocols*, Methods in Molecular Biology, vol. 2975, https://doi.org/10.1007/978-1-0716-4811-7_8,

cross-linked to the capillary wall using ultraviolet light, after which the matrix can be removed and replaced with antibodies and detection reagents. Advantages of this assay include (a) a reduction in the amount of sample required, allowing quantification of samples derived from cells or from limited amounts of patient tissue; (b) higher throughput, requiring approximately 2 hours of hands-on work to quantify up to 24 samples; and (c) increased reliability by avoiding technical pitfalls related to gel migration and transfer of such a large protein. Conversely, limitations of the capillary western immunoassay include the limited range of compatible antibodies and a generally higher consumable cost per sample than traditional western blots.

In this chapter, we provide a protocol for dystrophin quantification by capillary western immunoassay. This procedure is suitable for both mouse and human muscle samples, as well as cultured myotubes. The procedure begins with protein extraction and quantification of the protein concentration using the BCA assay, followed by sample preparation, electrophoresis and detection, and data analysis. The choice of antibody and protein concentration were informed by a series of optimization and validation studies, which also established the lower limit of detection of the assay, its linearity, and its reproducibility (Fig. 1). In these experiments, we found that the assay could detect dystrophin expression levels as low as 0.1% of normal and that its reproducibility and linearity compared favorably with traditional western blots.

2 Materials

2.1 Equipment

1. Cryostat.
2. Qiagen TissueLyser II bead mill.
3. 37 °C incubator.
4. Thermal cycler.
5. Microplate reader capable of measuring absorbance at 562 nM.
6. Centrifuge with a microplate rotor.
7. ProteinSimple Jess capillary western instrument.
8. Compass for Simple Western software.

2.2 Consumables

1. 5 mm steel beads. Beads may be washed and autoclaved for reuse.
2. 0.5 mL or 1 mL syringes with 30G needles.
3. Cell scrapers.
4. Clear, flat-bottom 96-well plate.
5. 96-well plate sealing film.

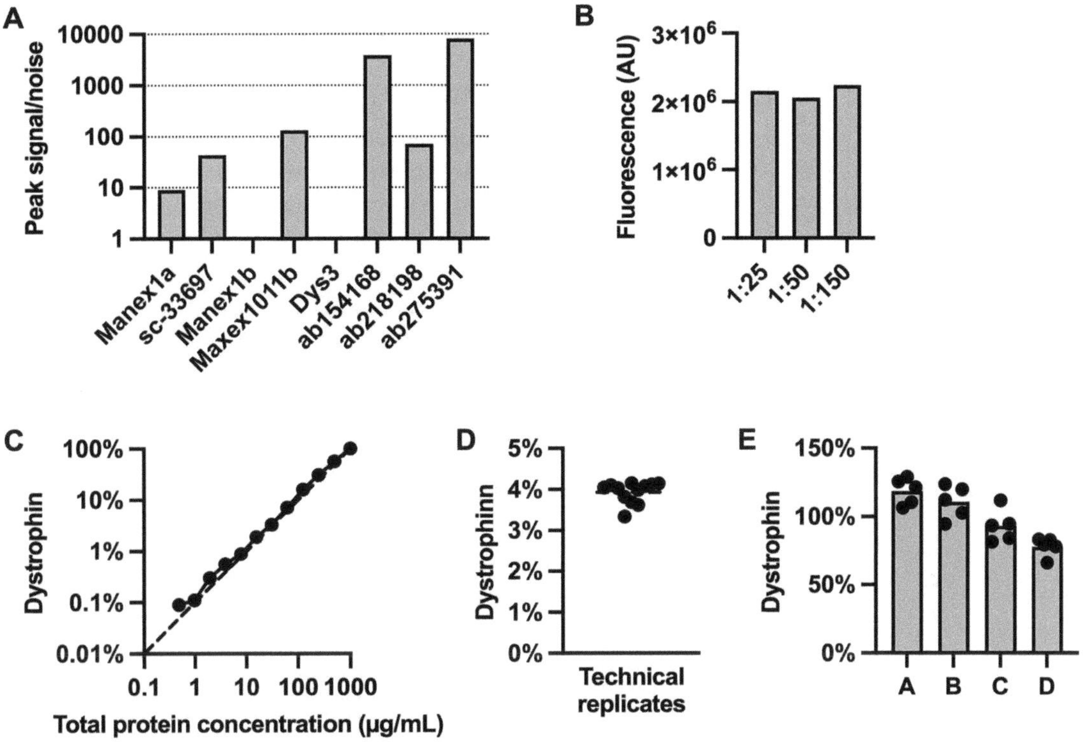

Fig. 1 Assay optimization and performance. (**a**) Two monoclonal rabbit C-terminal antibodies (ab154168 and ab275391) produced the largest peaks and highest signal to noise ratio. The ab275391 antibody however also produced a spurious peak at 203 kDa (not shown). (**b**) A 1:150 concentration of ab154168 was sufficient to maximize signal. (**c**) The assay's linear range was evaluated in serial dilutions of a C57bl/6 mouse muscle protein lysate. The lower limit of detection was approximately 0.1% of normal, and the quantification demonstrated good linearity at levels as low as 0.5% of normal. Similar results were observed with human muscle (not shown). (**d**) A coefficient of variation of 6.5% was observed between 11 technical replicates of a single sample processed concurrently. (**e**) The coefficient of variation was 11% when serial protein extractions and dystrophin quantifications were performed from the same healthy human control muscle samples

6. Deionized water.
7. 70% ethanol.
8. Phosphate-buffered saline (PBS).
9. RIPA buffer: Dilute 1 mL of 10× RIPA buffer in 9 mL of phosphate-buffered saline (PBS). Add one tablet of cOmplete Ultra, Mini Protease Inhibitor Cocktail and one tablet of cOmplete, Mini Protease Inhibitor Cocktail. Unused buffer may be stored at 4 °C for several weeks.
10. Pierce bicinchoninic acid (BCA) protein assay kit.
11. Bovine serum albumin standard set (0.125, 0.25, 0.50, 0.75, 1.0, 1.5, and 2.0 μg/μL).
12. C-terminal rabbit monoclonal dystrophin antibody (Abcam ab154168; *see* **Note 1**).

13. ProteinSimple 66–440 kDa separation module (13 or 25 capillaries per cartridge).
14. ProteinSimple anti-rabbit detection module.
15. 0.1× sample buffer: dilute 10 μL of 10× sample buffer from the 66–440 kDa separation module into 990 μL of deionized water. Unused buffer may be stored at room temperature for several weeks.

3 Methods

3.1 Protein Extraction from Muscle Tissue Using RIPA Buffer

1. Cut approximately 20 20 μm sections of frozen muscle from each tissue sample on a cryostat and collect them in a 2 mL round-bottom tube (*see* **Note 2**).
2. Add a 5 mm steel bead to each sample.
3. Add 150 μL of RIPA buffer to each sample.
4. Disrupt the sample in the bead mill for 2 min at 30 Hz.
5. Incubate on ice for 30 min.
6. Disrupt the sample in the bead mill for 1 min at 30 Hz.
7. Incubate on ice for 30 min.
8. Centrifuge at 20,000 g for 20 min at 4 °C.
9. Collect a total of approximately 90 μL of precipitate-free supernatant into 1.5 mL or 0.2 mL tubes. The samples may be stored at −80 °C until they are used (*see* **Note 3**).

3.2 Protein Extraction from Cultured Cells Using RIPA Buffer

Reagent volumes provided here are for each well in a 6-well plate. Volumes should be scaled accordingly for other plate sizes.

1. Aspirate the medium from each well, wash the cells with 1 mL of PBS, and then carefully aspirate all remaining PBS.
2. Add 80 μL of RIPA buffer to each well, and incubate at room temperature for 1 min.
3. Scrape the cells off the bottom of the well using a cell scraper.
4. Homogenize the cells by triturating with a syringe and 30G needle.
5. Transfer the lysate to a microcentrifuge tube and incubate on ice for 30 min.
6. Centrifuge the lysate at 14,000 g for 20 min at 4 °C.
7. Transfer the supernatant into a new 1.5 mL or 0.2 mL tube. The samples may be stored at −80 °C until they are used.

3.3 Measurement of Protein Concentration Using the BCA Assay

1. Thaw all protein lysates on ice. Keep the lysates on ice until they can be returned to a −80 °C freezer.
2. Prepare 20× diluted aliquots of protein lysates by diluting 2 μL of each protein lysate in 38 μL of RIPA buffer (*see* **Note 4**).
3. Load in triplicate 10 μL of each of the seven protein standards into rows A–G, columns 1–3, of a clear flat-bottom 96-well plate. Load 10 μL of RIPA buffer in row H, columns 1–3.
4. Load in triplicate 10 μL of each 20× diluted protein lysate into the remaining wells (columns 4–12). Up to 24 samples can be assayed in triplicate in a 96-well plate.
5. Prepare the Pierce BCA assay working reagent by mixing 50 parts of Reagent A with 1 part of Reagent B. The total amount of reagent (in mL) needed can be determined using the following formula: (8 + number of samples) × 0.66. Working reagent should be prepared immediately before use and any excess should be discarded.
6. Add 200 μL of working reagent to each well in the 96-well plate, and mix carefully (*see* **Note 5**).
7. Break any bubbles at the surface of the wells and cover the plate with a sealing film.
8. Incubate the plate at 37 °C for 30 min.
9. Measure absorbance in each well at 562 nM using a microplate reader.
10. Perform a linear regression on the absorbance values for each protein standard, and then use this equation to calculate the concentrations of each of the diluted samples. The resulting values should be multiplied by 20 to obtain the stock concentration of each lysate.

3.4 Capillary Western Immunoassay

1. Open the EZ-standard pack from the 66–440 kDa separation module. Briefly centrifuge the tubes, and then make a large opening in the foil sealing each vial using a hypodermic needle.
2. Reconstitute the 400 mM DTT by adding 40 μL of deionized water and gently pipetting.
3. Reconstitute the 5× fluorescent master mix by adding 20 μL of 10× sample buffer and 20 μL of 400 mM DTT and gently pipetting.
4. Reconstitute the biotinylated ladder by adding 20 μL of deionized water and gently pipetting.
5. In strips of 0.2 mL tubes, dilute an aliquot of each lysate to 1 μg/μL using 0.1× sample buffer (*see* **Notes 6–8**). A minimum of 5 μL of each diluted lysate should be prepared.
6. In a new series of PCR strips, combine 1 μL of 5× fluorescent master mix with 4 μL of each diluted lysate.

7. Vortex the samples, and then denature for 5 min at 95 °C in a thermal cycler. Briefly centrifuge to collect the liquid at the bottom of the tubes, and then place on ice.
8. Prepare luminol-peroxide solution by mixing 200 μL of luminol-S from the detection module with 200 μL of peroxide. Pipette to mix, and then place on ice. The solution should be prepared immediately before use and any excess should be discarded.
9. Dilute 2 μL of the primary antibody (C-terminal rabbit monoclonal dystrophin antibody) in 298 μL of antibody diluent 2 from the anti-rabbit detection module.
10. Load the samples, antibody diluent 2, biotinylated ladder, streptavidin-HRP, antibodies, and luminol-peroxide solution in the separation module plate according to the layout shown in the manufacturer's instructions (*see* **Note 9**).
11. Place the included cover over the plate and centrifuge for 5 min at 1000 g to dislodge bubbles from the bottom of the wells.
12. Add 500 μL of wash buffer from the 66–440 kDa separation module to the first three rows of large wells in the plate.
13. Carefully remove the foil covering the reagent wells at the bottom of the plate.
14. Break any bubbles still present in any of the wells (*see* **Note 10**).
15. Clean the Jess instrument door with a lint-free cloth and 70% ethanol.
16. Place the plate in the instrument, pressing down on each corner to ensure it is fully seated.
17. Place a capillary cartridge in the instrument and close the door.
18. Launch the Compass for Simple Western Software, then select File > New assay > Jess....
19. Select the chemiluminescence and 66-440 kDa options, along with the appropriate cartridge size (13 or 25 capillaries), then click OK.
20. Label the samples and antibodies on the plate diagram shown in the lower part of the screen. The electrophoresis and detection parameters should be left at their default values.
21. Click the start button to start the assay.

3.5 Capillary Western Immunoassay Data Analysis

1. Review the run summary to ensure there are no abnormalities in the current and voltage curves and that the video shows no air bubbles were aspirated through the capillaries.
2. Review the automated detection of the standard and ladder peaks, and adjust them as required.
3. Review the dystrophin signal in each capillary (Fig. 2).

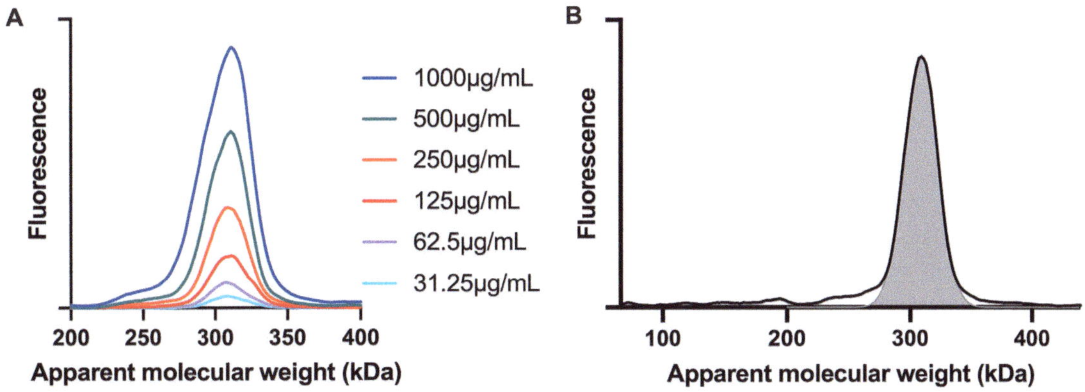

Fig. 2 Data analysis. (**a**) Full-length dystrophin migrates at an apparent molecular weight of 300–310 kDa on this platform. The dystrophin concentration in each sample is proportional to the area under the curve, as shown here with serial dilutions of a control sample. (**b**) A Gaussian fit is used to quantify dystrophin area under the curve

4. Click Edit > Analysis…, and select peak names in the left panel.
5. Add an analysis group, and then enter each dystrophin isoform (peak) that should be quantified, along with the molecular weight at which it migrates (*see* **Note 11**).
6. Select File > Export tables… to export the analysis results.
7. Dystrophin concentrations for each sample may be found in the *Capillary Peaks Area.txt* file and should be normalized to concurrently assayed controls. If lysates were loaded at different concentrations, the values should also be normalized accordingly.

4 Notes

1. Among the antibodies we tested, we have found this antibody to be the most suitable for capillary western immunoassay (Fig. 1). If a different antibody is used, its optimal concentration, signal-to-noise ratio, and dynamic range should be evaluated in a pilot experiment with a serial dilution of control muscle lysate prior to its use on experimental samples.
2. A single muscle fragment of approximately 5–10 mg may be used in place of sections. Smaller amounts of input material may also be used, in which case the amount of lysis buffer should be reduced accordingly.
3. Repeated freeze-thaw cycles should be avoided, and it is therefore recommended to divide the lysate into three to five aliquots.

4. The 20× dilution factor is suitable for measuring protein lysate concentrations between 2.5 μg/μL and 40 μg/μL, which is generally appropriate for muscle protein lysates. If the expected concentrations are outside this range, use a higher or lower dilution factor as needed.
5. Use of a multichannel pipette is recommended in order to fill the plate quickly, as the reaction begins immediately upon adding working reagent. Care should be taken to minimize bubbles created in the wells.
6. The present protocol quantifies dystrophin in a fixed amount of total protein, as measured by the BCA assay, without further normalization. If desired, the Simple Western platform allows experiment designs that additionally incorporate a total protein assay or total protein normalization. In our experience, we have found that normalization to the total protein assay does not improve the reproducibility of results, while combining the dystrophin and protein normalization assays limits the dynamic range of dystrophin quantification.
7. A loading concentration of 1 μg/μL maximizes the dynamic range of the assay and allows quantification of very low levels of dystrophin expression. The sensitivity and linearity of the assay using these parameters are shown in Fig. 1. The linearity of the assay may be improved with loading concentrations of 0.1–0.5 μg/μL, at the cost of sensitivity.
8. RIPA concentrations greater than 0.5× degrade the performance of the assay by broadening peaks on the electropherogram. If the samples have concentrations below 2 μg/mL, they should therefore be diluted at least 1:1 with 0.1× sample buffer.
9. Great care should be taken to avoid introducing bubbles at the bottom of the wells. Reverse pipetting should be used whenever possible. An electronic repeater pipette will accelerate loading of the antibodies and luminol-peroxide solution.
10. Bubbles are easiest to see by placing the plate on a solid color surface and looking at the plate from above.
11. Full-length (427 kDa) dystrophin isoforms migrate at an apparent molecular weight of 300–310 kDa on this platform. The molecular weights of shorter dystrophin isoforms (e.g., microdystrophin and internally truncated dystrophins resulting from deletions or exon skipping) can be empirically determined by examining the electropherograms and then entered into the analysis.

References

1. Mendell JR, Rodino-Klapac LR, Sahenk Z, Roush K, Bird L, Lowes LP, Alfano L, Gomez AM, Lewis S, Kota J, Malik V, Shontz K, Walker CM, Flanigan KM, Corridore M, Kean JR, Allen HD, Shilling C, Melia KR, Sazani P, Saoud JB, Kaye EM, Eteplirsen Study G (2013) Eteplirsen for the treatment of Duchenne muscular dystrophy. Ann Neurol 74(5):637–647. https://doi.org/10.1002/ana.23982
2. Clemens PR, Rao VK, Connolly AM, Harper AD, Mah JK, Smith EC, McDonald CM, Zaidman CM, Morgenroth LP, Osaki H, Satou Y, Yamashita T, Hoffman EP, Investigators CD (2020) Safety, tolerability, and efficacy of viltolarsen in boys with duchenne muscular dystrophy amenable to Exon 53 skipping: a phase 2 randomized clinical trial. JAMA Neurol 77(8):982–991. https://doi.org/10.1001/jamaneurol.2020.1264
3. Frank DE, Schnell FJ, Akana C, El-Husayni SH, Desjardins CA, Morgan J, Charleston JS, Sardone V, Domingos J, Dickson G, Straub V, Guglieri M, Mercuri E, Servais L, Muntoni F, Group S-NS (2020) Increased dystrophin production with golodirsen in patients with Duchenne muscular dystrophy. Neurology 94(21): e2270–e2282. https://doi.org/10.1212/WNL.0000000000009233
4. Zaidman CM, Proud CM, McDonald CM, Lehman KJ, Goedeker NL, Mason S, Murphy AP, Guridi M, Wang S, Reid C, Darton E, Wandel C, Lewis S, Malhotra J, Griffin DA, Potter RA, Rodino-Klapac LR, Mendell JR (2023) Delandistrogene moxeparvovec gene therapy in ambulatory patients (aged >/=4 to <8 years) with duchenne muscular dystrophy: 1-year interim results from study SRP-9001-103 (ENDEAVOR). Ann Neurol 94(5):955–968. https://doi.org/10.1002/ana.26755
5. Finkel RS, Flanigan KM, Wong B, Bonnemann C, Sampson J, Sweeney HL, Reha A, Northcutt VJ, Elfring G, Barth J, Peltz SW (2013) Phase 2a study of ataluren-mediated dystrophin production in patients with nonsense mutation Duchenne muscular dystrophy. PLoS One 8(12):e81302. https://doi.org/10.1371/journal.pone.0081302
6. Aartsma-Rus A, Arechavala-Gomeza V (2018) Why dystrophin quantification is key in the eteplirsen saga. Nat Rev Neurol 14(8): 454–456. https://doi.org/10.1038/s41582-018-0033-8
7. Unger EF, Califf RM (2017) Regarding "Eteplirsen for the treatment of Duchenne muscular dystrophy". Ann Neurol 81(1):162–164. https://doi.org/10.1002/ana.24842
8. Muntoni F, Straub V, Servais L, Mercuri E (2023) Letter to the editor: in response to P.R. Clemens et al., efficacy and safety of viltolarsen in boys with duchenne muscular dystrophy: results from the phase 2, open-label, 4-year extension study, and long-term functional efficacy and safety of viltolarsen in patients with duchenne muscular dystrophy. J Neuromuscul Dis 10(6):1151–1153. https://doi.org/10.3233/JND-239004
9. Koeks Z, Janson AA, Beekman C, Signorelli M, van Duyvenvoorde HA, van den Bergen JC, Hooijmans MT, Alleman I, Hegeman IM, Verschuuren J, JC VD, Spitali P, Datson NA, Niks EH (2021) Low dystrophin variability between muscles and stable expression over time in Becker muscular dystrophy using capillary Western immunoassay. Sci Rep 11(1): 5952. https://doi.org/10.1038/s41598-021-84863-w
10. Beekman C, Janson AA, Baghat A, van Deutekom JC, Datson NA (2018) Use of capillary Western immunoassay (Wes) for quantification of dystrophin levels in skeletal muscle of healthy controls and individuals with Becker and Duchenne muscular dystrophy. PLoS One 13(4):e0195850. https://doi.org/10.1371/journal.pone.0195850
11. Aartsma-Rus A, Morgan J, Lonkar P, Neubert H, Owens J, Binks M, Montolio M, Phadke R, Datson N, Van Deutekom J, Morris GE, Rao VA, Hoffman EP, Muntoni F, Arechavala-Gomeza V, Workshop P (2019) Report of a TREAT-NMD/world duchenne organisation meeting on dystrophin quantification methodology. J Neuromuscul Dis 6(1): 147–159. https://doi.org/10.3233/JND-180357
12. Stephenson AA, Nicolau S, Vetter TA, Dufresne GP, Frair EC, Sarff JE, Wheeler GL, Kelly BJ, White P, Flanigan KM (2023) CRISPR-Cas9 homology-independent targeted integration of exons 1-19 restores full-length dystrophin in mice. Mol Ther Methods Clin Dev 30:486–499. https://doi.org/10.1016/j.omtm.2023.08.009

Chapter 9

Quantification of Dystrophin in Human Muscle Biopsies by Mass Spectrometry

Emily H. Canessa and Yetrib Hathout

Abstract

With the increasing number of dystrophin replacement therapy trials, the need for reliable and accurate dystrophin quantification in muscle biopsies is becoming an important outcome measure. Quantification of dystrophin protein by mass spectrometry provides advantages over methods such as western blotting by offering improved precision, reproducibility, specificity, and a large quantifiable dynamic range. This protocol details a targeted mass spectrometry method for quantifying dystrophin in total muscle protein extracts using stable isotope-labeled dystrophin as a spike-in standard.

Key words DMD, Dystrophin, SILAC, In-gel digestion, SDS-PAGE, Quantification, Targeted mass spectrometry, Parallel reaction monitoring

1 Introduction

Dystrophin protein provides stability to the muscle sarcolemma by acting as a bridge between intracellular sarcomeric actin and the intramembrane dystroglycan complex, which binds to extracellular laminins [1]. Loss of dystrophin, such as in Duchenne muscular dystrophy (DMD), results in sarcolemma instability which in turn leads to muscle fiber degradation and progressive muscle wasting. Patients with DMD often lose ambulation by the age of 12 and have a lower life expectancy [2, 3]. Recently approved dystrophin replacement therapies [4, 5] have increased the need for accurate and reliable quantification methods of restored dystrophin in skeletal muscle [6]. Current methods such as western blotting and immunohistochemistry are commonly used but face challenges such as low lab-to-lab reproducibility, transfer variability, and over-saturation effects [7, 8]. Quantification of dystrophin by mass spectrometry [9, 10] allows excellent linearity over a wide dynamic range [11] with low run-to-run variability [12]. Given the low relative abundance of dystrophin in skeletal muscle (e.g.,

Kevin Flanigan and Stefan Nicolau (eds.), *Duchenne Muscular Dystrophy: Methods and Protocols*, Methods in Molecular Biology, vol. 2975, https://doi.org/10.1007/978-1-0716-4811-7_9,

0.002% of total proteins) [13], highly sensitive and precise methods are critical for accurate quantification. To achieve this goal, stable isotope-labeled human dystrophin protein is spiked as internal standard in patient muscle biopsy lysate before sample processing and analysis. The stable isotope-labeled dystrophin is produced by growing and differentiating human muscle cells in cell culture media where arginine (Arg) and lysine (Lys) are replaced by heavy $^{13}C_6$-Arg and $^{13}C_6$,$^{15}N_2$-Lys [14]. This method overcomes technical variability due to sample handling as well as provides accurate quantification with coefficients of variability (CVs) below 15% and a limit of quantification as low as 2% of the normal levels of dystrophin in healthy muscle [11].

This protocol details the reliable quantification of dystrophin protein by mass spectrometry [15]. Immortalized human myoblasts are cultured and differentiated using Stable Isotope Labeling by Amino Acid in Cell culture (SILAC) medium until all proteins are fully labeled with $^{13}C_6$-Arg and $^{13}C_6$,$^{15}N_2$-Lys. The cells are then harvested and lysed, and the lysate containing stable isotope-labeled dystrophin is used as an internal standard to spike into human muscle lysates. The experimental muscle lysates used as examples in this chapter were obtained from patients with Becker muscular dystrophy (BMD) [16], since they produce dystrophin, though generally at a lower percentage compared to normal. A calibration curve containing dystrophin percentages ranging from 0% to 100% is prepared by mixing muscle lysates from a DMD patient with muscle lysate from healthy controls at different ratios. Each specific ratio is spiked with the SILAC myotube lysate, and all samples are fractionated by SDS-PAGE. Gel bands containing dystrophin are excised and subjected to in-gel digestion. The resultant peptides are analyzed by LC-MS/MS, and a parallel reaction monitoring (PRM) [17] method is used to selectively target dystrophin and filamin-C tryptic peptides for MS/MS. The light/heavy ratios for each peptide are calculated and a calibration curve is used to determine the dystrophin percentages in each experimental sample. Filamin-C is used to normalize this percentage since it is a muscle specific protein [18] and can help to account for the fibrotic tissue common in DMD and BMD muscle biopsies. Thus, we have created a mass spectrometry-based method that can accurately quantify dystrophin protein in muscle biopsies.

2 Materials

2.1 SILAC Myotube Cell Culture

1. Immortalized normal human myoblasts: Obtained from a non-dystrophic person.
2. 0.4% porcine gelatin: Autoclave before use.
3. $^{13}C_6$-Arginine.

4. $^{13}C_6,^{15}N_2$-Lysine.
5. SILAC Complete Growth Media (S-CGM): DMEM for SILAC (without arginine or lysine) media supplemented with 1% penicillin-streptomycin, 15% FBS, and PromoCell growth supplement. Add $^{13}C_6$-arginine at 0.4 mM and $^{13}C_6,^{15}N_2$-lysine 0.78 mM concentrations. Filter media with a 0.2 μm filter unit before use.
6. SILAC Complete Differentiation Media (S-CDM): DMEM for SILAC (without arginine or lysine) media supplemented with 1% penicillin-streptomycin, 2% FBS, and PromoCell differentiation supplement. Add $^{13}C_6$-arginine at 0.4 mM and $^{13}C_6,^{15}N_2$-lysine 0.78 mM concentrations. Filter media with a 0.2 μm filter unit before use.
7. Dulbecco's phosphate buffered saline (DPBS).
8. Trypsin-EDTA for cell culture.

2.2 Sectioning and Lysis of Muscle Tissue Samples

1. Cell lysis buffer: Radio-Immunoprecipitation Assay (RIPA) buffer, 1× protease inhibitor.
2. Microtome.
3. Dry ice.
4. Water (HPLC grade).
5. 1.5 mL polypropylene tubes.
6. 1.5 mL Protein LoBind polypropylene tubes.
7. Polypropylene micro-pipette tips.
8. Muscle lysis buffer: 15 mM dithiothreitol, 0.1 mM Tris (pH 8.0), 10% sodium dodecyl sulfate (SDS), 1× protease inhibitor.
9. Pierce™ BCA Protein Assay Kit.
10. Thermoblock shaker.

2.3 Sample Preparation and SDS-PAGE Gel Electrophoresis

1. 3–8% Tris-Acetate Protein Gel (Biorad, Criterion™ XT Gel, 18 well).
2. HiMark™ Pre-stained Protein Standard.
3. SILAC myotube lysate.
4. Lithium dodecyl sulfate (LDS) sample buffer (4×).
5. Dithiothreitol (DTT).
6. Bio-Safe™ Coomassie Stain (Bio-Rad).
7. Single-edge razor blade.

2.4 In-Gel Digestion

1. Water (HPLC grade).
2. 25 mM ammonium bicarbonate (ABC): prepare with water.
3. Acetonitrile (ACN, HPLC grade).

4. 50 mM iodoacetamide (IAA): prepare with water.
5. 10 mM dithiothreitol (DTT): prepare with water.
6. 12.5 ng/μL trypsin (LC/MS proteomics grade): Solubilize a 20 μg vial of trypsin with 25 mM ABC. Keep on ice before use. Aliquot remaining trypsin in single-use volumes and store at −80 °C (*see* **Note 1**).
7. Vacuum centrifuge.

2.5 Quantification of Dystrophin by Mass Spectrometry

1. Formic acid (LC/MS proteomics grade).
2. Water bath ultrasonicator.
3. Polypropylene autosampler vials.
4. Dionex UltiMate 3000 RS UPLC.
5. Q Exactive™ HF-X Hybrid Quadrupole-Orbitrap™ Mass Spectrometer.
6. EASY-Spray™ LC Column (75 μm diameter, 500 mM length, pore size 100 Å, particle size 2 μm; Thermo Scientific).
7. Xcalibur software.
8. Proteome Discoverer software, version 2.2.
9. Skyline software, version 23.1.

3 Methods

Polypropylene tips and tubes should be used throughout all experiment steps. Other plastics can shed and contaminate samples with polymers that will interfere with peptide detection by mass spectrometry.

3.1 Stable Isotope Labeling by Amino Acid in Cell Culture (SILAC) of Human Myotube Cells

1. Seed 1×10^6 thawed immortalized normal human myoblast cells onto a 0.4% porcine gelatin coated T25 flask. Add 5 mL of S-CGM and allow cells to attach overnight. Transfer the cells into a new flask the next day.
2. Grow cells for five passages (~7 cell doublings) in S-CGM (*see* **Note 2**). Seed 5–10 T75 flasks at 30,000 cells/cm^2 depending on the need (*see* **Note 3**).
3. At 90% confluence (typically 1 day of growth), wash cells with DPBS and switch to S-CDM.
4. Differentiate cells for 12–15 days until they form multinucleated myotubes (*see* **Note 4**).
5. Detach myotubes using trypsin-EDTA and centrifuge at 1000 rpm for 5 min. Wash the cells twice with DPBS, and store the cell pellets at −80 °C in polypropylene tubes.

6. Add cell lysis buffer at three times the cell pellet volume (*see* **Note 5**). Let the cell pellet sit in the buffer on ice for 10 min. Vortex for 1 min and again let cell pellet sit on ice for 5 min.
7. Vortex for 1 min and centrifuge at 14,000 rpm at 4 °C for 10 min.
8. Collect the supernatant, and determine the protein concentration using the Pierce™ BCA Protein Assay Kit.
9. Store protein extract in 250 μg aliquots at −80 °C until needed.
10. Confirm the presence of dystrophin by western blot and confirm its labeling efficiency by mass spectrometry (*see* Subheadings 3.3, 3.4, and 3.5). A good SILAC lysate preparation should contain dystrophin with Arg and Lys that is >90% heavy labeled (*see* **Note 6**).

3.2 Sectioning and Lysis of Muscle Tissue Samples

1. Select muscle biopsies that will be used to generate the dystrophin calibration curve and experimental biopsies that will have their dystrophin content quantified. For the calibration curve, select a dystrophin-deficient muscle biopsy (typically from a DMD patient) that has been confirmed to have 0% dystrophin and any number of normal muscle biopsies from healthy donors (*see* **Note 7**).
2. Section human muscle tissue samples at 10 μm thickness using a microtome. Collect 50–75 cryosections per sample, and place them directly into pre-chilled polypropylene tubes.
3. Store sections at −80 °C until use.
4. Retrieve samples from the freezer when needed and store in dry ice until lysis. It is recommended to run the six-point calibration curve and ten experimental samples on the same 18-well gel.
5. To lyse muscle samples, individually add 200 μL of muscle lysis buffer (*see* **Note 8**) to a sample, and vortex the sample at high speed for 40 s. Immediately after, heat at 90 °C for 2 min, and then keep at room temperature (RT) (*see* **Note 9**). Proceed with lysis of all remaining samples the same way.
6. Centrifuge at 14,000 rpm for 10 min and collect the supernatant in a LoBind polypropylene tube.
7. Determine protein concentration using the Pierce™ BCA Protein Assay Kit.
8. Keep lysates at room temperature to prevent SDS precipitation, and use immediately.
9. When finished preparing SDS-PAGE gel, aliquot remaining sample, and flash freeze in liquid nitrogen. Store frozen lysates at −80 °C for backup analysis.

3.3 Sample Preparation and SDS-PAGE Gel Electrophoresis

1. For dystrophin quantification prepare a calibration curve using DMD muscle lysate and normal muscle lysate containing 100% dystrophin (*see* **Note 7**). The total amount of protein should be 30 μg of protein for each point of the curve as described in Table 1. Dilute the DMD lysate and normal lysate to 2 μg/μL using water. Each mixture containing a specific dystrophin percentage should be prepared in a separate tube. If protein concentrations in experimental samples are known to be lower than 30% of normal, a lower range calibration curve can be used (Table 2). For this curve dilute DMD and normal lysates to concentrations that allow for accurate pipetting before mixing the healthy and DMD muscle lysates.
2. Transfer 30 μg of each experimental sample to its own separate tube.
3. Spike each calibration point and experimental sample with 30 μg of SILAC myotube lysate.
4. Bring the volume of each sample to 20 μL using water, and add 1 μL of 200 mM DTT. Add 7.5 μL of 4× LDS sample buffer, and incubate samples at 70 °C for 5 min.

Table 1
Dystrophin calibration curve (0–100%)

% Dystrophin	DMD lysate (μg)	Normal lysate (μg)
0	30	0
20	24	6
40	18	12
60	12	18
80	6	24
100	0	30

Table 2
Dystrophin calibration curve (0–30%)

% Dystrophin	DMD lysate (μg)	Normal lysate (μg)
0	30	0
5	28.5	1.5
10	27	3
20	24	6
30	21	9
100	0	30

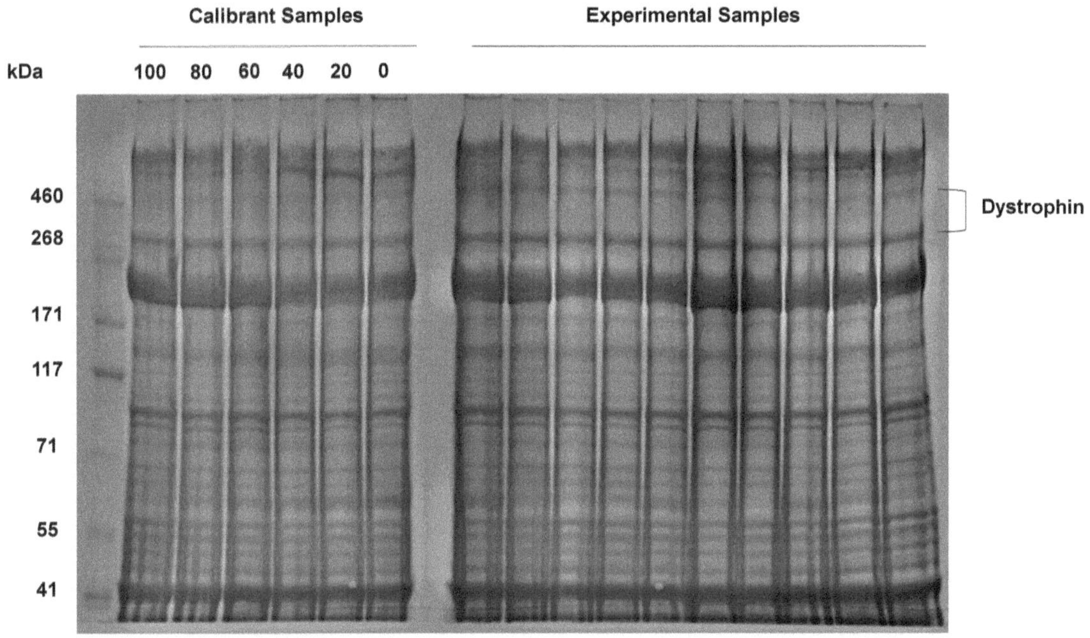

Fig. 1 SILAC-spiked muscle lysates ran for 1 h at 200 volts. Calibrant samples are DMD and normal muscle lysates combined at different ratios to create samples with 0–100% dystrophin. Experimental samples have an unknown amount of dystrophin that will be quantified. The 260–460 kDa region is indicated, showing where gel pieces containing dystrophin should be excised

5. Load the calibration and experimental samples onto a 3–8% Tris-acetate protein gel. Make sure to skip a well between sample types (Fig. 1) (*see* **Note 10**). Add 10 μL of HiMark™ Pre-stained Protein Standard to track the high-molecular-weight proteins in the gel.
6. Run the gel at 200 volts for 1 h to fractionate the proteins.
7. Stain the gel with Bio-Safe™ Coomassie Stain overnight at 4 ° C (*see* **Note 11**).

3.4 In-Gel Digestion

1. Destain the gel by rinsing the gel three times with water. Place the gel on a glass surface (*see* **Note 12**).
2. Using a single-edge razor blade, excise gel areas between 260 and 460 kDa from each lane (*see* Fig. 1) (*see* **Note 13**). Cut each band into small pieces of 1 mM^3 (*see* **Note 14**). Place the pieces into a 1.5 mL polypropylene tube filled with 100 μL of 25 mM ABC.
3. To remove Coomassie stain, add 100 μL of ACN to each tube, and let sit for 10 min. Vortex every 5 min to mix ABC and ACN (*see* **Note 15**). Remove supernatant and repeat ABC and ACN incubations two times. Remove supernatant and add 100 μL of ACN. Incubate for 5 min until gel pieces turn white and are fully dehydrated.

4. Add 100 μL of 10 mM DTT to cover gel pieces, and incubate for 30 min at 37 °C. Remove DTT and add 100 μL of 50 mM IAA (*see* **Note 16**). Incubate for 30 min at RT with tubes covered by aluminum foil to protect them from the light (*see* **Note 17**). Remove IAA.
5. Add 100 μL of ACN and incubate for 5 min. Remove ACN and add 100 μL of ABC. Add 100 μL of ACN and let sit for 10 min with vortexing every 5 min.
6. Remove supernatant and add 100 μL of ACN. Let sit for 5 min and replace supernatant with 100 μL of fresh ACN. Incubate for 5 min and remove supernatant.
7. Add 30 μL of 12.5 ng/μL trypsin to gel pieces and cover with 50 μL of 25 mM ABC. Incubate on ice for 15 min (*see* **Note 18**). Incubate at 37 °C overnight.
8. After 16–20 h of digestion, add 70 uL of 60% ACN in water with 0.1% formic acid to each tube. Place tubes containing gel pieces on a thermoblock shaker, and incubate for 5 min at 1000 rpm (at RT). Spin down and transfer the supernatant from each sample to its own new LoBind tube.
9. Repeat **step 9** two times and add the supernatant to same tube. Repeat incubation one last time with 50 uL of can, and add the supernatant to same tube.
10. Dry the samples in a vacuum centrifuge until completely dry, and store eluted peptides at −20 °C. The samples are now ready for mass spectrometry analysis.

3.5 Quantification of Dystrophin by Mass Spectrometry

1. Solubilize the 100% dystrophin sample in 15 μL of 0.1% FA. S-hake on thermoblock shaker for 5 min at 1000 rpm. Sonicate in a water bath ultrasonicator for 2 min. Centrifuge samples and transfer 13 μL into a polypropylene autosampler vial (*see* **Note 19**). Create a blank of 0.1% FA in a separate vial.
2. Inject 4 μL of the 100% dystrophin sample onto a Dionex UltiMate 3000 RS UPLC system tandem to a Q Exactive™ HF-X Hybrid Quadrupole-Orbitrap™ Mass Spectrometer. The samples are separated on an EASY-Spray™ LC Column at 35 °C (*see* **Note 20**).
3. Run the sample using a standard data-dependent acquisition (DDA) method that has been optimized for the mass spectrometer in use. The analysis should be done in positive mode. Suggested chromatography and mass spectrometry settings are detailed in Tables 3 and 4, respectively.
4. Analyze the .raw file using Proteome Discoverer and search against an updated human database. Include a tolerance of 10 ppm for precursor ions and 0.05 Da for transitions ions. Include as variable modifications oxidation (Methionine),

Table 3
Liquid chromatography method

Step	Time (min)	Flow (μL/min)	% A (0.1% FA)	% B (80% ACN, 0.1% FA)
1	0	0.3	98	2
2	8	0.3	98	2
3	47	0.3	60	40
4	49	0.3	5	95
5	53	0.3	5	95
6	55	0.3	98	2
7	70	0.3	98	2

Table 4
DDA method parameters

Parameter	Tune file	Full MS	dd-MS/MS	dd settings
Ion spray voltage	2 kV.			
Capillary temperature	350 °C.			
S-lens RF level	60			
Scan range (m/z)		375–1800		
Resolution		120,000	30,000	
Automatic gain control (AGC) target		3.00E+06	2.00E+05	
Maximum injection time (IT)		100 ms	50 ms	
Loop count			10	
Isolation window (m/z)			1.5	
Fixed first mass (m/z)			100	
(Normalized) collision energy			nce: 27	
Minimum AGC target				1.50E+03
Apex trigger				None
Charge exclusion				Unassigned, 1, 5–8, >8
Peptide match				Preferred
Exclude isotopes				On
Dynamic exclusion				20.0 s

Table 5
PRM peptides

Protein	Peptide	Expected RT (min)	Light precursor ion	Heavy precursor ion	Charge
Dystrophin	SYAYTQAAYVTTSDP TR	37	947.9445	950.9546	2
	VGNILQLGSK	39	514.8086	518.8157	2
	VLQEDLEQEQVR	34.5	743.3808	746.3909	2
	LFQKPANFEQR	31	459.9140	464.5921	3
	TAALQSATPVER	30	622.3357	625.3457	2
	TLNATGEEIIQQSSK	36.5	809.9178	813.9249	2
	SLEGSDDAVLLQR	39	701.8623	704.8723	2
Filamin-C	VTVLFAGQNIER	41	673.8750	676.8850	2
	GAGGQGQLDVR	27.5	529.2729	532.2830	2
	VYNVTYTVK	34	543.7951	547.8022	2

Table 6
An example inclusion list containing one dystrophin (VGNILQLGSK) and filamin-C (VYNVTYTVK) peptide

Mass [m/z]	CS [z]	Polarity	Start [min]	End [min]	(N)CE	Comment
514.808	2	Positive	37.00	41.00	27	VGNILQLGSK
518.815	2	Positive	37.00	41.00	27	VGNILQLGSK
543.795	2	Positive	39.30	43.30	27	VYNVTYTVK
547.802	2	Positive	39.30	43.30	27	VYNVTYTVK

carbamidomethylation (cysteine), N-terminal acetylation, $^{13}C_6$-arginine, and $^{13}C_6$,$^{15}N_2$-Lysine (*see* **Note 21**).

5. After analysis select peptides for dystrophin and filamin-C quantification. Suggested peptides are found in Table 5 (*see* **Note 22**). Select at least one peptide for each protein.
6. Create an inclusion list in the Xcalibur software Method Editor. Make a separate entry for the light and heavy precursor ion each peptide and create a retention time window ±2 min (*see* **Note 23**) of the observed retention time in the DDA analysis (*see* Table 6). Set up a PRM analysis as detailed in Table 7. All chromatography settings are kept the same as in the DDA method.
7. Solubilize all calibration and experimental samples as described in **steps 1–2**. Run these samples using the PRM method

Table 7
PRM method parameters

Parameter	Full MS	PRM
Scan range (m/z)	300–1000	
Resolution	60,000	60,000
AGC target	3.00E+06	5.00E+05
Maximum IT	100 ms	300 ms
Isolation window (m/z)		1.2
Fixed first mass (m/z)		100
(Normalized) collision energy		nce: 27

created in **step 6**. Run a Blank using the same method after each sample.

8. Analyze the .raw files from the PRM runs as described in **step 4**.
9. Copy the .raw files from the PRM analysis and the .msf files generated by Proteome Discoverer into a separate folder and create a new Skyline document in that folder. Add the inclusion list peptides to the "Targets" list, and "Import" the files to start a "Peptide search." Select the PRM workflow setting, and adjust the following parameters to match those from the PRM analysis (*see* Table 7).
10. Once the analysis finishes, select a peptide's light precursor ion, and click the plus sign on the left to see the transition ions. Manually adjust the "Peak Boundaries" of the peptide chromatogram to select the peak. Identify the top three ions for the peptide (*see* **Note 24**), and record the "Ratio to Standard" for each ion. The ratio for each peptide is calculated as the peak area of the *Light* peptide divided by the area of the *Heavy* peptide (*L/H*) (*see* **Note 25**) Average these ratios to get the final ratio for the peptide in that sample. Repeat this process for all samples (*see* Fig. 2).
11. Create a calibration curve using the ratios from the calibrant samples (*see* Fig. 2), and use this calibration curve to obtain the dystrophin percentage for each experimental sample (*see* Fig. 3). Normalize this percentage using the filamin-C ratio from each sample to get the final dystrophin percentage.

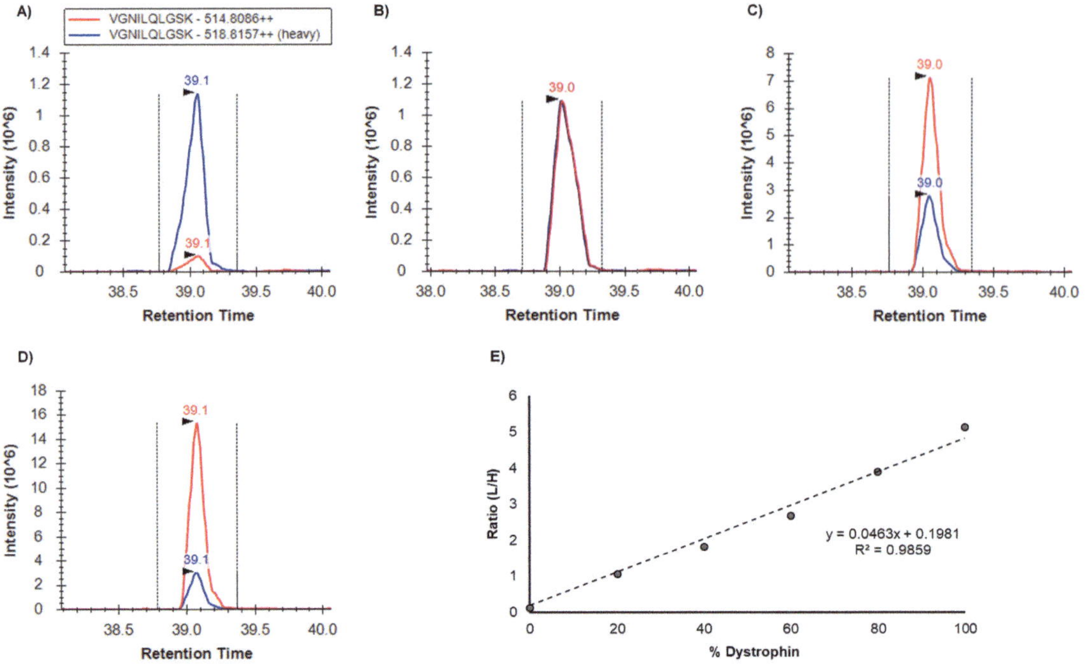

Fig. 2 Example skyline chromatograms of selected calibrant (**a**–**d**) points. Calibrant samples are in order: 0, 20, 60, and 100% dystrophin. The chromatogram indicates the light (red) and heavy (blue) form of each peptide. Ratios (L/H) are calculated from the chromatograms and used to generate the calibration curve. All six calibrant samples are used to create a calibration curve (**e**)

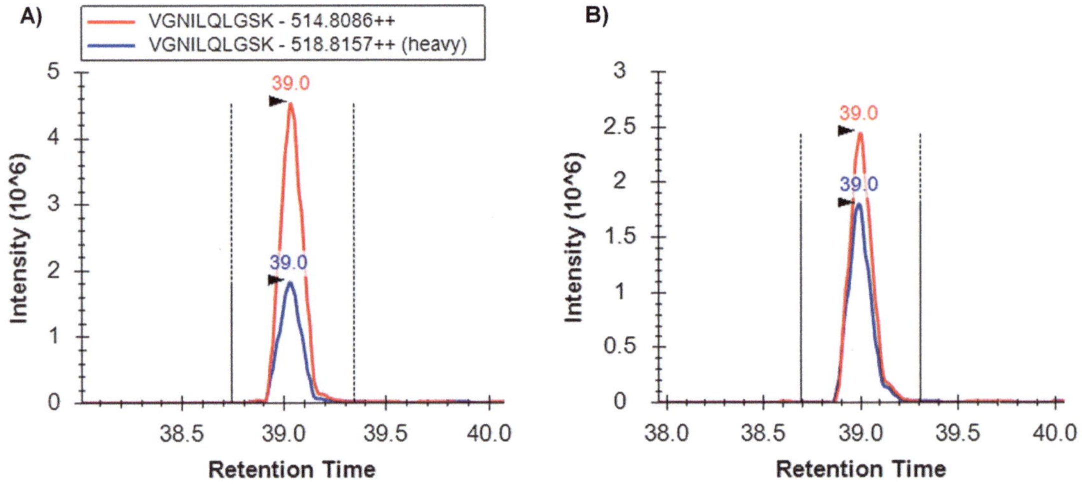

Fig. 3 Example skyline chromatograms of two experimental samples. The experimental samples have ratios of 2.5 (**a**) and 1.4 (**b**) and, by using the calibration curve from Fig. 2, are calculated to have dystrophin percentages of 49% and 25%. When normalized to filamin-C these sample have final quantification values of 83% and 70% dystrophin

4 Notes

1. Do not thaw and refreeze trypsin; doing so can reduce its efficiency [19].
2. Successive passaging is critical to ensure the cultured myotubes are fully labeled, as well as the full-length dystrophin they produce.
3. One T75 flask of lysed myotubes will generally produce about 3–5 mg of protein lysate.
4. Myotube cell lines can differentiate and make dystrophin at unique rates. Due to the high cost of heavy labeled amino acids, an experiment should be performed in non-SILAC media to determine when during differentiation the highest amount of dystrophin is produced. This can be determined easily by differentiating and harvesting cells at successive time-points over a 1- to 2-week time period and then performing a western blot to quantify dystrophin expression. Since the cells should grow the same way under SILAC and non-SILAC conditions, the SILAC cells should be differentiated until the maximum dystrophin expression timepoint is achieved.
5. Adding cell lysis buffer at this ratio will ensure the cell lysate is concentrated at 5–10 ug/μL. A high concentration of SILAC standard is needed when spiking muscle lysates for gel electrophoresis (*see* Subheading 3.3, **step 3**).
6. It is very hard to reach 100% labeling since a small amount of endogenous arginine and lysine are recycled by the cells.
7. The lysates from four muscle biopsies can be pooled together and used for generating the calibration curve. Doing so can reduce any bias caused by choosing only one sample as a normal standard. It is recommended to pool biopsy lysates from four different subjects, though this is flexible depending on sample availability. Repeat BCA protein quantification after pooling samples to determine the normal pool protein concentration.
8. It has been observed that higher SDS concentrations in lysis buffers allow for better solubilization of the cell membrane. This is especially true in tissues from DMD patients where fibrous and fatty tissue can constitute a larger percentage of the biopsy. It is recommended to use 10% SDS buffer for lysis, as higher concentrations can create interference when performing BCA. 5% SDS is acceptable for quantification as well, though will yield lower dystrophin amounts. To address this, all study samples and calibration samples should be prepared with the same SDS concentration.
9. Quick vortexing and heating is critical to preventing protein degradation.

10. Having a gap prevents the last calibration sample from spilling over into an experimental sample.
11. The gel can be incubated in stain for as little as 1 h, though overnight staining results in darker bands. This is helpful when later excising bands from the gel.
12. For this purpose, we use a bottom of a Pyrex dish. An orange sheet of paper can be placed underneath the glass dish to provide more contrast of the gel bands.
13. This range is where normal length dystrophin (427 kDa) and proteins that can be used for normalization such as filamin-C (290 kDa) fall. If quantifying truncated dystrophin or micro-dystrophin [4], adjust the cutting region as needed.
14. Smaller pieces will increase the surface area of the gel band, and improve digestion [15]. Make sure not to cut too small, as tiny gel fragments can block pipette tips. Reduce touching the gel pieces by hand to prevent keratin contamination. Handle gel pieces with razor blade and tips whenever possible.
15. ABC and ACN will separate over time, with ACN rising to the top. Vortexing also helps the gel absorb the incubating solution.
16. DTT reduces cysteine bonds, while IAA alkylates the cysteines to prevent the bonds from reforming. Breaking cysteine bonds fully dentures the proteins and improves protein digestion efficiency [20].
17. IAA is light-sensitive can degrade before all proteins are alkylated [20].
18. Make sure gel pieces become clear and fully absorb the trypsin.
19. Centrifuging helps remove any small particulates from the sample that may have been generated by the in-gel digestion procedure.
20. This protocol has been optimized for Thermo Scientific instruments and software. Other instruments can be used, though will require additional optimization.
21. Including heavy arginine and lysine allows for identification of peptides from the spiked SILAC myotubes.
22. If quantifying micro-dystrophin or dystrophin from a BMD biopsy or make sure to choose a peptide in a non-mutated exon.
23. Various factors such as temperature, humidity, and column age can affect retention time stability [21]. Increase the window if seeing large shifts in retention time. Consider changing to a new column as well.

24. The top three transition ions for peptide are the ions that are unique to that peptide and ionize well. Avoid y1 and b2 as these ion signals can have interference in a complex sample.
25. Because the same amount of SILAC lysate is spiked into each sample, the ratio of two different samples can be compared. As an example, sample A has a dystrophin ratio of 10 and sample B has a ratio of 1. Thus, sample A has 10× the amount of dystrophin as sample B.

References

1. Petrof B, Shrager J, Stedman HH et al (1993) Dystrophin protects the sarcolemma from stresses developed during muscle contraction. Proc Natl Acad Sci USA 90(8):3710–3714. https://doi.org/10.1073/pnas.90.8.3710
2. Nigro G, Comi LI, Politano L et al (1990) The incidence and evolution of cardiomyopathy in Duchenne muscular dystrophy. Int J Cardiol 26(3):271–277. https://doi.org/10.1016/0167-5273(90)90082-g
3. Mah JK, Korngut L, Dykeman J et al (2014) A systematic review and meta-analysis on the epidemiology of Duchenne and Becker muscular dystrophy. Neuromuscul Disord 24:482–491. https://doi.org/10.1016/j.nmd.2014.03.008
4. Mendell JR, Shieh PB, McDonald CM et al (2023) Expression of SRP-9001 dystrophin and stabilization of motor function up to 2 years post-treatment with delandistrogene moxeparvovec gene therapy in individuals with Duchenne muscular dystrophy. Front Cell Dev Biol. https://doi.org/10.3389/fcell.2023.1167762
5. Vasterling ME, Maitski RJ, Davis BA et al (2023) AMONDYS 45 (Casimersen), a novel antisense phosphorodiamidate morpholino oligomer: clinical considerations for treatment in Duchenne muscular dystrophy. Cureus 15(12):e51237. https://doi.org/10.7759/cureus.51237
6. Anthony K, Arechavala-Gomeza V, Taylor LE et al (2014) Dystrophin quantification biological and translational research implications. Neurology 83:2062–2069. https://doi.org/10.1212/WNL.0000000000001025
7. Bass JJ, Wilkinson DJ, Rankin D et al (2017) An overview of technical considerations for western blotting applications to physiological research. Scand J Med Sci Sports 27(1):4–25. https://doi.org/10.1111/sms.12702
8. Anthony K, Arechavala-Gomeza V, Taylor LE et al (2014) Dystrophin quantification: biological and translational research implications. Neurology 83(22):2062–2069. https://doi.org/10.1212/WNL.0000000000001025
9. Farrokhi V, Walsh J, Palandra J et al (2022) Dystrophin and mini-dystrophin quantification by mass spectrometry in skeletal muscle for gene therapy development in Duchenne muscular dystrophy. Gene Ther 9(10–11):608–615. https://doi.org/10.1038/s41434-021-00300-7
10. Tominari T, Takatoya M, Matsubara T et al (2024) Establishment of a triple quadrupole HPLC-MS quantitation method for dystrophin protein in mouse and human skeletal muscle. Int J Mol Sci 25(1):303. https://doi.org/10.3390/ijms25010303
11. Canessa EH, Goswami MV, Alayi TD et al (2020) Absolute quantification of dystrophin protein in human muscle biopsies using parallel reaction monitoring (PRM). J Mass Spectrom 55(2):e4437. https://doi.org/10.1002/jms.4437
12. Chen X, Wei S, Ji Y et al (2015) Quantitative proteomics using SILAC: principles, applications, and developments. Proteomics 15(18):3175–3192. https://doi.org/10.1002/pmic.201500108
13. Chelly J, Kaplan JC, Maire P et al (1988) Transcription of the dystrophin gene in human muscle and non-muscle tissue. Nature 333(6176):858–860. https://doi.org/10.1038/333858a0
14. Ong SE, Mann M (2006) A practical recipe for stable isotope labeling by amino acids in cell culture (SILAC). Nat Protoc 1(6):2650–2660. https://doi.org/10.1038/nprot.2006.427
15. Shevchenko A et al (2006) In-gel digestion for mass spectrometric characterization of proteins and proteomes. Nat Protoc 1(6):2856–2860. https://doi.org/10.1038/nprot.2006.468
16. Worton R (1995) Muscular dystrophies: diseases of the dystrophin-glycoprotein complex. Science 270(5237):755–756. https://doi.org/10.1126/science.270.5237.755

17. Rauniyar N (2015) Parallel reaction monitoring: a targeted experiment performed using high resolution and high mass accuracy mass spectrometry. Int J Mol Sci 16(12): 28566–28581. https://doi.org/10.3390/ijms161226120
18. Han S, Cui C, Zhao X et al (2021) Filamin C regulates skeletal muscle atrophy by stabilizing dishevelled-2 to inhibit autophagy and mitophagy. Mol Ther Nucleic Acids. https://doi.org/10.1016/j.omtn.2021.11.022
19. Vatić S, Mirković N, Milošević JR et al (2021) Trypsin activity and freeze-thaw stability in the presence of ions and non-ionic surfactants. J Biosci Bioeng 3:234–240. https://doi.org/10.1016/j.jbiosc.2020.10.010
20. Sechi S, Chait B (1998) Modification of cysteine residues by alkylation. A tool in peptide mapping and protein identification. Anal Chem 70(24):5150–5158. https://doi.org/10.1021/ac9806005
21. Dolan J (2002) Temperature selectivity in reversed-phase high performance liquid chromatography. J Chromatogr A 965(1–2): 195–205. https://doi.org/10.1016/s0021-9673(01)01321-8

Chapter 10

Isolation and Culture of Primary Myoblasts from Humans and Mice

Felipe de Souza Leite and Emanuela Gussoni

Abstract

Primary myoblasts are generated from committed stem cells, responsible for repairing skeletal muscle fibers upon injury. They are a valuable resource for scientific research in muscle disease and drug screening and hold promise for cell-based therapies. However, skeletal muscle biopsies contain a plethora of non-myogenic cells, exacerbated in diseased muscle tissues. Thus, purification methods of the myogenic cell population are necessary. This protocol describes techniques for dissociating cells from human and mouse skeletal muscle biopsies and enrichment for a highly myogenic population using magnetic isolation for mouse muscles and by fluorescence-activated cell sorting (FACS) using cell surface antibodies for humans. We also describe methods to expand myoblasts in culture and assess myogenicity.

Key words Skeletal muscle, Tissue dissociation, Magnetic enrichment, Fluorescence-activated cell sorting (FACS), Cell culture

1 Introduction

Myogenic cells derived from skeletal muscle are often used to study cell division and regeneration potential in normal and diseased muscle [1–3]. Mouse models are available for many types of muscle disorders, including muscular dystrophies and myopathies [4–10]. These models have been instrumental in studying some aspects of human disease pathogenesis and for testing new promising therapies [11, 12]. While many aspects of human disease are often conserved in animal models, some exceptions have also been noted, and parallel studies of human and mouse cells can be necessary to determine similarities and differences [13]. Skeletal muscles are mostly composed of mature myofibers; however, many cell types are also contained, including immune cells, vascular cells, myogenic stem cells, and fibro-adipogenic progenitors. These cell types cooperate to maintain healthy muscle homeostasis and can often be altered in diseased muscle. Thus, isolation of primary cells from

Kevin Flanigan and Stefan Nicolau (eds.), *Duchenne Muscular Dystrophy: Methods and Protocols*, Methods in Molecular Biology, vol. 2975, https://doi.org/10.1007/978-1-0716-4811-7_10,

mouse models and from human tissue can be essential for downstream studies on pathogenesis and/or for translational studies. Importantly, while tissue dissociation of mouse and human muscles can be achieved using essentially identical experimental steps, cell surface antigens for the enrichment of myogenic cells are species-specific [14–18], and different strategies must be implemented for selection of myogenic progenitors from mouse or human tissues [17, 19, 20].

This chapter describes the common experimental steps for dissociating mononuclear cell populations from mouse or human muscles. Next, individual purification protocols are included for mouse and human cells. A magnetic enrichment protocol is used for isolation of myogenic progenitors from dissociated mouse tissue [17, 19, 20]. This method is a negative selection strategy, where non-myogenic cells (immune cells, endothelial cells and fibro-adipogenic progenitors are retained in the magnetic column while myogenic cells are collected in the flow-through portion. For purification of human myogenic cells, a FACS-based protocol is used via the cell surface markers CD56 and CD82 [14, 21].

2 Materials

2.1 Dissociation of Primary Cells (Human and Mouse)

1. Disposable scalpels with retractable stainless steel blade size #10.
2. Sterile 10 cm tissue culture-treated plastic dishes.
3. Assorted sterile pipettes (5, 10, and 25 mL).
4. Sterile 0.22 μm polyethersulfone (PES) low protein binding membrane filters (500 mL).
5. Sterile conical centrifuge tubes (15 and 50 mL).
6. (Optional) gentleMACS C Tubes.
7. (Optional) GentleMACS Octo Dissociator with heaters, or tissue culture incubator at 37 °C.
8. Sterile nylon cell strainers (100 μm and 40 μm pore sizes).
9. 10× Hank's Balanced Saline Solution (HBSS) or 10× phosphate base salt (PBS), free of calcium chloride, magnesium chloride, and magnesium sulfate. Dilute 10× stock to 1× with double-distilled water, and filter-sterilize using a 0.22 μm sterile filter unit. This solution can be stored at 4 °C or room temperature.
10. Complete growth medium: High glucose (4.5 g) Dulbecco's Modified Eagle's Medium (DMEM), 20% fetal bovine serum (FBS), 1% Penicillin-Streptomycin-Glutamine (PSG 100× stock). Mix 395 mL of high glucose DMEM with 100 mL of FBS (*see* **Note 1**) and 5 mL of 100× PSG. Sterilize by filtering

the solution through a 500 mL 0.22 μm PES filter unit. Store at 4 °C and use within 1 month.

11. Sterile HEPES buffered saline solution without phenol red.
12. 1 M calcium chloride solution: Dissolve 1.47 g calcium chloride dihydrate powder in 10 mL of double-distilled water. Store at 4 °C.
13. Dispase II stock solution at 2.4 U/mL: Dissolve 1 g dispase II powder in 100 mL HEPES buffered saline. Check the total units for the specific dispase lot, and add the appropriate volume of DMEM to reach the 2.4 U/mL concentration. Filter-sterilize the solution through a PES 500 mL filter; aliquot into 15 mL conical tubes (10 mL/tube), and store aliquots at −20 °C.
14. Collagenase D stock solution: dissolve 2.5 g of collagenase D powder in 250 mL of 1× HBSS solution supplemented with 1.25 mL of 1 M $CaCl_2$. Then, sterilize it by filtering it through a PES filter unit. The filtered solution can be dispensed in 15 mL conical tubes (10 mL/tube) and stored at −20 °C.
15. (Optional) Sterile freezing medium: Prepare the desired volume of freezing medium with 90% FBS and 10% dimethyl sulfoxide (DMSO), and immediately store on ice. Store unused amounts of sterile freezing medium at 4 °C for up to 4 weeks.
16. Sterile red blood cell lysis solution (such as Qiagen RBC Lysis solution). Store the solution at room temperature.
17. Sterile CryoTube™ vials (1.8 mL).
18. Benchtop centrifuge.
19. Hemocytometer.
20. Sterile laminar flow biosafety cabinet.
21. −150 °C freezer, liquid nitrogen storage tank.
22. Humidified 5% CO_2 incubator set to 37 °C.

2.2 Magnetic Enrichment of Mouse Myogenic Cells

1. MACS LS depletion columns (Miltenyi).
2. Satellite cell isolation cocktail, mouse (Miltenyi).
3. Cell debris removal solution (Miltenyi).
4. Miltenyi magnet (such as Quadro MACS) and magnet stand.
5. Sterile 1× PBS 0.5% bovine serum albumin (BSA) 2 mM EDTA (abbreviated PBS + BSA + EDTA). For 500 mL solution, dilute 50 mL of 10× PBS (no Ca++ or Mg++) to 1× using distilled water. Add 2.5 g BSA (cell culture grade) and 0.37 g EDTA. Sterilize the solution by filtering it and store at 4 °C until used.
6. 15 mL collection tubes.

7. 5 mL sterile pipettes.
8. Benchtop centrifuge.

2.3 FACS-Sorting Human Cells

1. Antibodies (all stored at 4 °C and protected from light): APC anti-CD56 antibody, Clone HCD56 (BioLegend); PE anti-CD82 antibody, Clone ASL-24 (BioLegend).
2. Calcein blue solution: Resuspend an entire 1 mg vial in 200 μL of dimethyl sulfoxide (DMSO). Aliquot in 25 μL aliquots and store them at −20 °C (stock). Use 0.5 μL stock calcein per 10^6 cells.
3. Sterile 2% fetal bovine serum (FBS) in 1× PBS: Add 10 mL FBS to 500 mL 1× PBS. Sterilize by filtering the solution through a 500 mL 0.22 μm PES filter unit. Store at 4 °C.
4. Sterile laminar flow biosafety cabinet.
5. Benchtop centrifuge.
6. Inverted microscope.
7. Hemocytometer.
8. FACS 5 mL round-bottom tubes.
9. Cell sorter within a BSL-2 biosafety cabinet.

2.4 In Vitro Culture of Myoblasts

1. Sterile 50 mL conical centrifuge tubes.
2. Sterile 10 cm tissue culture-treated plastic dishes.
3. Sterile 0.22 μm PES filter, 50 mL volume.
4. Sterile 1× DPBS.
5. 0.1% gelatin: Add 0.5 g gelatin to 500 mL of double-distilled water, do not shake the bottle. Sterilize the solution by autoclaving for 20 min and store at 4 °C.
6. iMatrix-511 silk laminin coating reagent.
7. TrypLE™ Express Dissociation Enzyme with Phenol Red.
8. Sterile complete growth medium (Subheading 2.1, **item 10**).
9. Differentiation medium (50 mL): Mix 48.5 mL of low glucose Dulbecco's Modified Eagle's Medium (DMEM) with 1 mL of horse serum (HS) and 0.5 mL of 100× Penicillin-Streptomycin-Glutamine (PSG). Sterilize by filtering the solution through a 150 mL 0.22 μm PES filter unit. Store at 4 °C and use within 1 month.
10. Sterile laminar flow biosafety cabinet.
11. Water bath set to 37 °C.
12. Humidified 5% CO_2 incubator set to 37 °C.
13. Benchtop centrifuge.

14. Inverted microscope.
15. Hemocytometer.

3 Methods

3.1 Dissociation of Mouse or Human Primary Cells (See Note 2)

Institutional Review Boards (IRB) must approve the collection of human skeletal muscles before any procedure.

1. All steps in this protocol should be performed in a sterile laminar flow biosafety cabinet using proper sterile tissue culture techniques (Fig. 1a). Pre-weigh one 10 cm tissue culture plate with its lid (W1) for each sample to be dissociated.
2. Place the tissue sample to be dissociated in a second (not pre-weighed) 10 cm tissue culture plate. Then, use sterile scalpels to dissociate connective tissue from muscle tissue (Fig. 1b). Add as many drops of sterile 1× HBSS as necessary to the tissue to prevent muscle tissue from drying out.

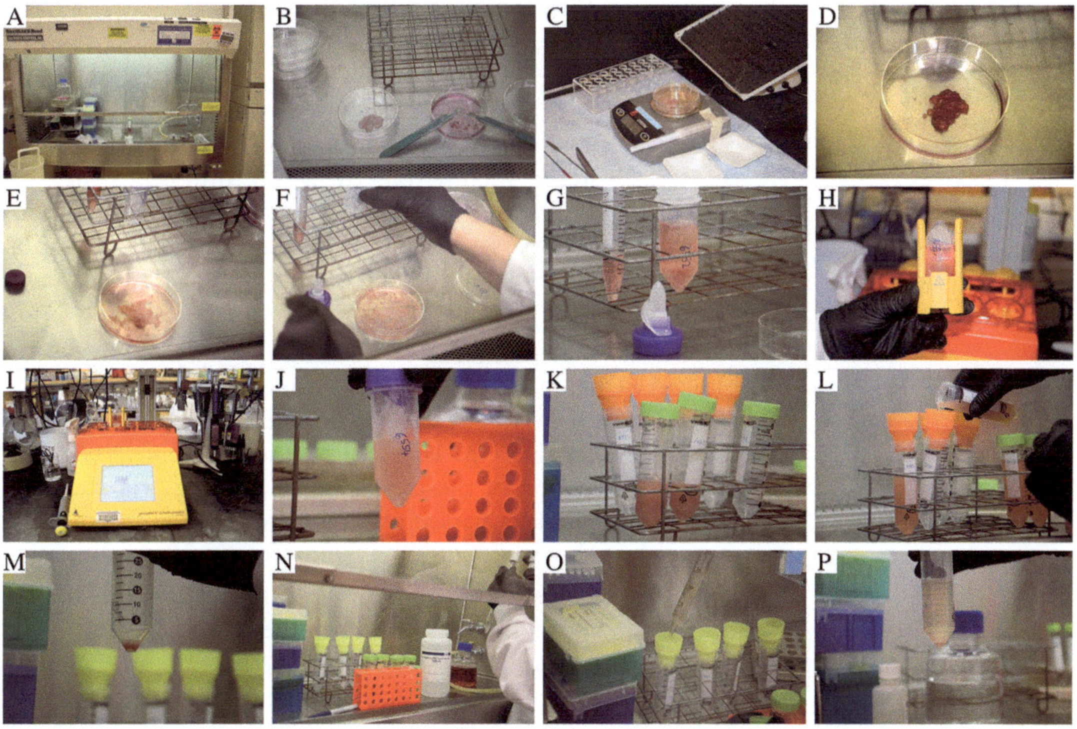

Fig. 1 Steps involved in the dissociation of primary mouse and human cells. (**a**) Pre-weigh sterile plates; (**b**) transfer clean muscle tissue to pre-weighed plate; (**c**) weigh plate again with sample to determine the amount of enzymes necessary for digestion; (**d**–**f**) Mince tissue with scalpels before adding the enzymes; (**g**–**i**) transfer sample to C tube for dissociation on Gentle MACS Dissociator; (**j**–**l**) Filter dissociated samples on 100 μm filters and dilute enzymes with media; (**m**–**p**) steps to lyse red cells in the samples, filter through 30–40 μm filters, and prepare samples for magnetic enrichment or for FACS sorting

3. Place muscle tissue in the pre-weighed plate, covered it with its original lid, and weigh the plate again (W2, Fig. 1c). Subtract W1 from W2 to calculate the amount of muscle tissue to be dissociated (W2 − W1 = g tissue).
4. Thaw frozen dispase II and collagenase D aliquots in a 37 °C water bath. Each solution will be added at a volume of 3.5 mL per gram of muscle tissue to be dissociated. Thaw only the amounts of collagenase D and dispase II necessary for dissociation. However, excesses can be refrozen once and reused.
5. Mince the muscle sample using sterile scalpels until the tissue resembles a fine paste. Add a few drops of sterile 1× HBSS during mincing to prevent exposed tissue from drying out. Tissue should always appear moist but with no excess liquid (Fig. 1d).
6. After finely mincing tissue (Fig. 1e), add equal amounts of the thawed dispase II and collagenase D solutions (digestion solution). The final digestion solution concentration will be 5 mg/mL for collagenase D and 1.2 U/mL for dispase II.
7. If no gentleMACS Dissociator is available, pipette the minced tissue up and down a few times in the digestion solution using a sterile wide-bore 25 mL pipette and incubate plate in a humidified 5% CO_2 incubator set to 37 °C for 15 min. After 15 min, pipette the tissue in the digestion solution up and down through a sterile 25 mL pipette a few times and incubate again for 15 min. Repeat this step once or twice more, until the slurry easily passes through a sterile 5 mL pipette and all tissue chunks are dissolved. The total digestion time will range between 45 and 75 min.
8. (Optional) If gentleMACS Dissociator is available, transfer the minced tissue in the digestion enzymes in the gentleMACS C tube (1 tube per sample to be dissociated) (Fig. 1f, g). Place the tube in the heater (Fig. 1h), and position the assembled tube and heater in one of the dissociator slots (Fig. 1i). Start the skeletal muscle dissociation program pre-saved in the machine memory (it should countdown from 1 h and 30 min).
9. Once tissue digestion is complete (Fig. 1j), add two volumes of complete growth medium (based on the total volume of digestion solution) to the digested slurry (Fig. 1k). Filter the mixed solution through a 100 μm cell strainer placed over a 50 mL conical tube (Fig. 1l). Change cell strainer if it appears clogged. Pellet cells by centrifuging for 10 min at 510 × g at room temperature (Fig. 1m).
10. Resuspend the pellet in 3–5 mL of red blood cell lysis solution. Wait for 2 min at room temperature to ensure red cells are lysed, and then add 20–25 mL of complete growth medium. Invert the tube a few times, and then filter the solution

through a 40 μm cell strainer placed over a new 50 mL conical tube (Fig. 1n, o).

11. Count cells using a hemocytometer, and then pellet the cells by centrifuging for 10 min at 510 × g at room temperature (Fig. 1p). Proceed to the next purification steps described for mouse (3.2) or human (3.3) myogenic cell isolation. If FACS sorting is not possible on the same days as isolation of human cells, freeze human cells at a concentration of 1×10^7 cells/mL in ice-cold freezing medium. Store cryovials at −80 °C overnight, and then transfer them to −150 °C where they can be permanently stored until necessary. Cell freezing is not required if all reagents and FACS equipment are immediately available. In this case, proceed to **step 5** of Subheading 3.3.

3.2 Purification of Mouse Satellite Cells/Myoblasts Using Magnetic Enrichment

This protocol continues immediately from Subheading 3.1, **step 11** (end of tissue dissociation and cell counting, Fig. 2a) and proceeds with elimination of cell debris prior to magnetic column selection.

1. Centrifuge filtered sample for 10 min at 510 × g, aspirate off liquid just leaving the pellet, add 6.2 mL of cold 1× PBS to each tube (sample). Transfer each sample resuspended in 6.2 mL PBS to a new 15 mL tube.
2. Add 1.8 mL of debris removal solution, mix by pipetting up and down five times; overlay gently with 4 mL of cold 1× PBS.
3. Centrifuge at 2800 × g for 10 min at room temperature; three layers should form. The middle layer should be a thin line of debris. Aspirate off the top two layers (all material above the debris and the debris layer).
4. Fill sample to 15 mL with cold PBS, invert three times and centrifuge at 910 × g for 10 min; aspirate off supernatant, making sure to get as much liquid out as possible without disturbing the clean cell pellet.
5. Resuspend cell pellet in 80 μL of PBS + BSA + EDTA per gram of muscle (calculated in Subheading 3.1, **step 3**). Add 20 μL of mouse satellite cell isolation kit per gram of muscle (Fig. 2b). Mix by flicking tube. Incubate at 4 °C for 15 min.
6. Prepare magnet and place LS columns within the magnet slots with a collection tube (15 mL) underneath (Fig. 2c), and rinse LS column with 3 mL PBS + BSA + EDTA.
7. At the end of incubation (**step 5**), bring volume to 500 μL final using PBS + BSA + EDTA, and load the sample to the rinsed LS column (Fig. 2c).
8. Wait for all the liquid to enter the top part of the column and rinse the column 3 times with 1 mL PBS + BSA + EDTA.

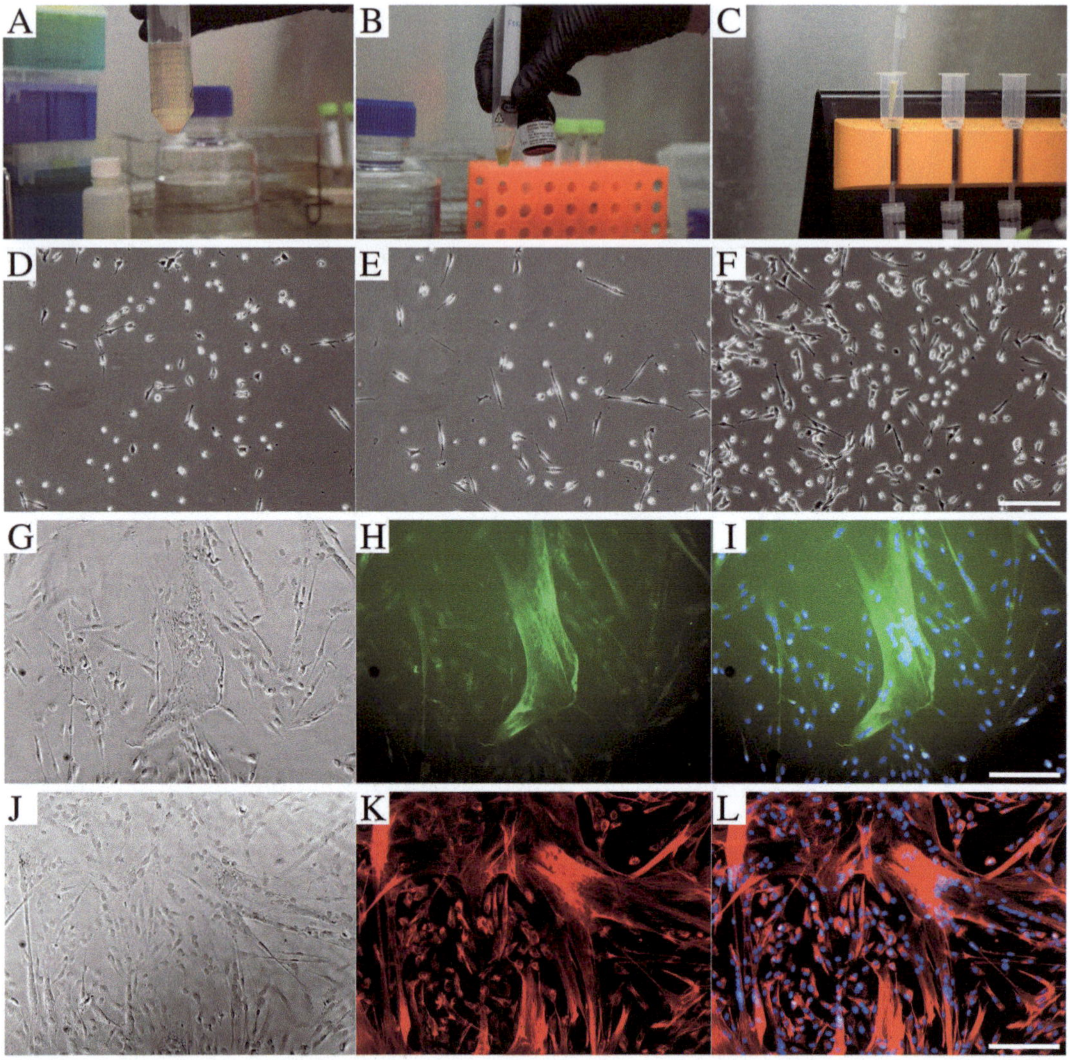

Fig. 2 (**a**) Cell debris removal and (**b**) labeling of mouse cells with antibodies conjugated with magnetic beads; (**c**) isolation of myogenic cells via collection of flow-through (negative selection) of samples. Unwanted cells are retained by the magnet within the column, while myogenic cells pass through the column and are collected in the flow-through tubes; (**d**, **e**) examples of mouse cells 48 h and (**f**) 72 h after plating post column enrichment; (**g**–**i**) example of differentiating human cells showing formation of myotubes in vitro. (**g**) Bright-field image; (**h**) myotube stained in green with myosin heavy chain antibody (MF20); (**i**) merged image with nuclei stained in blue with DAPI. (**j**–**l**): Brightfield image of human myotubes (**j**); (**k**) immunostaining with the myogenic marker desmin (red) shows both myotubes and mononuclear myogenic cells; (**l**) merged image showing DAPI-positive nuclei (blue) within the desmin-positive myotubes and myoblasts. Scale bars are 100 μm

9. When all liquid is eluted, spin flowthrough containing the myogenic cells 10 min at 510 × g.
10. Add complete growth media to pelleted cells and place in cell culture incubator.

3.3 Purification of Human Myoblasts Using FACS

All steps in this protocol should be performed in a sterile laminar flow hood or BSL-2 conditions using sterile tissue culture technique. If starting from frozen cells, a detailed protocol for defrosting the cells and then proceed to FACS sorting has been previously published [21].

1. Pre-warm complete growth medium and 1× DPBS in a 37 °C water bath. Place the 1× PBS 2% FBS on ice.
2. Centrifuge the 50 mL conical tubes containing the cells described in Subheading 3.1, **step 11** at 510 × g at 4 °C for 10 min to pellet the cells.
3. Remove the supernatant and resuspend the cells in 10 mL 2% FBS/PBS.
4. Determine the cell concentration using a hemocytometer or other cell counting device (*see* **Note 3**).
5. For FACS controls, use 5 mL round–bottom test tubes and set aside 2.5×10^5 cells in 500 μL 2% FBS/PBS for each of the following controls: "Unstained" control; Calcein blue single-color control (live cells); CD56 single color control; CD82 single color control (*see* **Note 4**).
6. Pipette the "unstained" control sample through the strainer cap of a 5 mL round-bottom test tube. Keep on ice.
7. Primary antibody incubation: add CD56 and CD82 antibodies to the single-color control tubes and to the sample to be sorted at a concentration of 5 μL per 1×10^6 cells (or as recommended by the manufacturer).
8. Add Calcein blue at a concentration of 0.5 μL per 1×10^6 cells to the single-color control and to the sample to be sorted. Gently mix and place on ice protected from the light for 30 min.
9. After this incubation, wash the cells once by adding cold 2% FBS/PBS to a final volume of 15 mL. Centrifuge the cells for 10 min at 510 × g at 4 °C.
10. Resuspend the CD56 and CD82 single color controls in 500 μL of 2% FBS/PBS, and pipette through the strainer cap of a 5 mL round bottom test tube. Store on ice in the dark.
11. Resuspend the CD56/CD82/Calcein blue stained cells in 1 mL of 2% FBS/PBS, and pipette through the strainer cap of a 5 mL round bottom test tube. Store on ice in the dark.
12. Prepare collection tube for $CD56^+CD82^+$ sorted cells by pipetting 500 μL of growth medium into a new tube. Store on ice.
13. Bring all tubes to FACS sorter core. The cell sorter operator will run unstained, single-color samples and the stained sample before drawing sorting gates, see **Note 5.**

3.4 In Vitro Culture of Myoblasts

All steps in this protocol should be performed in a sterile laminar flow hood using sterile tissue culture technique.

1. For human cells: coat sterile 10 cm tissue culture-treated plates with 10 mL 0.1% gelatin for 1 h in a humidified 5% CO_2 incubator set to 37 °C, and then remove the gelatin solution by aspiration. Let plates dry briefly in the biosafety cabinets and replace lid.
2. For mouse cells: coat sterile plates using iMatrix-511 diluted in cold 1× PBS (48 μL iMatrix-511 in 9 mL PBS) overnight at 4 ° C. Remove coating prior to plating cells. Cells can be added immediately without drying the plates.
3. Pre-warm complete growth medium in a water bath set to 37 ° C.
4. Resuspend sorted CD56/CD82 double-positive cells at 0.5–1 × 10^6 cells/10 mL complete growth medium, and plate on coated gelatin plates. Gently and evenly distribute cells on plates, and then place in a 5% CO_2 incubator set to 37 °C. Sorted cells will be small and have a bright, rounded appearance and should attach within 1 day post-sorting (Fig. 2d).
5. Propagate the cells to 60–75% confluency (*see* **Note 6**). This should take approximately 2–3 days; however, if necessary, replace the medium with fresh growth medium every 2 days until the plate is at 60–75% confluency (Fig. 2e shows an example of low confluency, and Fig. 2f shows optimum confluency).
6. To passage cells, coat sterile tissue culture-treated plates with 0.1% gelatin as in Subheading 3.4, **step 1** for human cells or **step 2** for mouse cells. Pre-warm the following in a water bath set to 37 °C: 1× DPBS, TrypLE™ Express dissociation enzyme, and complete growth medium.
7. Remove medium from plate by aspiration, and wash the cells twice with 10 mL (10 cm plate) 1× DPBS. Remove DPBS by aspiration. Repeat this wash for a total of three times.
8. Pipette 2 mL TrypLE™ Express onto the plate and incubate in a humidified 37 °C CO_2 incubator for 2–3 min. Gently remove the cells from the plate by pipetting up and down a few times before transferring cells into a sterile conical tube. Wash any remaining cells from the surface of the plate with additional complete growth medium.
9. Centrifuge the cells at 510 × g at room temperature for 10 min.
10. Resuspend the cells in 10 mL fresh complete growth medium.

11. Determine the cell concentration using a hemocytometer, and plate the cells at 0.5–1 × 10^6 cells in 10 mL complete growth medium/10 cm plate.
12. Cells should be passaged every 2–3 days and should not be grown past 75% confluency (Fig. 2f). Cells will spontaneously differentiate into myotubes when reaching higher confluency. Figure 2g–l show differentiated myotubes.

4 Notes

1. FBS varies considerably between companies and even lot to lot from the same company. Therefore, several different FBS samples should be tested using the in vitro methods described in Subheading 3.4 to determine which lot/company works best.
2. Institutional review and animal protocol approval are required prior to collection and processing of human or mouse tissue. All personnel handling biological tissues must receive appropriate biosafety and human and mouse subject education training.
3. There will be many floating, live cells in the culture, which is normal for dissociated human skeletal muscle. It is also likely that there will be small clumps of cells in the culture, and the number of clumps will vary. These clumps will be filtered out prior to cell sorting. Additionally, the dissociation process results in a large amount of debris in addition to cells. This will make the culture appear "dirty" (i.e., little black specks, etc.), but again, this is normal and should not be considered contamination. This debris will be removed during the FACS sample preparation process.
4. Some FACS machines may require tubes that are different in diameter/size from the tube specified in this protocol. Check in advance that the tubes fit in the FACS machine. Calcein blue is a cell viability dye and is used in this protocol to discriminate live from dead cells/debris during the FACS.
5. Determine optimal excitation voltages and compensation values using the "no stain" and single-color controls. Determine the live cell population gating for Calcein blue-positive cells. Determine the double-positive $CD56^+/CD82^+$ cell population and proceed to sorting, as this fraction contains myogenic cells.
6. Cells should never reach 100% confluency when proliferating, as they will begin to differentiate and fuse on contact. The high serum growth medium will lower in serum concentration over time and will not be able to prevent fusion.

Acknowledgments

This work is supported by a grant from the Muscular Dystrophy Association #1058516 (EG) and by the National Institute of Arthritis and Musculoskeletal and Skin Diseases of the National Institutes of Health under Award Number 1R01AR069582 (EG). The content is solely the responsibility of the authors and does not necessarily represent the official views of the National Institutes of Health. This protocol was modified from previous work [14, 21].

References

1. Shadrin IY, Khodabukus A, Bursac N (2016) Striated muscle function, regeneration, and repair. Cell Mol Life Sci 73:4175–4202. https://doi.org/10.1007/s00018-016-2285-z
2. Khodabukus A, Prabhu N, Wang J, Bursac N (2018) In vitro tissue-engineered skeletal muscle models for studying muscle physiology and disease. Adv Healthc Mater 7:e1701498. https://doi.org/10.1002/adhm.201701498
3. Romagnoli C, Iantomasi T, Brandi ML (2021) Available in vitro models for human satellite cells from skeletal muscle. Int J Mol Sci 22. https://doi.org/10.3390/ijms222413221
4. Sicinski P, Geng Y, Ryder-Cook AS et al (1989) The molecular basis of muscular dystrophy in the mdx mouse: a point mutation. Science (1979) 244:1578–1580
5. Danko I, Chapman V, Wolff JA (1992) The frequency of revertants in mdx mouse genetic models for duchenne muscular dystrophy. Pediatr Res 32:128–131. https://doi.org/10.1203/00006450-199207000-00025
6. Nance JR, Dowling JJ, Gibbs EM, Bönnemann CG (2012) Congenital myopathies: an update. Curr Neurol Neurosci Rep 12:165–174. https://doi.org/10.1007/s11910-012-0255-x
7. van Putten M, Lloyd EM, de Greef JC et al (2020) Mouse models for muscular dystrophies: an overview. Dis Model Mech 13. https://doi.org/10.1242/dmm.043562
8. Konishi R, Ichimura Y, Okiyama N (2023) Murine models of idiopathic inflammatory myopathy. Immunol Med 46:9–14. https://doi.org/10.1080/25785826.2022.2137968
9. Lawal TA, Wires ES, Terry NL et al (2020) Preclinical model systems of ryanodine receptor 1-related myopathies and malignant hyperthermia: a comprehensive scoping review of works published 1990-2019. Orphanet J Rare Dis 15:113. https://doi.org/10.1186/s13023-020-01384-x
10. Fusto A, Moyle LA, Gilbert PM, Pegoraro E (2019) Cored in the act: the use of models to understand core myopathies. Dis Model Mech 12. https://doi.org/10.1242/dmm.041368
11. Heydemann A, Ceco E, Lim JE et al (2009) Latent TGF-beta-binding protein 4 modifies muscular dystrophy in mice. J Clin Invest 119:3703–3712. https://doi.org/10.1172/JCI39845
12. Bogdanovich S, Krag TOB, Barton ER et al (2002) Functional improvement of dystrophic muscle by myostatin blockade. Nature 420: 418–421. https://doi.org/10.1038/nature01154
13. Sztretye M, Szabó L, Dobrosi N et al (2020) From mice to humans: an overview of the potentials and limitations of current transgenic mouse models of major muscular dystrophies and congenital myopathies. Int J Mol Sci 21. https://doi.org/10.3390/ijms21238935
14. Alexander MS, Rozkalne A, Colletta A et al (2016) CD82 is a marker for prospective isolation of human muscle satellite cells and is linked to muscular dystrophies. Cell Stem Cell 19:800–807. https://doi.org/10.1016/j.stem.2016.08.006
15. Striedinger K, Barruet E, Pomerantz JH (2021) Purification and preservation of satellite cells from human skeletal muscle. STAR Protoc 2:100302. https://doi.org/10.1016/j.xpro.2021.100302
16. Maesner CC, Almada AE, Wagers AJ (2016) Established cell surface markers efficiently isolate highly overlapping populations of skeletal muscle satellite cells by fluorescence-activated cell sorting. Skelet Muscle 6:35. https://doi.org/10.1186/s13395-016-0106-6
17. Sacco A, Doyonnas R, Kraft P et al (2008) Self-renewal and expansion of single transplanted muscle stem cells. Nature 456:502–506. https://doi.org/10.1038/nature07384
18. Bentzinger CF, von Maltzahn J, Dumont NA et al (2014) Wnt7a stimulates myogenic stem

cell motility and engraftment resulting in improved muscle strength. J Cell Biol 205: 97–111. https://doi.org/10.1083/jcb.201310035

19. Motohashi N, Asakura Y, Asakura A (2014) Isolation, culture, and transplantation of muscle satellite cells. J Vis Exp. https://doi.org/10.3791/50846

20. Hernández-Torres F, Rodríguez-Outeiriño L, Aránega A (2020) Isolation and culture of quiescent skeletal muscle satellite cells. Methods Mol Biol 2155:141–150. https://doi.org/10.1007/978-1-0716-0655-1_12

21. Pakula A, Spinazzola JM, Gussoni E (2019) Purification of myogenic progenitors from human muscle using fluorescence-activated cell sorting (FACS). Methods Mol Biol 1889: 1–15. https://doi.org/10.1007/978-1-4939-8897-6_1

Chapter 11

Isolation of Immune Cells from Skeletal Muscles for Flow Cytometry

Zachary M. Howard, Jeovanna Lowe, and Jill A. Rafael-Fortney

Abstract

Inflammation is a characteristic of skeletal muscle wound healing and of many muscle diseases. Analysis and isolation of the immune cell populations infiltrating skeletal muscles and their molecular signatures allow optimal use of anti-inflammatory treatments and identification of novel therapeutic targets. Here we describe a protocol for unbiased isolation of skeletal muscle immune cells. This technique allows for downstream assessment of immune cell populations or their isolation for downstream molecular or cellular assays from individual skeletal muscles during normal wound healing or disease.

Key words Myeloid cells, Macrophages, T-cells, Diaphragm, Quadriceps, Skeletal muscles

1 Introduction

Density-mediated cell separation techniques are used routinely to isolate and analyze immune cells from the blood or lymphoid tissues, using commercially available compounds such as lympholyte or histopaque. Several of the original studies of skeletal muscle immune cells also used these density-mediated cell separation techniques to remove tissue debris and enrich mononuclear cells [1, 2]. However, studies of immune cell isolation from other tissues supported that although density-mediated mononuclear cell isolation works well for peripheral blood mononuclear cells (PBMCs), altered cell-size and granularity of tissue-infiltrated immune cells results in significant loss of leukocyte yield [3]. A direct isolation technique was described for isolating immune cells from single mouse hearts, and we previously adapted this method for individual skeletal muscles [4–6]. When comparing density-mediated isolation techniques with the direct isolation method described here, we showed that 38.6% and 71.3% of CD45$^+$ leukocytes from diseased muscles were lost by isolating cells at the gradient interface of histopaque and lympholyte, respectively [4]. Furthermore, 80–90%

Kevin Flanigan and Stefan Nicolau (eds.), *Duchenne Muscular Dystrophy: Methods and Protocols*, Methods in Molecular Biology, vol. 2975, https://doi.org/10.1007/978-1-0716-4811-7_11,

of F4/80^{Hi} CD68^{+} macrophages, a highly transcriptionally active population in diseased muscles, did not localize at the gradient interface for either histopaque or lympholyte and were found in the pellets at the bottom of the tubes suggesting significant under-representation of this population in studies using density-mediated immune cell isolations from skeletal muscles [4].

The method outlined here has demonstrated vast increases in total immune cells in muscles after damage or in diseased muscles compared to normal muscles at homeostasis and the ability to isolate subpopulations of immune cells from individual muscle types allowing comparisons between muscles affected differentially by a particular disease [4, 7, 8]. We also include methods for isolation of blood and spleen cells for instances where a comparison of muscle immune infiltrate with circulating or splenic immune cells is warranted.

2 Materials

1. Digestion buffer: Dissolve 10 mg (21 u) of collagenase P and 50 μL (50 u) of RQ1 DNase in 50 mL of Dulbecco's Modified Eagle Medium (DMEM). Make immediately prior to experiment, store on ice until use, and prewarm immediately before use.
2. Flow buffer: autoMACS Running Buffer (store at 4 °C).
3. DPBS, no calcium, no magnesium (store at 4 °C).
4. 1% Paraformaldehyde (PFA) diluted in DPBS: Prepare this buffer wearing a lab coat, gloves, and safety glasses. Prepare this buffer in a chemical fume hood while wearing a lab coat, gloves, and safety glasses. Add 9 mL of PBS to 15 mL Falcon tube, and then add 0.1 g paraformaldehyde. Heat the tube in a 37 °C incubator or water bath for approximately 15 min and vortex. Add a drop of 1 M NaOH and invert the tube several times to mix. If the paraformaldehyde is not dissolved, add another drop of NaOH, and repeat until it is fully dissolved. The PFA will only dissolve when it reaches the correct pH. Store the working solution at 4 °C for up to 1 week. Extra aliquots can be stored long term at −20 °C.
5. Fetal bovine serum (FBS).
6. Dissection tools (autoclaved).
7. 70 μm cell strainers.
8. 40 μm cell strainers.
9. 35 mm non-treated cell culture dishes.
10. Goldenrod animal lancet, 4 mm sterile.
11. BD Microtainer dipotassium EDTA blood collection tube.

12. 1× RBC lysis buffer (Thermo Fisher).
13. Sterile gauze.
14. Four 50 mL Falcon high clarity polypropylene (PP) centrifuge tubes, conical bottom, per animal.
15. One 15 mL Falcon high clarity PP centrifuge tube per animal.
16. 3 mL syringes.
17. 1 glass plate.

3 Methods

3.1 The Day Prior to Starting the Experiment

1. Chill at 4 °C: Flow buffer, DPBS, FBS (thawed from −20 °C or −80 °C).
2. If required, make 1% PFA.
3. Autoclave dissection tools.
4. Label for each mouse a 35 mm non-treated cell culture dish and EDTA microtainer tubes for blood.
5. Label four 50 mL tubes for each mouse for each muscle tissue being isolated: skeletal muscle digestion tube, skeletal muscle 70 μm filtration, skeletal muscle 40 μm filtration, and splenocyte isolation tube.

   Label one 15 mL tube for each mouse for blood collection.

3.2 The Morning of the Experiment, Prepare

1. Mouse dissection setup.
2. 70% ethanol spray bottle.
3. Serological pipets: 5, 10, 25 mL (many).
4. Ice bucket.
5. 37 °C water bath.
6. 40 μm and 70 μm cell strainers.
7. Prepare digestion buffer and store on ice, then prewarm to 37 °C immediately before use.

3.3 Submandibular Blood Collection

1. Securely scruff the mouse close to the head and puncture the facial vein using a sterile 4 mm lancet (*see* **Note 1** and Fig. 1a).
2. Collect blood (≤ 100 μL) in an EDTA blood tube.
3. Flick the tube following collection and store on ice temporarily.
4. Loosen the scruff on mouse and apply sterile gauze to stop excess bleeding.

3.4 Spleen Collection

1. Sacrifice mice using IACUC-approved methods of euthanasia.
2. Spray the mouse with 70% ethanol prior to dissection, and then remove the spleen quickly without perforating.
3. Place the spleen in cold PBS in the 35 mm dish on ice (*see* **Note 2**).

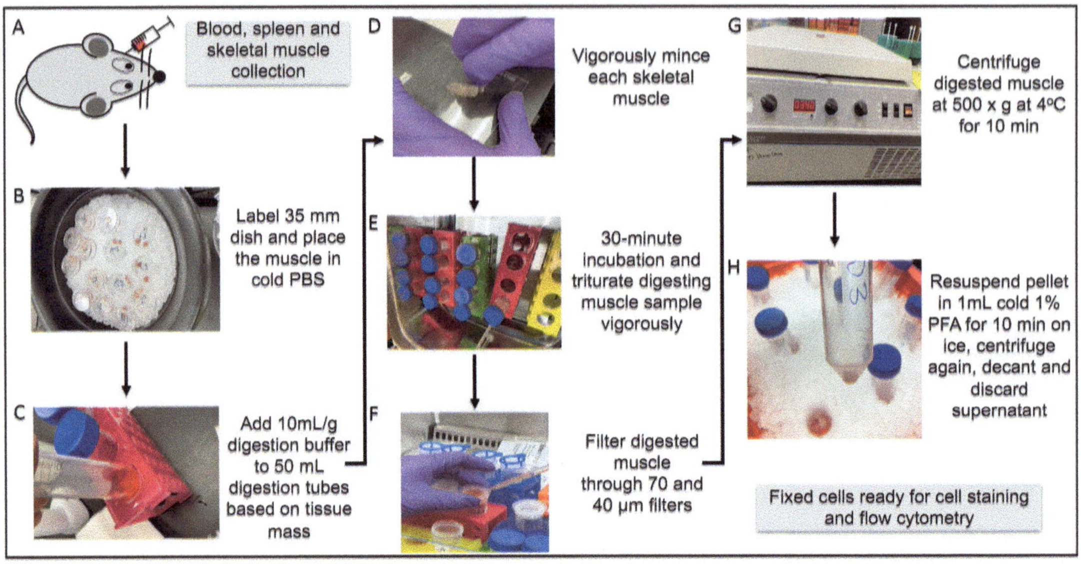

Fig. 1 Photos of key steps in skeletal muscle processing for immune cell isolation. (**a**) Blood, spleen, and skeletal muscle are collected. (**b**) Skeletal muscles are submerged in cold PBS in labeled 35 mm dishes on ice. (**c**) Digestion buffer is added to labeled digestion tubes based on tissue mass. (**d**) Muscle tissues are then vigorously minced on a glass plate with a razor blade. (**e**) Digested muscle samples are triturated vigorously followed by a 30-min incubation period at 37 °C and (**f**) then filtered through 70 and 40 μm filters to obtain isolated cells. (**g**) The collected flow through is centrifuged at 500 × g at 4 °C for 10 min. (**h**) The resulting pellet is resuspended in 1 mL of cold 1% PFA and incubated for 10 min on ice. Subsequently, the cells are centrifuged again, the supernatant is decanted and discarded, leaving the fixed cells ready for flow cytometry procedures. This protocol enables the efficient isolation and fixation of immune cells, facilitating their subsequent analysis and characterization

3.5 Skeletal Muscle Collection

1. Dissect the diaphragm and quadriceps separately, or other muscles of interest, carefully removing adipose tissue and avoiding lymph nodes.
2. Place tissues separately in the cold PBS in the 35 mm dish on ice (Fig. 1b).
3. Blot each tissue on a paper towel, weigh the tissue and record its weight. Return to PBS to prevent drying out. Repeat for each tissue (*see* **Note 2**).

3.6 Blood Processing

1. Transfer at least 50 μL of blood to a 15 mL tube on ice.
2. Add 2 mL of RBC Lysis Buffer and incubate for 5 min at room temperature.
3. Add 2 mL of cold flow buffer to quench lysis, centrifuge at 400 × g for 8 min at 4 °C.
4. Carefully decant the supernatant and discard.
5. Resuspend the pellet in 1 mL of cold 1% PFA for 10 min on ice.
6. Add 5 mL of cold flow buffer to quench the fixative.

7. Centrifuge at 400 × g for 8 min at 4 °C.
8. Aspirate fluid off the pellet and resuspend in 200 μL of cold flow buffer.
9. Measure the exact volume and record it (*see* **Note 3**); fixed cells can be stored at 4 °C for up to 2 weeks before flow cytometry.

3.7 Spleen Processing

1. Add 10 mL of cold flow buffer to a 50 mL tube on ice; place a 40 μm cell strainer on the tube.
2. Transfer spleens from the 35 mm dish to the 40 μm cell strainer.
3. Dissociate the spleens on the cell strainer using the flat side of a plunger from a 3 mL syringe.
4. Wash the filter with 5 mL of cold flow buffer.
5. Centrifuge at 400 × g for 8 min at 4 °C.
6. Carefully decant the supernatant and discard.
7. Resuspend the pellet in 5 mL of RBC Lysis Buffer and incubate for 5 min at room temperature.
8. Add 5 mL of cold flow buffer to quench lysis, centrifuge at 400 × g for 8 min at 4 °C.
9. Carefully decant the supernatant and discard.
10. Resuspend the pellet in 1 mL of cold 1% PFA for 10 min on ice.
11. Add 5 mL of cold flow buffer to quench the fixative.
12. Centrifuge at 400 × g for 8 min at 4 °C. Carefully decant the supernatant and discard. Approximately 400 μL of flow buffer should remain in tube after decanting.
13. Resuspend the pellet by flicking the bottom of the 50 mL tube.
14. Measure the exact volume and record it (*see* **Notes 3** and **4**); fixed cells can be stored at 4 °C for up to 2 weeks before flow cytometry.

3.8 Skeletal Muscle Processing

1. Add 10 mL/g of digestion buffer to 50 mL digestion tubes based on tissue mass (Fig. 1c).
2. Vigorously mince each skeletal muscle sample separately with a razor blade on a glass plate until it reaches a paste-like consistency (*see* **Note 5**, Fig. 1d). To safeguard the cellular structure and enhance process efficiency, it is crucial to conduct the mincing rapidly, particularly when utilizing a laminar flow hood for flow sorting, to mitigate the risk of tissue drying (*see* **Note 2**).
3. Transfer the minced tissue to a 50 mL digestion tube.

4. If digesting tissues from many mice, place each minced tissue in the 50 mL tube with the digestion buffer at room temperature while mincing the others.
5. Place the tubes in a 37 °C water bath and set a timer for 30 min.
6. During the 30-min incubation:
   (a) Coat a 5 mL pipet with FBS for each sample by pipetting an aliquot of FBS into the pipet and expelling.
   (b) Every 10 min over the course of 30 min, vigorously triturate (pipet up and down) (Fig. 1e) the muscle sample being digested using a 5 mL pipet coated with FBS 10–15 times until the digested muscle solution easily passes through the tip of the pipet.
7. After the 30-min digestion is complete, pour the digested muscle through a 70 μm cell strainer (Fig. 1f), and collect the flow through in a new 50 mL tube on ice (*see* **Note 6**).
8. Directly following 70 μm filtration, add an equal volume of cold flow buffer to the tubes with digested muscle to quench the collagenase reaction.
9. Wash the filter with 3 mL cold flow buffer, and collect the flow through in the same collection tubes for each sample.
10. Filter the digested muscle through a 40 μm cell strainer (Fig. 1f), and collect the flow through into a new 50 mL tube.
11. Centrifuge the digested muscle (Fig. 1g) at 500 × g at 4 °C for 10 min.
12. Carefully decant the supernatant and discard.
13. Fresh, unfixed cell samples are required for flow sorting. If this is desired, adjust the remaining volume in the tube to 400 μL with flow buffer, and proceed immediately to the staining process.
14. Resuspend the pellet in 1 mL of cold 1% PFA, and vortex quickly to ensure the PFA mixes with the cells. Incubate for 10 min on ice (Fig. 1h).
15. Add 5 mL of cold flow buffer to quench the fixative.
16. Centrifuge at 500 × g at 4 °C, low brake, for 8 min.
17. Carefully decant the supernatant and discard. Approximately 400 μL of flow buffer should remain in tube after decanting (*see* **Notes 3, 4**, and 7).
18. Measure the exact volume and record it (*see* **Note 8**).
19. Vortex briefly to avoid cell clumping.
20. Fixed cells can be stored at 4 °C for up to 2 weeks before flow cytometry (see Table 1 for a suggested myeloid flow panel).

Table 1
Suggested antibody panel for myeloid cell staining with isotype controls

Cell type or marker	Antibody	Supplier, catalog number
Myeloid staining panel		
Immune cells	PE-Cy7_CD45	Thermo Fisher #25-0451-82
Myeloid	APC_CD11b	BioLegend, 101212
Neutrophil	APC-Fire750_LY6G	BioLegend # 127652
Monocyte	eFluor450_LY6C	Thermo Fisher #48-5932-82
Macrophage	BV605 _CD64	BioLegend #139323
CD206	PerCP-eFluor710_CD206	Thermo Fisher #46-5866-82
CCR2	FITC_CCR2	BioLegend #150608
Isotype control antibodies	Isotype PE-Cy7 Isotype APC Isotype APC-Fire750 Isotype eFluor450 Isotpe BV605 Isotype PerCP-eFluor710 Isotype FITC	Thermo Fisher #25-04031-81 BioLegend #400611 BioLegend #400567 Thermo Fisher #48-4321-82 BioLegend #400161 Thermo Fisher #46-4321-80 BioLegend # 400605

4 Notes

1. Cardiac puncture can be used for collecting blood in terminal experiments if the heart is not being collected for other analyses. Facial vein collection can be done longitudinally in live mice.
2. A laminar flow hood should be used for sample processing if unfixed cells will be sorted and used for downstream applications, particularly in cell culture experiments. However, sample processing can be done on the bench when fixing the sample for flow cytometry analysis to avoid tissues drying out from air flow in a laminar flow hood.
3. The reason for measuring the volume of sample remaining after white blood cell, splenocyte, or skeletal muscle immune cell isolation is to normalize flow cell counts. For example, if 400 μL are obtained at the end of the skeletal muscle isolation, a 100 μL aliquot could be used for flow cytometry and the cell counts could be normalized by multiplying by 4.
4. For splenocyte and skeletal muscle fixation, the effective concentration of PFA will be less than 1% because of the remaining volume in the tube after decanting. This is sufficient for fixation and will minimally affect the integrity of cell surface markers for flow cytometry.

models. These models work well for validating drugs, but they are not appropriate for large screening because of the lengthy and expensive drug development process [1]. Zebrafish and dystrophin-deficient *Caenorhabditis elegans* models have been used as an alternative, but hit compounds discovered using these organisms have not effectively translated into effective DMD treatments [2]. While gene-editing technologies like CRISPR/Cas9 have facilitated the creation of animal models [3], it remains a challenging and costly endeavor.

In contrast, patient-derived cell lines offer a more accessible option, covering a wide spectrum of mutations in diseases such as DMD. The establishment of cell models is pivotal for developing personalized therapies tailored to the specific mutations in these pathologies. Cellular models derived directly from human tissues are valuable for modeling various human genetic disorders. They offer the advantage of retaining the original genomic context and, in many instances, replicating the molecular and cellular features observed in patients. In the context of neuromuscular disorders, muscle biopsies have served as a crucial source of human myoblasts, aiding in unraveling pathological mechanisms, and serving as essential tools for in vitro testing of drugs and gene therapies. While obtaining myoblasts from muscle biopsies is relatively straightforward, their culture and maintenance present challenges due to limited proliferation rates and replicative senescence in vitro [4]. To overcome these limitations, myoblasts can be immortalized through the insertion of human telomerase (hTERT) and/or cyclin-dependent kinase 4 (CDK4) genes, preserving skeletal muscle characteristics [5–7]. However, the dependence on muscle biopsy, a surgical procedure with drawbacks for patients with advanced muscle degeneration, remains a constraint, as the muscle often has a significant proportion of fibrotic and/or adipose tissue, necessitating cell purification before immortalization [5]. In contrast, skin biopsies are more accessible and less invasive, providing an alternative route for obtaining primary fibroblasts in vitro. Although fibroblasts are not primarily affected by mutations causing DMD, they can be transdifferentiated into myoblasts through the introduction of the *MyoD* gene, a myogenic regulatory transcription factor [8].

Among the investigated personalized therapies, exon-skipping strategies show promise for various muscular dystrophies [9, 10]. This approach involves producing a shorter yet functional protein by concealing the exon definition from the spliceosome, thus excluding the exon from the final messenger RNA in order to restore the reading frame. This technology, which has led to 4 FDA-approved drugs for DMD, is explored in this protocol through methods to transfect myoblasts with two exon-skipping technologies: antisense oligonucleotides (AON) and U7 small nuclear RNAs (U7snRNA) delivered by adeno-associated virus

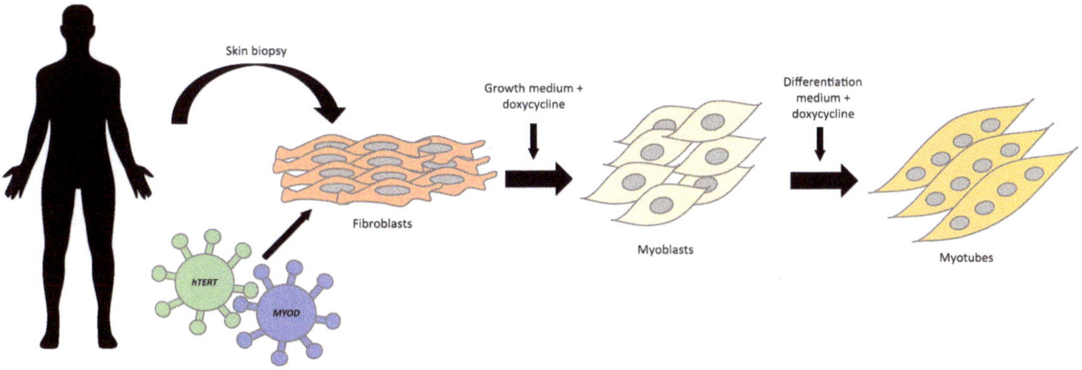

Fig. 1 Representation of fibroblasts transdifferentiating into myogenic cells. Human skin biopsies are collected to establish primary fibroblast cultures. Initially, the fibroblasts are transduced with the hTERT gene and subsequently with the MyoD gene, using lentiviral vectors. Following antibiotic selection of the infected cells, myoblast conversion is initiated by introducing doxycycline, CHIR, and DAPT into the myoblast growth medium. Over a span of 2–4 days, the cells undergo elongation and align in a parallel orientation. Upon transitioning to a differentiation medium, the myoblasts fuse with each other, giving rise to multinucleated myotubes. (Figure adapted from Almeida et al. [19])

(AAV) vectors. AON transfection serves as a tool for the initial screening of sequences designed to promote exon skipping [11], but it is only transiently active. In contrast, modified U7snRNAs in combination with AAV vectors can be used to achieve sustained expression of antisense sequences [12]. U7snRNA plays a role in histone mRNA processing. It can be modified to selectively bind proteins, directing its integration into the spliceosome and facilitating the delivery of antisense sequences [13]. This approach overcomes limitations associated with AONs, resulting in continuous AON expression and improved tissue transduction efficiency [14–16]. This chapter outlines the protocol for obtaining transdifferentiated myoblasts, starting from the lentiviral production to the generation of differentiated myotubes (refer to Fig. 1 for a concise overview of the method).

2 Materials

2.1 Cell Culture

1. Plastic/disposable supplies: 100 mM dishes, 75 cm^2 and 175 cm^2 flasks, scalpels, and serological pipets.
2. Hemocytometer.
3. Mr. Frosty™ freezing container.
4. 1× phosphate buffered saline (PBS).
5. 0.25% trypsin-EDTA, phenol red.
6. Hygromycin B.
7. Puromycin.

8. HEK293T growth medium: combine DMEM (high glucose (4.5 g/L), GlutaMAX, 100 mM sodium pyruvate) with 10% fetal bovine serum and 1% antibiotic-antimycotic. Store at 4 °C. Warm up to 37 °C before use.
9. Fibroblast growth medium: combine DMEM (high glucose, GlutaMAX, sodium pyruvate) with 20% fetal bovine serum and 1% antibiotic-antimycotic. Store at 4 °C. Warm up to 37 °C before use.
10. Doxycycline 1000× stock solution: dissolve 8 mg of doxycycline hydrochloride in 1 mL ultra-pure water. Filter through a 0.22 μm syringe filter. Aliquot in PCR tubes. Store at −20 °C, protected from light, and do not re-freeze after thawing. Add doxycycline to the medium just before media changes to prevent degradation, as it quickly degrades when diluted and stored at 4 °C.
11. CHIR-99021: reconstitute in DMSO for a 10 mM stock solution.
12. DAPT: reconstitute in DMSO for a 10 mM stock solution.
13. Myoblast growth medium: mix basal medium with supplements (Skeletal Muscle Cell Growth Medium, Promocell). Store at 4 °C for up to 2 months. Warm up to 37 °C before use, and add doxycycline (8 μg/mL final concentration), CHIR (5 μM), and DAPT (10 μM). Prepare enough volume for a single use.
14. Myotube differentiation medium: mix basal medium with supplements (Skeletal Muscle Cell Differentiation Medium, Promocell). Store at 4 °C for up to 2 months. Warm up to 37 °C before use and add doxycycline (8 μg/mL final concentration), CHIR (5 μM), and DAPT (10 μM). Prepare enough volume for a single use.
15. Freezing medium: mix 90% fibroblast growth medium with 10% DMSO.
16. Matrigel matrix.
17. HEK293T cells.

2.2 Lentivirus Production

1. 2.5 M $CaCl_2$ solution: dissolve 11 g of calcium chloride hexahydrate in 20 mL of deionized water. Filter the solution through a 0.22 um filter. Store at 4 °C.
2. 2× HEPES buffer (pH 7.05–7.1).
3. Nuclease-free water.
4. DNA plasmids diluted to 1 μg/mL: MDL-helper, REV-helper, CMV-VSVG, hTERT, and MyoD. Lentivirus production plasmids are obtained from the group that published Chaouch et al. (2009)s paper, as detailed in Aure et al. (2007) [17] for

the hTERT plasmid and Barde et al. (2006) [18] for the Tet-on system used in MyoD plasmid design. These plasmids are acquired through a Material Transfer Agreement with Genethon, France (contact Dr. Vincent Mouly at vincent.mouly@upmc.fr). Briefly, hTERT consists of hTERT variant 1 driven by a CMV promoter, while puromycin is driven by a PGK promoter. The MyoD plasmid contains MyoD variant 1 driven by a CMV promoter under rtTA2 repressor control, along with hygromycin selection expressed by the SV40 promoter.

5. Beckman Coulter Optima XE-100 ultracentrifuge and SW 28 rotor.
6. Beckman Coulter 38.5 mL open-top Thinwall polypropylene ultracentrifuge tubes, 25 × 89 mM.
7. TNE buffer (Tris-HCl, NaCl, EDTA).
8. qPCR Lentivirus Titer Kit (Applied Biological Materials).

2.3 Immunostaining

1. 4% PFA: dissolve 4 g of paraformaldehyde in 80 mL PBS, add 200 μL of 5 M NaOH, adjust pH to 7.4 with HCl, and complete the volume with PBS to 100 mL.
2. Permeabilization solution: 0.5% TritonX-100 in 1× PBS.
3. Blocking solution: 10% goat serum in 1× PBS.
4. Anti-myosin heavy chain antibody (MF20, Developmental Studies Hybridoma Bank).
5. Goat anti-mouse antibody, Alexa Fluor 488.

2.4 Antisense Oligonucleotide Transfection

1. 2'o-methyl-phosphorothioate antisense oligonucleotide.
2. OptiMEM reduced serum medium.
3. Lipofectamine 3000 Transfection Reagent.

2.5 AAV Transduction

1. AAV1 vector encoding a modified U7snRNA.

3 Methods

Carry out all procedures under a biological safety cabinet level 2.

3.1 Lentivirus Production

1. Culture HEK293T cells until confluent (80–90%) in 175 cm^2 flasks.
2. Change medium on the cells 2 h before transfection (25 mL complete DMEM with 10% FBS for 175 cm^2 flasks).
3. Warm 2.5 M $CaCl_2$ and 2× HEPES buffer to room temperature.

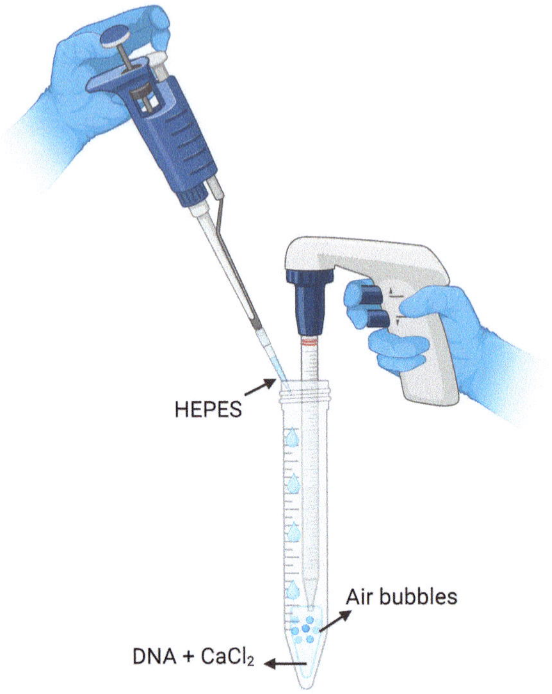

Fig. 2 $CaCl_2$ DNA precipitation for transfection. Schematic representation showing how to add HEPES buffer to the DNA-$CaCl_2$ solution. Make sure to constantly create bubbles in the solution. (Created with BioRender.com)

4. In a 15 mL conical tube, prepare the DNA (quantities calculated for a 175 cm^2 confluent flask and DNA diluted to 1 μg/mL): mix 17.6 μg MDL-helper, 6.85 μg REV-helper, 9.1 μg CMV-VSVG, and 27.3 μg hTERT or MyoD plasmids. Complete the volume to 1080 μL with nuclease-free water and mix well. Add 120 μL 2.5 M $CaCl_2$ (final volume is 1.2 mL) and vortex briefly to mix.
5. Use a 2 mL pipette to make bubbles in the DNA solution. At the same time, add 1.2 mL 2× HEPES buffer dropwise (Fig. 2). Let sit for 10 min, no longer than 15 min. The solution may or not become cloudy.
6. Collect the medium from the 175 cm^2 flask and transfer it to a 50 mL conical tube. Add the DNA-$CaCl_2$-HEPES solution to the medium, mix, and transfer the medium back to the flask with cells.
7. Incubate at 37 °C overnight. The next morning, replace the medium with fresh medium and incubate the cells at 37 °C overnight.
8. The following day, to harvest the lentivirus particles, collect the medium (~25 mL/flask) and filter using a 0.22 μm filter. Add fresh medium to the cells and return the flask to the incubator.

Split the filtered medium into ultracentrifuge tubes. Balance the tubes to a difference no higher than 0.01 g.

9. Ultracentrifuge at 22,500 rpm (~66,000 g), for 2 h, at 4 °C.
10. Carefully aspirate the supernatant and resuspend the virus pellet (which may or may not be visible) in 200 μL TNE buffer. Cover the ultracentrifuge tube with parafilm and place it inside a 50 mL conical tube and store it at 4 °C.
11. Repeat **steps 8–10** for another 2 days (total of 3 harvests), adding the filtered medium to the same ultracentrifuge tube to combine and concentrate the lentivirus solution.
12. After the third ultracentrifugation, resuspend the lentivirus pellet in 600 μL TNE buffer. Cover the ultracentrifuge tube with parafilm, and place it inside a 50 mL conical tube, and store it at 4 °C for 24 h to allow complete resuspension of the lentivirus.
13. Titer the lentivirus concentration using the qPCR Lentivirus Titer Kit, according to manufacturer instructions.
14. Make 20–25 μL aliquots of lentivirus, and flash freeze them in liquid nitrogen. Store at −80 °C for long-term.

3.2 Initiation of Dermal Fibroblasts Culture

1. Add 10 mL of fibroblast medium in 15 mL conical tubes for skin biopsy transportation. The material should be kept at 4 °C until processing. Preferentially, the skin biopsy is processed on the same day, but it can be kept at 4 °C until the next day (*see* **Note 1**).
2. Aspirate all the liquid from the tube and wash the biopsy with 10 mL of 1× PBS (at room temperature) three times. After the third wash, keep the biopsy in PBS.
3. Transfer the skin biopsy to a 10 cm^2 dish with PBS to not dry out the sample.
4. Chop the biopsy in small fragments using sterile scalpels. The fragments should be as small as possible (~1–2 mM^3) for proper adherence to the plate.
5. Transfer the skin fragments one by one using a pipette. To facilitate the transfer and not let the fragment stick to the pipette tip, aspirate some PBS together, and drop it into a clean 10 cm^2 dish. Distribute 10–12 fragments per dish, spaced approximately 1 cm apart.
6. Carefully, remove the excess of PBS around each fragment.
7. Place the lid on the dish leaving an opening to allow the rest of PBS to dry. This should take from 5 to 20 min. However, it is important that the fragments do not dehydrate to maintain the viability of the cells.

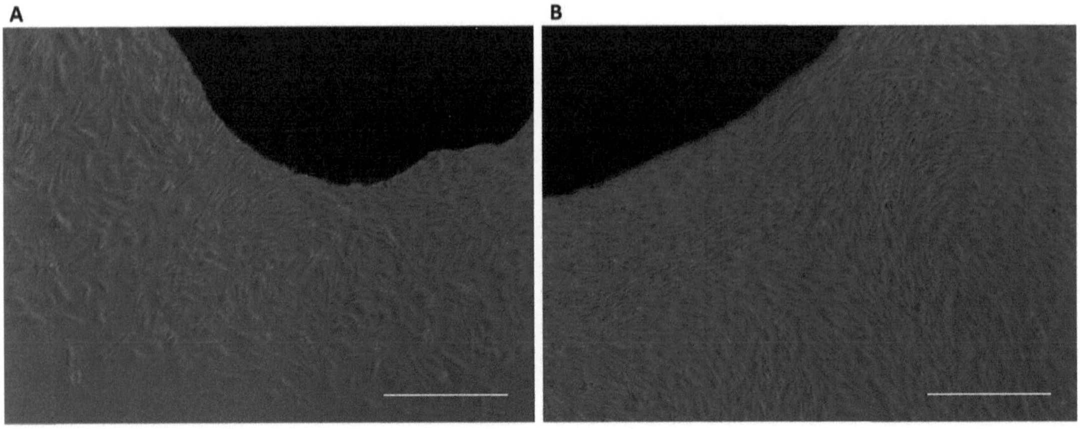

Fig. 3 Skin fragments in culture. (**a**) Fibroblasts emerge from the skin fragment within a week. (**b**) When the fibroblasts surrounding the skin fragment, it is time to transfer them to a new surface for further proliferation. Scale bar: 50 μm. (Figure adapted from Almeida et al. [19])

8. Incline the dish at 45° and carefully add 10 mL of fibroblast medium to the bottom of the dish. Slowly lower the dish to spread the medium, without lifting the fragments.
9. Transfer the dishes to the incubator (37 °C, 5% CO_2). Monitor for the appearance of the first cells from each fragment every other day during the first week. It is not necessary to replace the medium during this period.
10. Within the first week, the first fibroblasts should emerge from the skin fragments (Fig. 3). When there is an accumulation of cells around the fragment, they need to be spread. For that, aspirate the medium, wash with PBS, and add 1 mL of 0.25% trypsin. Incubate at 37 °C for 5 min or until all cells are lifted. Add 10 mL of fibroblast growth medium to collect the cells and transfer it to a new dish (*see* **Note 2**).
11. Once the cells are confluent (80–100%), transfer them to 75 cm^2 flasks for proliferation of larger amounts of cells.

3.3 Cryopreservation of Primary Fibroblast Lines

1. When the 75 cm^2 flask reaches confluency, rinse it with 10 mL of 1× PBS, and then remove the PBS.
2. Add 2 mL of 0.25% trypsin to the cell surface, and place the flask in the incubator for 5 min. Check under the microscope for cell detachment. If needed, return the flask to the incubator for an additional 5 min.
3. Once cells are detached, add 7 mL of fibroblast medium to the flask and resuspend the cells by pipetting up and down. Transfer the cells into a 50 mL conical tube.
4. Prepare a 100 μL aliquot of trypan blue, and mix it with 100 μL of the sample being cryopreserved. Load the mixture onto the

hemocytometer for cell counting. Count cells in four different hemocytometer fields under the microscope. Calculate the cell concentration according to the conversion factor provided by the manufacturer.

5. Centrifuge the conical tubes at 300 × g for 10 min at either room temperature or 4 °C.
6. Remove the medium, and resuspend the cells in an appropriate volume of freezing medium, to a concentration of 1 million cells per mL. Homogenize the solution by pipetting up and down and distribute 1 mL into each cryovial labeled with the cell line identification numbers, passage number, number of cells, and date (*see* **Note 3**).
7. Place the vials into the freezing container and allow them to freeze at a rate of 1 °C/min in a −80 °C freezer overnight.
8. The next day, transfer the vials to a liquid nitrogen tank or a − 150 °C freezer for long-term storage.

3.4 Establishment of Fibromyoblast (FM) Cell Line

1. Begin by seeding primary fibroblasts at approximately 30% confluency in two wells of a 12-well plate (2×10^4 cells/well) to achieve around 50% confluency the following day.
2. For lentiviral transduction, introduce 2–5 × 10^9 viral genome particles (vg) of hTERT-puromycin lentivirus in 400 μL of fibroblast medium into one well, and add only 400 μL of fibroblast medium to the second well. The next day, supplement with 1 mL of medium.
3. One or 2 days later, transfer the cells to a 6-well plate, and allow them to grow until reaching 60–70% confluency.
4. Supplement the fibroblast medium with 1 μg/mL of puromycin and add 2 mL to each well.
5. Keep the cells under selection until all cells in the control well are dead (up to 5 days), changing the media every 2–3 days. Passage the cells from the 6-well plate into two 10 cm^2 dishes for further proliferation.
6. Freeze vials of fibroblasts after selection, labeled as F(hTer).
7. Seed hTERT-expressing fibroblasts (F(hTer)) at about 30% confluency in fibroblast medium in three wells of a 12-well plate, aiming for about 50% confluency the next day.
8. For lentivirus transduction, mix 2–5 × 10^9 vg of MyoD-hygromycin B lentivirus in 400 μL of fibroblast medium per well and add to two wells; add 400 μL of fibroblast medium to the third well as a non-transduced negative control. Add 1 mL of medium the next day.
9. One or two days later, transfer the cells to a 6-well plate and allow them to grow until 60–70% confluency.

10. Supplement the fibroblast growth medium with hygromycin B (400 μg/mL), and add 2 mL to each well.
11. Keep the cells under selection until all cells in the control well are dead (up to 5 days), changing the media every 2–3 days.
12. Freeze vials of fibroblasts after selection, labeled as FM followed by the cell identification number/name.

3.5 Trans-differentiation Protocol

1. Plate transduced FM cells onto 10 cm^2 dishes or multi-well plates (depending on the purpose of the experiment) at 30–40% confluency (*see* **Note 4**).
2. To induce myoblast transformation, when fibroblasts reach 70–80% confluency (as depicted in Fig. 4a), wash the cell surface with 1× PBS and supplement with fresh myoblast medium containing 8 μg/mL doxycycline, 10 mM DAPT, and 5 mM CHIR (*see* **Notes 5** and **6**).
3. Two to three days later, cells will reach 90–95% confluency, and their morphology will have altered (Fig. 4b). Wash the cell surface with 1× PBS, and add fresh differentiation medium containing 8 μg/mL doxycycline, 10 mM DAPT, and 5 mM CHIR.
4. Continue changing the medium every 2–3 days without passaging until myotubes are formed (confirm via morphology) (Fig. 4c).
5. Seven to ten days post-initiation of myotube differentiation, cells should be fully differentiated and may begin to detach or die. Before detachment occurs, harvest myotubes for further analysis (*see* **Notes** 7 and **8**) (Fig. 5).
6. To harvest myotubes, collect the medium and transfer it to a 50 mL conical tube. The medium may contain detached myotubes.

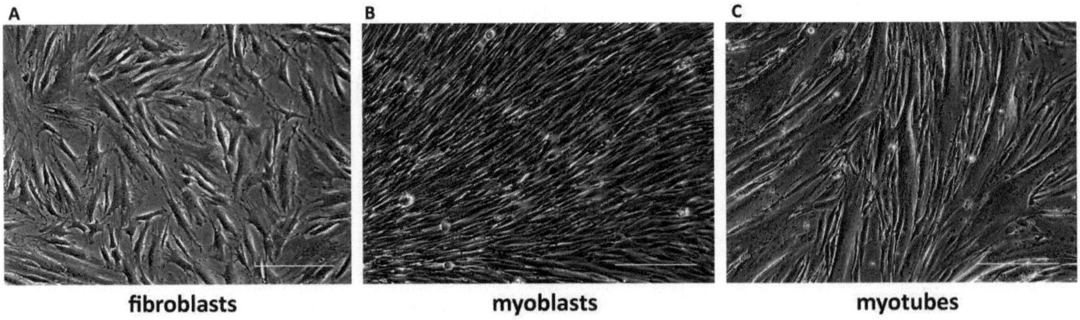

Fig. 4 Fibroblasts transdifferentiation. (**a**) 70% confluent fibroblasts ready for myoblast conversion. (**b**) Within 2–4 days the cells become elongated. (**c**) After 7 days in myotube medium, multinucleated myotubes can be observed. Scale bar: 50 μm. (Figure adapted from Almeida et al. [19])

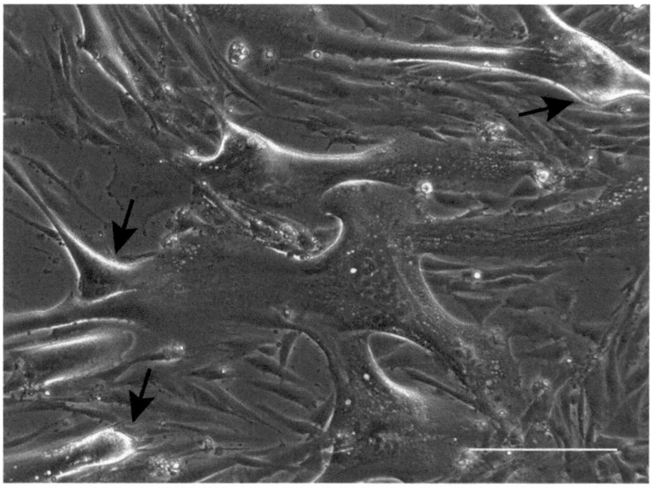

Fig. 5 Detaching myotubes. Upon maturation, the myotubes may start to detach as evidenced by white and bright edges (arrows). Scale bar: 50 μm. (Figure adapted from Almeida et al. [19])

7. Rinse the myotubes with 5 mL 1× PBS and transfer the PBS to the 50 mL tube. Some myotubes may detached by just rinsing.
8. Add 1 mL (for a 10 cm dish) of 0.25% trypsin to the cell surface. Incubate the dish in the incubator for 5 min. Check under the microscope for cell detachment. If not detached, return the dish to the incubator for an additional 5 min.
9. Once the cells are detached, transfer 10 mL of the collected medium and PBS to the dish and pipette up and down to resuspend the cells. Transfer the cells back into the 50 mL conical tube.
10. Centrifuge at 1200 × g for 7 min at 4 °C.
11. Carefully remove the supernatant without disturbing the pellet. Add 1 mL 1× PBS to resuspend the cells and transfer them to a 1.5 mL tube.
12. Centrifuge at 1200 × g for 7 min at 4 °C.
13. Carefully remove the supernatant without disturbing the pellet. Store the pellets at −80 °C until further processing.

3.6 Myosin Heavy Chain Immunostaining of Differentiated Myotubes

For immunostaining, cultivate cells on glass coverslips or chamber slides, following the instructions above.

1. Upon full differentiation of myotubes, remove the medium, and delicately rinse the slides with 1× PBS. Aspirate the PBS.
2. Add 4% PFA (500 μL per well in a 12-well plate), and incubate at room temperature for 10 min. Aspirate the PFA.
3. Rinse with 1 mL 1× PBS.

4. Permeabilize with permeabilization solution (300 μL/well in a 12-well plate) for 10 min, and then aspirate the solution.
5. Block with 300 μL/well of blocking solution for 10 min, and then aspirate the solution.
6. Apply the primary antibody, diluted 1:50 in 300 μL of blocking solution, for 2 h at room temperature with gentle rocking.
7. Rinse three times with 1 mL/well of 1× PBS for 5 min, with gentle rocking.
8. Incubate with the secondary antibody, diluted 1:500 in 300 μL of blocking solution, for 1 h at room temperature with gentle shaking. Cover the plate with aluminum foil during this incubation.
9. Rinse three times with 1 mL/well of 1× PBS for 5 min, with gentle shaking.
10. Incubate with DAPI diluted in 1× PBS for 10 min. Rinse three times with 1 mL/well of PBS.
11. Apply a drop of mounting medium to a glass slide. Remove the coverslip with forceps and place it face down on the drop of mounting medium.
12. Invert the slide onto a paper towel and gently press to eliminate bubbles and excess mounting medium.
13. Seal the slides with nail polish and store at 4 °C until imaging. Myotubes derived from DMD patient cells show good myogenic fusion and express myosin heavy chain protein (Fig. 6).

3.7 Antisense Oligonucleotide Transfection

The following procedure is designed for transfecting a 6-well plate. Adjust the volumes as needed for smaller or larger wells. The transfection occurs in confluent fibroblasts when the cells are prepared for the differentiation step.

1. Remove the fibroblast growth medium and wash the cells with 1 mL of 1× PBS.
2. Add 500 μL/well of OptiMEM medium and incubate at 37 °C for 1 h.
3. Dilute the antisense oligonucleotide (AON) in 100 μL of OptiMEM per well to the desired final concentration (e.g., 50 nM, 100 nM, 200 nM, 500 nM). Incubate at room temperature for 5 min (*see* **Note 9**).
4. Dilute the lipofectamine in OptiMEM, to a final volume of 100 μL per well. 1 μL of lipofectamine should be used for every μg of AON. Incubate at room temperature for 5 min.
5. Mix the diluted lipofectamine with the diluted AON. Gently pipette and incubate for 20 min at room temperature to facilitate complex formation. Avoid air bubbles.

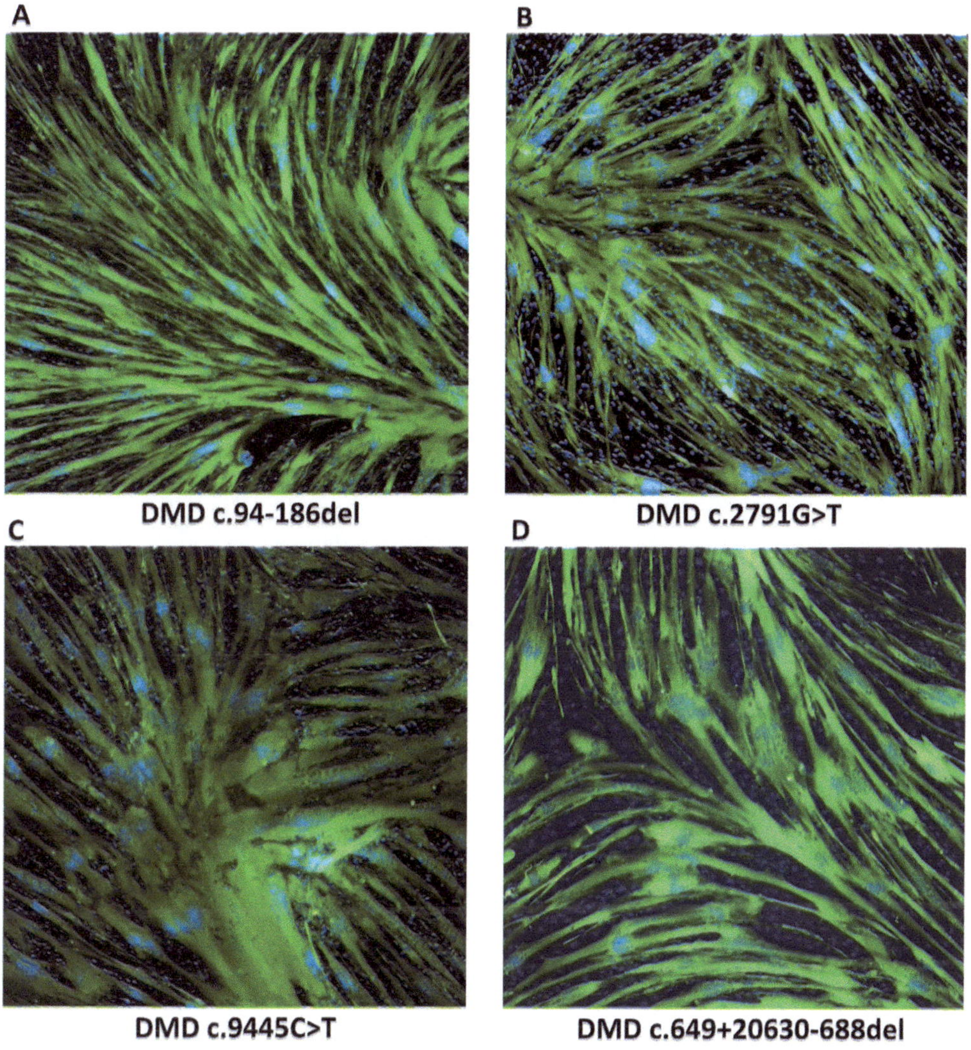

Fig. 6 Immunofluorescence of fused myotubes. Immunostaining of myosin heavy chain in myotubes derived from DMD patients with various mutations. Scale bar: 100 μm

6. Add 200 μL of the lipofectamine and AON mix to each well. Incubate the cells overnight at 37 °C in 5% CO_2.
7. On the following day, remove the transfection mix, and add 2 mL of warm myoblast growth medium supplemented with doxycycline, DAPT, and CHIR.
8. Collect cells at least 3 days later for RNA extraction or 7–21 days for protein analysis (*see* **Note 10**) (Fig. 7).

3.8 AAV1-U7 Transduction

This procedure was optimized for 6-well plates. Adjust the volumes proportionally based on the culture surface area. The transduction occurs in confluent fibroblasts prepared for the differentiation step.

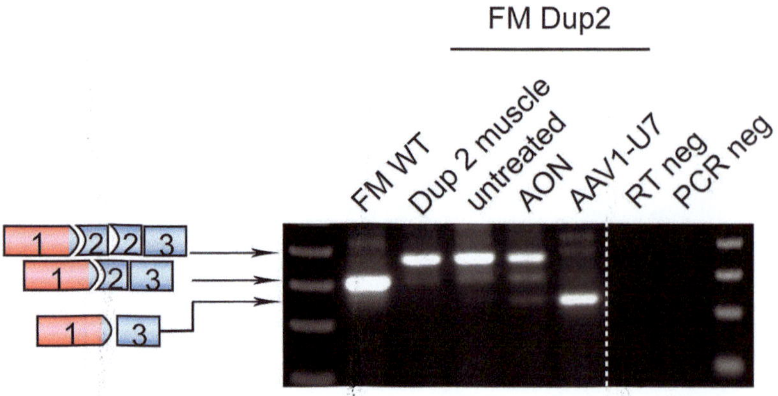

Fig. 7 Representative RT-PCR showing *DMD* exon skipping in FM cells. DMD expression by RT-PCR. FM cells with DMD exon 2 duplication. RNA from muscle biopsy was used as control to show that FM cells express the mutated transcript. Treatment with AON led to partial skipping of exon 2 duplication, while treatment with AAV1-U7 results in improved skipping. (Figure adapted from Almeida et al. [19])

AAV1 has the most effective transduction capacity in cultured myoblasts.

1. Remove the fibroblast growth medium and wash the cells with 1 mL of 1× PBS.
2. Dilute 0.5–1 × 10^{11} viral particles of AAV1-U7 in 700 μL of warm myoblast growth medium supplemented with doxycycline, DAPT, and CHIR (*see* **Note 11**).
3. Homogeneously add the viral mix to the well by dropping.
4. On the following day, supplement with 1.3 mL of warm myoblast growth medium supplemented with doxycycline, DAPT, and CHIR.
5. Collect cells at least 3 days later for RNA extraction or between 7 and 21 days for protein analysis (Fig. 7).

4 Notes

1. It is better to use the skin biopsy within 24–36 h to avoid potential growth of contamination. In addition, the prompt processing of the skin biopsy increases the likelihood of obtaining viable fibroblasts.
2. If the number of cells is low, use less medium to collect the cells and transfer them to a smaller dish or 6-well plates. It is important to not let the cells too spaced, because the closer contact enhances proliferation.

3. Passage numbering: P1 is defined when the initial fibroblasts derived from the skin biopsy are moved to a new flask to undergo proliferation. Managing the passage number of fibroblast cultures is vital due to their limited proliferative capacity, as they can enter replicative senescence after numerous passages. Therefore, it is advisable to maintain a stock of fibroblasts at a low passage number and transform cells at the earliest possible passage.
4. For immunostaining purposes, seed cells onto glass coverslips or chamber slides coated with Matrigel. Matrigel should be diluted at a ratio of 1:10 in DMEM, applied to the well in sufficient quantity to cover the surface, and allowed to sit at room temperature for 1 h. Matrigel should then be aspirated immediately before cell seeding.
5. If the cells are below 70% or above 80% confluency, myogenic differentiation may be compromised.
6. CHIR is an activator of the Wnt signaling pathway, which upregulates myogenic regulatory factors. DAPT is an inhibitor of Notch1 signaling pathway, which is involved in the differentiation and maturation of myoblasts. While the addition of CHIR and DAPT helps the differentiation process, it is also possible to perform experiments without these. However, differentiation may take longer and the myotubes may be less mature.
7. The timeframe for myotube formation varies based on the cell line. Mutations in muscle-related proteins may affect myogenic potential. Bright appearance and whitening at the borders of myotubes indicate detachment initiation (refer to Fig. 4).
8. The differentiation capacity of FM cells persists up to passage 25–27, particularly in wild-type controls. Some DMD cell lines maintain myogenic potential beyond P20, while a myotonic dystrophy type 1 (DM1) cell line shows reduced myogenicity after P8. The retention of myogenic differentiation capacity varies by cell line but generally remains up to P20–25.
9. This protocol is optimized for 2'omethyl-phosphorothioate AONs.
10. The duration of differentiation required to detect RNA and/or protein expression may vary based on the gene of interest or the cell line. In the case of DMD, mRNA detection is possible within 3 days, while dystrophin protein detection necessitates at least 7 days. This timeframe may vary depending on the cell line. High concentrations of AON and transfection reagent can impact transdifferentiation.
11. Viral concentration was determined using qPCR. The amount of virus utilized may vary depending on the quantification method and should be determined in a preliminary experiment using a reporter transgene.

Acknowledgments

We would like to thank Dr. Vincent Mouly for sharing his knowledge in the past regarding the model. This work has been supported by the US National Institutes of Health National Institute of Neurological Disorders and Stroke (R01 NS043264 (Kevin Flanigan and Robert Weiss)) and the US National Institutes of Health National Institute of Arthritis and Musculoskeletal and Skin Diseases (NIAMS) (P50 AR070604-01 (Kevin Flanigan, Kim McBride, and Nicolas Wein). This work was also supported by internal discretionary funds. IRB number: IRB #: IRB10-00358/ CR00005138 and IBCSC#: IBS00000123.

References

1. Szabo M, Akusjärvi SS, Saxena A et al (2017) Cell and small animal models for phenotypic drug discovery. Drug Des Devel Ther 11: 1957–1967. https://doi.org/10.2147/DDDT.S129447
2. Kawahara G, Kunkel LM (2013) Zebrafish based small molecule screens for novel DMD drugs. Drug Discov Today Technol 10:e91–e96. https://doi.org/10.1016/J.DDTEC.2012.03.001
3. Zarei A, Razban V, Hosseini SE, Tabei SMB (2019) Creating cell and animal models of human disease by genome editing using CRISPR/Cas9. J Gene Med 21:e3082. https://doi.org/10.1002/jgm.3082
4. Bigot A, Jacquemin V, Debacq-Chainiaux F et al (2008) Replicative aging down-regulates the myogenic regulatory factors in human myoblasts. Biol Cell 100:189–199. https://doi.org/10.1042/bc20070085
5. Mamchaoui K, Trollet C, Bigot A et al (2011) Immortalized pathological human myoblasts: towards a universal tool for the study of neuromuscular disorders. Skelet Muscle 1. https://doi.org/10.1186/2044-5040-1-34
6. Pantic B, Borgia D, Giunco S et al (2016) Reliable and versatile immortal muscle cell models from healthy and myotonic dystrophy type 1 primary human myoblasts. Exp Cell Res 342:39–51. https://doi.org/10.1016/j.yexcr.2016.02.013
7. Thorley M, Duguez S, Mazza EMC et al (2016) Skeletal muscle characteristics are preserved in hTERT/cdk4 human myogenic cell lines. Skelet Muscle 6:43. https://doi.org/10.1186/s13395-016-0115-5
8. Chaouch S, Mouly V, Goyenvalle A et al (2009) Immortalized skin fibroblasts expressing conditional MyoD as a renewable and reliable source of converted human muscle cells to assess therapeutic strategies for muscular dystrophies: validation of an exon-skipping approach to restore dystrophin in duchen. Hum Gene Ther 20:784–790. https://doi.org/10.1089/hum.2008.163
9. Echevarría L, Aupy P, Goyenvalle A (2018) Exon-skipping advances for Duchenne muscular dystrophy. Hum Mol Genet 27:R163–R172
10. Hwang J, Yokota T (2019) Recent advancements in exon-skipping therapies using antisense oligonucleotides and genome editing for the treatment of various muscular dystrophies. Expert Rev Mol Med 21:e5
11. Wein N, Vulin A, Findlay AR et al (2017) Efficient skipping of single exon duplications in DMD patient-derived cell lines using an antisense oligonucleotide approach. J Neuromuscul Dis 4:199–207. https://doi.org/10.3233/JND-170233
12. De Angelis FG, Sthandier O, Berarducci B et al (2002) Chimeric snRNA molecules carrying antisense sequences against the splice junctions of exon 51 of the dystrophin pre-mRNA induce exon skipping and restoration of a dystrophin synthesis in Δ48-50 DMD cells. Proc Natl Acad Sci USA 99:9456–9461. https://doi.org/10.1073/pnas.142302299
13. Gorman L, Suter D, Emerick V et al (1998) Stable alteration of pre-mRNA splicing patterns by modified U7 small nuclear RNAs. Proc Natl Acad Sci USA 95:4929–4934. https://doi.org/10.1073/pnas.95.9.4929
14. Wein N, Vulin A, Falzarano MS et al (2014) Translation from a DMD exon 5 IRES results in a functional dystrophin isoform that attenuates dystrophinopathy in humans and mice.

Nat Med 20:992–1000. https://doi.org/10.1038/nm.3628

15. Findlay AR, Wein N, Kaminoh Y et al (2015) Clinical phenotypes as predictors of the outcome of skipping around DMD exon 45. Ann Neurol 77. https://doi.org/10.1002/ana.24365
16. Wein N, Avril A, Bartoli M et al (2010) Efficient bypass of mutations in dysferlin deficient patient cells by antisense-induced exon skipping. Hum Mutat 31:136–142
17. Auré K, Mamchaoui K, Frachon P et al (2007) Impact on oxidative phosphorylation of immortalization with the telomerase gene. Neuromuscul Disord 17:368–375. https://doi.org/10.1016/j.nmd.2007.01.019
18. Barde I, Zanta-Boussif MA, Paisant S et al (2006) Efficient control of gene expression in the hematopoietic system using a single Tet-on inducible lentiviral vector. Mol Ther 13:382–390. https://doi.org/10.1016/j.ymthe.2005.09.012
19. Almeida CF, Frair EC, Huang N et al (2021) Direct reprogramming of human fibroblasts into myoblasts to investigate therapies for neuromuscular disorders. J Vis Exp 2021. https://doi.org/10.3791/61991

Chapter 13

Culture and Electrophysiological Analysis of Patient-Specific iPSCs Using Microelectrode Array Dishes

Bayardo I. Garay and Rita C. R. Perlingeiro

Abstract

Microelectrode arrays offer an exciting opportunity to probe and characterize the functional status of dystrophin-deficient muscle tissue vis-à-vis patient-specific iPSC-derived skeletal muscle (SkM) and cardiac muscle (CM). Here, we present a method to culture and analyze baseline electrophysiological profiles of Duchenne muscular dystrophy (DMD) iPSC-SkM and iPSC-CM using MEA dishes.

Key words Patient-specific iPSC, Duchenne muscular dystrophy, Skeletal muscle, Cardiac muscle, Microelectrode array, Electrophysiology, Dystrophin, Field potentials, Conduction velocity

1 Introduction

The discovery of human-induced pluripotent stem cells (hiPSCs) in 2007 [1] has been a complete paradigm shift in our efforts to model and understand the pathophysiology of muscular dystrophies, like Duchenne muscular dystrophy (DMD). Given the vast heterogeneity of mutations and disease severity in DMD, patient-specific iPSCs have become an exciting tool in which to probe for functional differences between large numbers of affected and unaffected control iPSC-derived myocytes. We and others have developed robust methods for the generation of hiPSCs into more mature skeletal muscle (SkM) [2–4] and cardiomyocytes (CM) [5, 6]. Once generated in the dish, these patient-derived iPSC-SkM and iPSC-CMs can then be assessed for functional status using technologies such as microelectrode arrays (MEA, Fig. 1a), which can measure changes in the extracellular membrane potential.

In 1972, Thomas, Jr. and colleagues reported for the first time the detection of extracellular membrane potentials from dissociated embryonic chick heart tissue using MEAs [7]. They detected peak-to-peak amplitudes, i.e., field potentials, of up to 2.5 mV after a few

Kevin Flanigan and Stefan Nicolau (eds.), *Duchenne Muscular Dystrophy: Methods and Protocols*, Methods in Molecular Biology, vol. 2975, https://doi.org/10.1007/978-1-0716-4811-7_13,

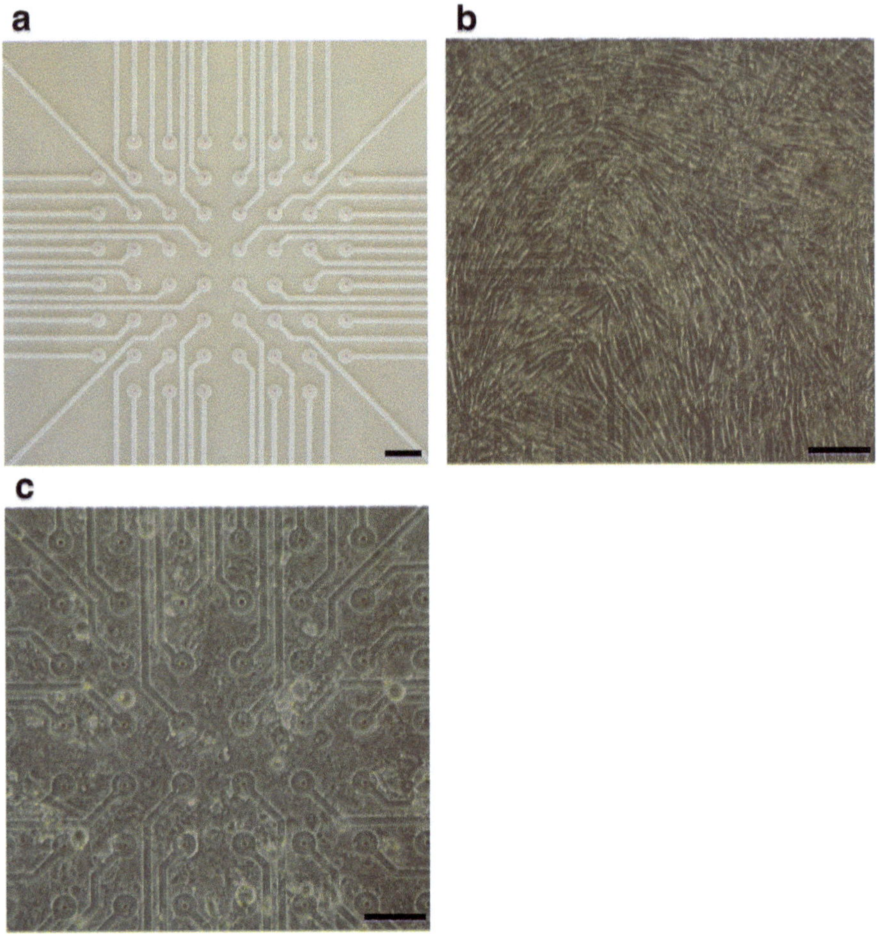

Fig. 1 MEA system for functional characterization of DMD patient-derived iPSC-SkM and iPSC-CMs. (**a**) Representative MEA dish with 59-recording electrode configuration. Representative DMD patient-derived (**b**) skeletal muscle and (**c**) cardiac muscle cultured on MEA dish. Scale bar = 100 μm

hours in culture. More recently, Eisen and colleagues have used this technology to probe for functional differences between DMD and unaffected control iPSC-CMs [8]. Although there are some reports of primary skeletal muscle cultures being analyzed via MEA [9, 10], their application to DMD patient-derived iPSC-SkM remains largely untested. Certainly, the technology has improved over the past few decades with new designs that emphasize the simultaneous recording from multiple chambers while maintaining temperature and CO_2 concentrations constant. However, one of the main barriers for the use of this technology continues to be the high degree of technical expertise needed to ensure successful recordings. Overcoming these challenges are particularly important for contracting myocytes, which tend to easily detach from their substrate after a few days. Here, we present a more detailed protocol to obtain well-adherent DMD patient-specific iPSC-SkM (Fig. 1b) and iPSC-CM

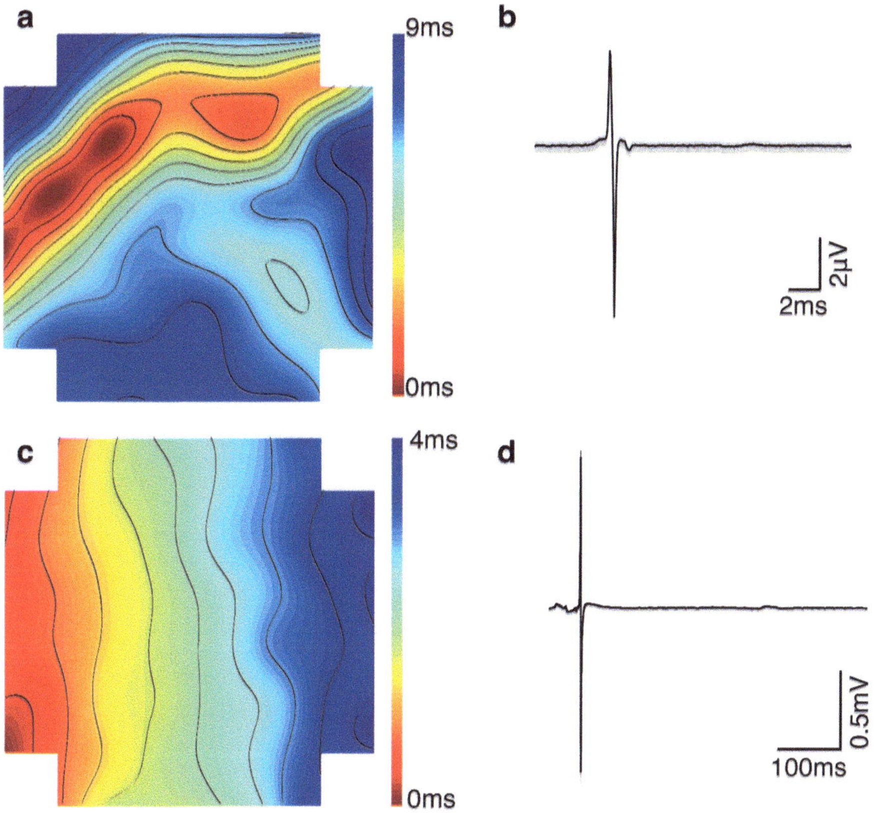

Fig. 2 Electrophysiological properties of DMD patient-derived iPSC-SkM and iPSC-CMs using MEA technology. Representative wave propagation maps showing the spread of action potentials in DMD patient-derived (**a**) iPSC-SkM and (**c**) iPSC-CMs. Representative field potential traces in (**b**) iPSC-SkM and (**d**) iPSC-CMs. Trace shown as average ± SEM

(Fig. 1c) that have the prerequisite maturation status to measure their electrophysiological properties using MEA dishes, including their conduction velocity with propagation maps (Fig. 2a, c) and field potential amplitude and duration (Fig. 2b, d).

2 Materials

Unless otherwise specified, all media preparation occurs under sterile and aseptic conditions inside a biosafety level-2 (BSL2) cabinet.

1. Microelectrode arrays: 60MEA100/10iR-ITO-gr (Multichannel Systems, Harvard Bioscience, Inc). Each 12-mM-tall dish has the following technical configuration: 8 × 8 electrode grid with 59 recording titanium nitride (TiN) electrodes and 1 silicon nitride (SiN) isolator with an electrode spacing of 100 μm and an electrode diameter of 10 μm (*see* **Note 1**).

2. Gas-permeable MEA-MEM covers, O-rings, loading tool (Multichannel Systems, Harvard Bioscience, Inc).
3. Compatible MEA system: MEA1060 system or MEA2100 system with integrated blanking or non-blanking amplifier, temperature control, and real-time detection with feedback using either PCI MC-card or USB-ME system (Multichannel Systems, Harvard Bioscience, Inc.; *see* **Note 2**).
4. 10 cm tissue culture dish for transport and incubation.
5. Cardio2D software (Multichannel systems, Harvard Bioscience, Inc).
6. Cardio2D+ software (Multichannel systems, Harvard Bioscience, Inc).
7. Microsoft Excel.
8. GraphPad Prism (GraphPad).
9. Matrigel® solution: 160 μg/mL Growth Factor Reduced (GFR) Matrigel® (Phenol-red free, LDEV-free, Corning) in KnockOut Dulbecco's Modified Eagle Medium (DMEM), and store at 4 °C (*see* **Note 3**).
10. 0.25% trypsin-ethylenediaminetetraacetic acid (EDTA).
11. 1× Dulbecco's Phosphate-Buffered Saline (DPBS) without calcium and magnesium.
12. ROCK inhibitor (Y-27632).
13. H-1152 dihydrochloride.
14. Recombinant Human FGF-basic (154 a.a.) (Preprotech).
15. Cardiac quenching media (RPMI1620): sterile-filtered 1× RPMI1640 with L-glutamine, 20% fetal bovine serum (FBS).
16. Hemocytometer.
17. Tabletop centrifuge.
18. Autoclaved DNAse-free and RNAse-free 1.5 mL-microcentrifuge tubes.
19. 0.4% trypan blue solution.
20. Myogenic medium (S1): Iscove's Modified Dulbecco's Medium (IMDM) without glutamax, 15% FBS, 10% heat-inactivated horse serum (HS), 1% KnockOut Serum Replacement (KOSR), 4.5 mM α-monothioglycerol (MP Biochemicals), 1× GlutaMAX supplement, 50 μg/mL L-ascorbic acid, and 5 U/mL penicillin-streptomycin (Pen/Strep).
21. Skeletal muscle terminal differentiation medium (S2): KnockOut DMEM supplemented with 20% KOSR, GlutaMax, Pen/-Strep, and nonessential amino acids.
22. Skeletal muscle maturation cocktail (S3): S2 supplemented with 10 μM TGFβ inhibitor (SB431542), 10 μM

dDexamethasone (Cayman Chemical), 10 μM Forskolin (Cayman Chemical), and 10 μM γ-secretase inhibitor (DAPT, Selleckchem).

23. Cardiac maintenance media (C1): RPMI1640 with L-glutamine, 1× serum-free B27 supplement with insulin.
24. Cardiac muscle maturation cocktail (C2): C1 media supplemented with 10 μM mirdametinib (PD0325901, Cayman Chemical) and 5 μM SB203580 (Cayman Chemical).
25. Day-16 human iPSC myogenic progenitors as described in [2, 3, 11].
26. Day-27 human iPSC-CMs as described in [6].
27. Cleaning MEA dish: 1% Tergazyme™ Enzyme-Active Powered Detergent (Alconox) in distilled water.

3 Methods

Carry out cell culture procedures using sterile and aseptic techniques at room temperature inside a biosafety level-2 (BSL2) cabinet. Always wear personal protective equipment when handing human stem cells and their derivatives. Dispose of all waste materials following your institution's guidelines.

3.1 MEA Dish Preparation

1. Start by placing the MEA dish in 1% Tergazyme™ solution in distilled water overnight under gentle rocking at room temperature (RT). The following day perform three sequential washes in distilled water under gentle rocking for 15–20 min each time at RT. Transfer MEA dish(es) onto a 10 cm culture dish inside the BSL-2 cabinet, add 1.5 mL of sterile molecular grade water to each dish, and incubate under UV light overnight inside the tissue culture BSL-2 cabinet (*see* **Note 4**).
2. Gently and carefully, without touching the center of the dish, aspirate the water and add 1.0 mL of 160 μg/mL GFR-Matrigel® for a final concentration of ~45 μg/cm^2 (*see* **Note 5**). Transfer to a sterile 10 cm dish and incubate at 37 °C for at least 30 min.
3. Before seeding cells onto the MEA dish, wash the dish three times with 1× DPBS.

3.2 Skeletal Muscle Preparation

1. Obtain day-16 human iPSC myogenic progenitors as described in [2, 3, 11]. Dissociate the myogenic progenitors using 0.25% trypsin-EDTA for 3–5 min, and transfer to a 50 mL conical tube containing 2× volume in S1 media. For example, 10 mL S1 media for every 5 mL 0.25% trypsin-EDTA. Singularize the cells by pipetting up and down with a serological pipette.

2. Centrifuge for 5 min at 400 *g* at RT. Carefully aspirate the supernatant, and resuspend in 5 mL of S1 media supplemented with 10 μM ROCK inhibitor and 10 ng/mL bFGF. Prepare to count cells using a 1:4 dilution ratio. For example, 75 μL 0.4% trypan blue solution and 25 μL cell suspension. Transfer 1,000,000 myogenic progenitors resuspended in 500 μL S1 media supplemented with 10 μM ROCK inhibitor and 10 ng/mL bFGF onto the center of the MEA dish and place inside the incubator at 37 °C and 5% CO_2 for 1 h (*see* **Note 6**). After 1 h, add an additional 1.0 mL of S1 media with ROCK and bFGF to the MEA dish to avoid evaporative loss overnight (*see* **Note 7**).
3. On the following day, carefully aspirate the media, and wash once with 1.0 mL of 1× DPBS (*see* **Note 8**). Then, add 1.5 mL of S3 media for terminal skeletal muscle differentiation and maturation, and place back inside the incubator for 2 days. On day 3, wash once with 1.0 mL of 1× DPBS, and add fresh S3 media. Keep inside the incubator for an additional 3 days. Terminally differentiated and mature skeletal muscle fibers should be visible on days 5–7 (*see* Fig. 1b).
4. On the day of data acquisition, cover the MEA dish with a MEA-MEM ring inside the BSL-2 cabinet, and take to the MEA system at your disposal.
5. Set the surface temperature of the platform at 37 °C before placing the MEA dish onto the MEA system (*see* **Note 9**). Set up the Cardio2D software for data acquisition with the following parameters: Sample rate 50 kHz, Input Range ± 4096 mV. Click on viability to ensure there is a reliable signal (*see* **Note 10**). Record 120 s of data (*see* **Note 11**).
6. Export the data and analyze using the Cardio2D+ software (*see* **Note 12**).
7. Because the MEA dish is covered using the MEA-MEM ring, the cultures will remain sterile. Place back inside the incubator at 37 °C after removing the ring inside the BSL-2 cabinet, and repeat readings as needed. If keeping the cultures beyond day 7, we recommend carefully changing the media using S2 media.

3.3 Cardiac Muscle Preparation

1. Obtain day-27 human iPSC-CMs as described in [6]. Dissociate the cells using 0.25% trypsin-EDTA for 20 min, and singularize first by pipetting up and down with a P1000. Then, transfer to a 50 mL conical tube containing 3× volume in RPMI1620 supplemented with 10 μM ROCK inhibitor, and pipette up and down to mix and fully quench the reaction (*see* **Note 13**). For example, 30 mL RPMI1620 media for every 10 mL 0.25% trypsin-EDTA.

2. Centrifuge at 300 *g* for 5 min at RT. Carefully aspirate the supernatant, and resuspend in 5 mL of C2 media supplemented with 10 μM ROCK inhibitor and 2 μM H-1152. Prepare to count cells using a 1:4 dilution ratio, as described above. Transfer 500,000 hiPSC-CMs resuspended in 500 μL C2 media supplemented with 10 μM ROCK inhibitor and 2 μM H-1152 onto the center of the MEA dish, and place inside the incubator at 37 °C and 5% CO_2 for 1 h (*see* **Note 6**). After 1 h, add an additional 1.0 mL of C2 media supplemented with 10 μM ROCK inhibitor and 2 μM H-1152 to the MEA dish to avoid evaporative loss overnight (*see* **Note** 7). Spontaneous beating should be observed within 12 h.
3. On the following day, carefully aspirate the media, and wash once with 1.0 mL of 1× DPBS (*see* **Note 8**). Then, add 1.5 mL of C2 media alone, and place back inside the incubator until day 30.
4. On the day of data acquisition, cover the MEA dish with a MEA-MEM ring inside the BSL-2 cabinet, and take to the MEA system at your disposal.
5. Set the surface temperature of the platform at 37 °C before placing the MEA dish onto the MEA system (see **Note 9**). Set up the Cardio2D software for data acquisition with the following parameters: sample rate 50 kHz, input range ± 4096 mV. Click on viability to ensure there is a reliable signal (*see* **Note 14**). Record ≥300 beats of data (*see* **Note 15**).
6. Export the data and analyze using the Cardio2D+ software (*see* **Note 16**).
7. Because the MEA dish is covered using the MEA-MEM ring, the cultures will remain sterile. Place back inside the incubator at 37 °C after removing the ring inside the BSL-2 cabinet, and repeat readings as needed.

3.4 MEA Maintenance

1. Clean the MEA dish as outlined above after every experiment to preserve the integrity of the electrodes and the longevity of the dish itself.
2. When not in use, add sterile water, cover with parafilm, and keep at 4 °C. Replace the water at least once a month. If done correctly, the MEA dishes should last for at least 50 experiments.

4 Notes

1. In the event that electrical pacing is indicated for the experiment, we recommend using an electrode diameter of 30 μm instead of 10 μm as this will result in lower voltages needed to

successfully pace the cultures while still retaining cell viability. The indications for electrical stimulation depend on the question being explored, so a detailed exploration is beyond the scope of this chapter. However, one example includes the addition of sympathomimetic or sympatholytic compounds to a stepwise frequency stimulation protocol to assess the response to stress in iPSC-CMs with intact or impaired calcium homeostasis, as in DMD. Another example could be the addition of cholinergic agonists or antagonists to a high-frequency low-voltage training protocol of iPSC-SkM. Because the MEA dishes are compatible with microscopy techniques, addition of calcium-sensitive dyes opens another layer of complexity that can be further explored.

2. Each MEA system has its benefits and disadvantages and the choice of the system to use depends on the investigator. The MEA1060 system, which we employed in our experiments, utilizes a single-chamber system, while the MEA2100 can allow up to four-chamber simultaneous recordings. For initial optimization and standardization of workflow, we recommend a single-chamber approach since it offers the greatest amount of flexibility. Newer MEA systems have a built-in blanking amplifier. However, whether the amplifier has a blanking or non-blanking function is not consequential since the raw data undergoes rigorous unbiased filtering during analysis with the Cardio2D+ software. Of note, data acquisition using the PCI MC-card is no longer compatible with Windows 10, so we recommend using systems that have a USB capability.
3. The Matrigel® solution can be used for up to 2 weeks if kept at 4 °C. Always use ice when handling Matrigel® as it tends to polymerize at temperatures >8–10 °C. If you observe a precipitate floating in the solution, discard immediately, and prepare a new solution from a fresh aliquot.
4. We prefer sterilizing under UV light overnight, but we have also gotten successful cultures with 3–4 h of UV exposure.
5. Once polymerized, the MEA dishes can be kept inside the incubator for at least 1–2 days, but we recommend using them within 1–2 h. We have tested other matrices compatible with cell culture such as gelatin and human fibronectin at various concentrations, but these matrices result in a faster rate of cell detachment compared to GFR-Matrigel®. Similar matrices like hESC-Matrigel®, which provide a dilution ratio have a higher degree of experimental variability since it is harder to know the specific concentration of matrix being used.
6. Make sure to distribute the cells evenly after seeding. This can be accomplished by gently rocking the dish in the North, South, then East, and West directions before placing inside the incubator.

7. On the first time following this protocol, we recommend using two to three MEA dishes per cell line to ensure reproducibility. Once established, we always perform at least three independent biological replicates per cell line per cell type, where a biological replicate is a new experiment starting from the human iPSC stage. When adding media to the MEA dish, use a P1000 pipette tip to dispense the volume *slowly* to one of the borders of the dish at an angle of 15–30°, while making sure not to aim directly on top of the cells as this could disturb their distribution.
8. When aspirating media from the MEA dish, use a glass Pasteur pipette or other equally long suction device. It is also possible to use a P1000 pipette to remove the volume, but exercise caution to not touch the cells or the electrodes.
9. Adding 100% ethanol to the heating plate before placing the MEA dish allows better contact and reduces noise signal during data acquisition. Although it temporarily decreases the temperature, it should still come back to 37 °C within 1 min.
10. Before assessing the cell "viability" using the Cardio2D software for skeletal muscle cells, adjust the default threshold of what is detected by setting the analyzer to the following: minimum rise time 100 μs, maximum rise time 300 μs, signal-dependent ≥5 standard deviations (SD), detection dead time 5 ms, and min/max search ±1 ms. These parameters are not final and can be changed during data analysis with the Cardio2D+ software.
11. Start with 120 s worth of data and adjust as needed based on research needs. However, given the difference in action potential duration between SkM (2–4 ms) and CM (200–700 ms), there is a risk of oversampling SkM data with recordings >2 min.
12. Optimal settings for analysis will need to be empirically determined. We have provided an initial set of filtering criteria that works well for our cultures, but many factors can influence the approach—how well the cultures attached to the dish, how well differentiated the cultures are, expression of ion channels, amount of spontaneous twitching observed, temperature, amount of time that the cultures remain outside of the incubator, etc. Specifically for skeletal muscle, we find that setting a stringent signal-dependent threshold (≥5 SD) is helpful to start, but if there is too much signal passing through, then increasing the threshold can be helpful. Setting the signal-dependent threshold to the maximum (50SD) can result in no signal being detected.

13. Cardiac cells are exquisitely sensitive to shearing forces and tend to die quickly when not handled properly. We recommend using the minimum amount of pipetting needed to singularize the cells before quenching. However, not singularizing the cells enough will result in inaccurate cell counting and inappropriate attachment of cells downstream. Therefore, great care must be taken to properly singularize them. If 20 min of trypsinization is insufficient to quickly singularize the cells within the first few attempts, place back inside the incubator for an additional 5–10 min. Do not keep in trypsin longer than 40 min. We have found that transferring the hiPSC-CMs into a quenching media containing an anti-apoptotic agent, such as ROCK, greatly improves viability.
14. Before assessing the cell "viability" using the Cardio2D software for cardiac muscle, adjust the default threshold of what is detected by setting the analyzer to the following: minimum rise time 100 μs, maximum rise time 1 ms, signal-dependent ≥25 SD (ideally 50), detection dead time 50 ms, min/max search ±10 ms. These parameters are not final and can be changed during data analysis with the Cardio2D+ software.
15. Unlike skeletal muscle, it is much easier to regulate the heart rate of spontaneously beating hiPSC-CMs, namely, via electrical pacing and temperature. The disadvantage of electrical pacing on MEAs is that we lose the field potential amplitude data because it is overpowered by the pacing signal (typically, biphasic pulses of 1–2 mV amplitude). The amplitude of the field potential is correlated to the expression levels and amount of open sodium-voltage gated ion channels ($Na_v1.5$), which is an important measure of functionality in the context of DMD [12, 13]. Electrical pacing is often used because the same beat frequency is needed for all the samples, which allows comparisons of conduction velocities between groups. Generally, lower ambient temperatures will result in a slower beat rate. Regardless of experimental design, we recommend ensuring that each cardiac culture has the same beat rate when recording a session to allow for downstream comparisons of field potential amplitude, duration, and conduction velocity. Another option is to regulate the beating rate with temperature. We have been successful ensuring rates of 60 bpm ± 5 bpm using surface temperatures of 35–37 °C. If the cardiac cultures are beating too fast (>100 bpm), then we recommend changing the media inside the MEA dish with C1 media after day 30 and returning to the incubator for at least 1–2 h before reattempting recording.

Many parameters for analysis will need to be empirically determined. Determining the field potential duration will depend on beating rate of the culture, with faster rates having a shorter duration, and the functional deficit intrinsic to the cells. For example, patients with DMD often present with prolonged QTc intervals compared to unaffected controls.

Acknowledgments

This work was supported by National Institutes of Health NIAMS grants R21 AR079236, R01 AR078571, AR078624, and AR081882 (R.C.R.P.). B.I.G. was supported by NHLBI F30 HL151138 and NIGMS T32 GM008244. We thank Dr. Samuel Dudley, Jr., and Dr. Hong Liu for helpful discussions about the MEA technology.

Disclosure R.C.R.P. is cofounder and holds equity in Myogenica.

References

1. Takahashi K, Tanabe K, Ohnuki M, Narita M, Ichisaka T, Tomoda K, Yamanaka S (2007) Induction of pluripotent stem cells from adult human fibroblasts by defined factors. Cell 131(5):861–872. https://doi.org/10.1016/j.cell.2007.11.019
2. Selvaraj S, Mondragon-Gonzalez R, Xu B, Magli A, Kim H, Lainé J, Kiley J, Mckee H, Rinaldi F, Aho J, Tabti N, Shen W, Perlingeiro RC (2019) Screening identifies small molecules that enhance the maturation of human pluripotent stem cell-derived myotubes. elife 8. https://doi.org/10.7554/eLife.47970
3. Mondragon-Gonzalez R, Selvaraj S, Perlingeiro RCR (2023) In vitro maturation of human pluripotent stem cell-derived myotubes. Methods Mol Biol 2640:129–142. https://doi.org/10.1007/978-1-0716-3036-5_10
4. Darabi R, Arpke RW, Irion S, Dimos JT, Grskovic M, Kyba M, Perlingeiro RC (2012) Human ES- and iPS-derived myogenic progenitors restore DYSTROPHIN and improve contractility upon transplantation in dystrophic mice. Cell Stem Cell 10(5):610–619. https://doi.org/10.1016/j.stem.2012.02.015
5. Lian X, Hsiao C, Wilson G, Zhu K, Hazeltine LB, Azarin SM, Raval KK, Zhang J, Kamp TJ, Palecek SP (2012) Robust cardiomyocyte differentiation from human pluripotent stem cells via temporal modulation of canonical Wnt signaling. Proc Natl Acad Sci USA 109(27): E1848–E1857. https://doi.org/10.1073/pnas.1200250109
6. Garay BI, Givens S, Abreu P, Liu M, Yücel D, Baik J, Stanis N, Rothermel TM, Magli A, Abrahante JE, Goloviznina NA, Soliman HAN, Dhoke NR, Kyba M, Alford PW, Dudley SC, van Berlo JH, Ogle B, Perlingeiro RRC (2022) Dual inhibition of MAPK and PI3K/AKT pathways enhances maturation of human iPSC-derived cardiomyocytes. Stem Cell Rep. https://doi.org/10.1016/j.stemcr.2022.07.003
7. Thomas CA, Springer PA, Loeb GE, Berwald-Netter Y, Okun LM (1972) A miniature microelectrode array to monitor the bioelectric activity of cultured cells. Exp Cell Res 74(1):61–66. https://doi.org/10.1016/0014-4827(72)90481-8
8. Eisen B, Ben Jehuda R, Cuttitta AJ, Mekies LN, Shemer Y, Baskin P, Reiter I, Willi L, Freimark D, Gherghiceanu M, Monserrat L, Scherr M, Hilfiker-Kleiner D, Arad M, Michele DE, Binah O (2019) Electrophysiological abnormalities in induced pluripotent stem cell-derived cardiomyocytes generated from Duchenne muscular dystrophy patients. J Cell Mol Med 23(3):2125–2135. https://doi.org/10.1111/jcmm.14124
9. Rabieh N, Ojovan SM, Shmoel N, Erez H, Maydan E, Spira ME (2016) On-chip, multisite extracellular and intracellular recordings from primary cultured skeletal myotubes. Sci

Rep 6:36498. https://doi.org/10.1038/srep36498
10. Lewandowska MK, Bogatikov E, Hierlemann AR, Punga AR (2018) Long-term high-density extracellular recordings enable studies of muscle cell physiology. Front Physiol 9:1424. https://doi.org/10.3389/fphys.2018.01424
11. Kim H, Selvaraj S, Kiley J, Azzag K, Garay BI, Perlingeiro RCR (2021) Genomic Safe Harbor expression of PAX7 for the generation of engraftable myogenic progenitors. Stem Cell Rep 16(1):10–19. https://doi.org/10.1016/j.stemcr.2020.11.001
12. Gavillet B, Rougier JS, Domenighetti AA, Behar R, Boixel C, Ruchat P, Lehr HA, Pedrazzini T, Abriel H (2006) Cardiac sodium channel Nav1.5 is regulated by a multiprotein complex composed of syntrophins and dystrophin. Circ Res 99(4):407–414. https://doi.org/10.1161/01.RES.0000237466.13252.5e
13. Hesse M, Kondo CS, Clark RB, Su L, Allen FL, Geary-Joo CT, Kunnathu S, Severson DL, Nygren A, Giles WR, Cross JC (2007) Dilated cardiomyopathy is associated with reduced expression of the cardiac sodium channel Scn5a. Cardiovasc Res 75(3):498–509. https://doi.org/10.1016/j.cardiores.2007.04.009

Chapter 14

Canine Model of Duchenne Muscular Dystrophy

James Teixeira, Matthew J. Burke, Roland W. Herzog, and Dongsheng Duan

Abstract

Dystrophin-deficient canines are the best-characterized large animal model for Duchenne muscular dystrophy (DMD). Unlike dystrophin-null mice, affected dogs display many phenotypical features of human patients. Furthermore, dogs' body size and body weight are much closer to humans than those of mice's. These features make affected canines an excellent model system for studying pathogenic mechanisms of the disease and for testing novel experimental therapeutics. The golden retriever muscular dystrophy (GRMD) dog is the first canine DMD model with a confirmed null mutation in the *DMD* gene. Since then, dystrophin-deficient dogs have been identified in many different breeds. Research colonies have been established worldwide. Here, we describe our experience in generating and maintaining affected dogs. Adeno-associated virus (AAV) vectors have been used to restore dystrophin expression by gene replacement or gene repair therapies. We provide protocols for intramuscular AAV injection in adult canines and systemic delivery of AAV to neonatal and young adult canines. We also include a muscle biopsy protocol for periodic monitoring of gene transfer before the terminal study.

Key words Duchenne muscular dystrophy, DMD, Dystrophin, Canine model, Dog breeding, Dog care, Adeno-associated virus, AAV, Intramuscular injection, Systemic gene delivery, Biopsy

1 Introduction

Duchenne muscular dystrophy (DMD) is an X-linked progressive muscle-wasting disease [1]. It is caused by mutations in the *DMD* gene. Dystrophin, the protein product of the *DMD* gene, is essential for muscle membrane integrity and muscle signaling. Loss of dystrophin leads to muscle degeneration, inflammation, and fibro-fatty replacement. Numerous naturally occurring or experimentally engineered DMD models have been established [2, 3]. These include nonmammalian models (*Caenorhabditis elegans*, *Drosophila melanogaster*, and zebrafish), rodent models (mouse and rat), and medium-size to large-size animal models (cat, rabbit, dog, pig, and

Authors James Teixeira and Matthew J. Burke have equally contributed to this chapter.

Kevin Flanigan and Stefan Nicolau (eds.), *Duchenne Muscular Dystrophy: Methods and Protocols*, Methods in Molecular Biology, vol. 2975, https://doi.org/10.1007/978-1-0716-4811-7_14,

monkey). Among these, the canine model is particularly attractive for translational studies because of its clinical similarity and large body size [4–6].

The golden retriever muscular dystrophy (GRMD) dog is the most widely used canine DMD model [7]. In GRMD dogs, a point mutation in intron 6 leads to erroneous splicing of exon 6 to exon 8 and the disruption of the reading frame of the resulting transcript [8]. GRMD colonies have been established in Australia, Brazil, France, and the USA. The GRMD mutation has also been crossed to the Beagle background, and a colony was established in Japan [9]. Many important findings have been made with this model. These include, but are not limited to, the demonstration of antisense oligonucleotide-mediated exon-skipping therapy [10], adeno-associated virus (AAV)-mediated microdystrophin gene therapy [11, 12], AAV-mediated exon-skipping therapy [13], and the discovery of the involvement of the Notch signaling pathway in DMD pathogenesis [14].

The DE50-MD dog is a recently established canine DMD model [15]. It was initially discovered in Cavalier King Charles spaniels and subsequently crossed to the beagle background [16]. A colony of DE50-MD dogs has been established in the UK. DE-50 dogs carry a point mutation in the major mutation hotspot of the *DMD* gene. Specifically, a point mutation in intron 50 leads to the exclusion of exon 50 in the resulting transcript [15]. The phenotype of the DE50-MD model has been described in a series of recent publications [17–22]. Notably, a study performed in the DE50-MD model showed successful restoration of dystrophin using the CRISPR editing technology [23].

Given the outbreeding nature of human patients, we have intentionally established a mixed-breed (golden retriever, Labrador retriever, beagle, spaniel, and Welsh corgi) DMD dog colony to better model the disease. Affected dogs in our colony carry three different mutations including a point mutation in intron 6 (same as the GRMD model), an intronic insertion in intron 13 (originally identified in the Welsh corgi muscular dystrophy model, WCMD) [24], and an intronic insertion in intron 19 [25]. The dystrophic phenotype of our colony has been characterized in multiple publications [26–35]. Using this model, we demonstrated for the first time that AAV microdystrophin gene therapy can ameliorate histological and physiological defects in a dystrophic large mammal [36]. We also provided proof-of-principle for dual AAV reconstitution of a larger synthetic mini-dystrophin gene in the canine model [37]. Importantly, we established the first evidence that intravenous AAV delivery can efficiently transduce dystrophic muscles in ~2-month-old affected dogs, thus, paving the way for systemic AAV microdystrophin trials [38, 39]. Most recently, we showed that the Cas9-induced cytotoxic T-lymphocyte response is a critical hurdle for translating CRISPR editing therapy to DMD patients [40].

Over the years, we have developed a set of protocols to study histological, molecular, biochemical, immunological, and physiological changes in the canine DMD model. These protocols were published recently [41–45]. In this chapter, we present detailed protocols for the generation and care of affected dogs. We also present protocols for intramuscular injection and systemic delivery of AAV vectors to affected dogs and the muscle biopsy protocol to examine AAV-injected muscles.

2 Materials

2.1 Generation of Affected Dogs

1. Personal protective equipment (PPE), including disposable gowns and/or dedicated scrubs, shoe covers or dedicated shoes, safety glasses or splash goggles, and gloves.
2. Female carrier canines (*see* **Note 1**).
3. Semen from normal or affected canines (*see* **Note 2**).
4. Software for calculating genetic diversity (PedScope, Tenset Technologies, UK).
5. Transcervical insemination endoscope (Karl Storz, Tuttlingen, Germany).
6. Transcervical insemination catheter (Artisan Medical, Medford, NY, USA).
7. GE Logiq E ultrasound machine (GE Healthcare, Chicago, IL, USA) (*see* **Note 3**).
8. Radiograph machine (*see* **Note 3**).
9. Digital thermometer (ReliOn Inc., Spokane Valley, WA, USA) (*see* **Note 4**).
10. Stethoscope (3 M Littmann, St. Paul, MN, USA).
11. Whelping kennel. This includes a whelping box (The Jonart Whelping Box, Ada, MI, USA), white blankets, and an infrared heating lamp (Erin's Edge Dog Supply, Richfield, WI, USA).
12. Animal hair clipper with a #10 blade (Oster Professional Products, Sunbeam Products Inc., Boca Raton, FL, USA).
13. Nail polish (Sally Hansen, New York, NY, USA).
14. PetAg Esbilac puppy milk replacer (PETCO Animal Supplies, San Diego, CA, USA).
15. 8 French feeding tube (Jorgensen Laboratories, Loveland, CO, USA).
16. 10 mL syringe (Becton, Dickinson, and Company, Franklin Lakes, NJ, USA).
17. Sterile scissors (Fine Science Tools, Inc. Foster City, CA, USA).

18. Sterile #2 Suture (Surgical Specialties Corp., Wyomissing, PA, USA).
19. 1.5 mL Microcentrifuge tube (Sigma-Aldrich Inc., St. Louis, MO, USA).
20. Bulb syringe (Dynarex, Orangeburg, NY, USA).
21. Digital gram scale (Zieis, Apple Valley, MN, USA).
22. Color-coded BD® vacutainer blood collection tubes (Becton, Dickinson, and Company, Franklin Lakes, NJ, USA).
23. Safety blood collection set (23G × ¾ inch or 21G × ¾ inch butterfly needle attached to a 12″ infusion line with a Luer Lock connector) (Greiner Bio-One, Kremsmünster, Austria).
24. BD Vacutainer one-use blood collection tube holder (Becton, Dickinson, and Company, Franklin Lakes, NJ, USA).
25. Tris(hydroxymethyl)aminomethane (Tris) (Sigma-Aldrich, St. Louis, MO, USA).
26. Ethylenediaminetetraacetic acid (EDTA) (Sigma-Aldrich, St. Louis, MO, USA).
27. Sodium chloride (NaCl) (Sigma-Aldrich, St. Louis, MO, USA).
28. Sodium dodecyl sulfate (SDS) (Sigma-Aldrich, St. Louis, MO, USA).
29. Proteinase K (Research Products International, Mount Prospect, IL, USA).
30. Screw-top tubes (2 mL) with O-ring cap (Thermo Fisher Scientific, Hampton, NH, USA).
31. Rotary hybridization oven (Thermo Fisher Scientific, Waltham, MA, USA).
32. Model 5417C Eppendorf centrifuge (Eppendorf, Hamburg, Germany).
33. Microcentrifuge tube (Thermo Fisher Scientific, Hampton, NH, USA).
34. Isopropyl alcohol (isopropanol) (Sigma-Aldrich, St. Louis, MO, USA).
35. Ethyl alcohol (ethanol) (Sigma-Aldrich, St. Louis, MO, USA).
36. Incubator (Thermo Fisher Scientific, Waltham, MA, USA).
37. PCR primers for the golden retriever muscular dystrophy dog (GRMD) genotyping. Forward primer, CTTAAGGAAT GATGGGCATGGG; Reverse primer, ATGCATAGTTTCT CT<u>T</u>TCATGC.
38. PCR primers for Welsh corgi muscular dystrophy dog (WCMD) genotyping. Forward primer, AGCTTTATTAGCC-CAAGGTTGTTC; reverse primer for normal dogs,

GCAATGGCCTATCAGTACAGTCC; reverse primer for affected dogs, TGAAGCAAATGGTGGGTATTTG.

39. 5× Green GoTaq Flexi PCR kit (Promega; Madison, WI, USA). The kit contains 5× Green GoTaq Flexi buffer and GoTaq DNA polymerase.
40. Eppendorf Mastercycler personal PCR machine (Eppendorf).
41. *Sau*96I restriction enzyme (New England Biolab, Ipswich, MA, USA).

2.2 Caring for Affected Dogs

1. PPE including disposable gowns and/or dedicated scrubs, safety glasses or splash goggles, head cover, shoe covers, surgical mask, and gloves.
2. Stethoscope (3 M Littmann, St. Paul, MN, USA).
3. Thermometer (ReliOn Inc., Spokane Valley, WA, USA).
4. Digital gram scale (Zieis, Apple Valley, MN, USA).
5. Digital kilogram scale (OptimaScale, Rancho Cucamonga, CA, USA).
6. Color-coded BD® vacutainer blood collection tubes: (Becton, Dickinson, and Company, Franklin Lakes, NJ, USA).
7. Safety blood collection set (23G × ¾ inch or 21G × ¾ inch butterfly needle attached to a 12″ infusion line with a Luer Lock connector) (Greiner Bio-One, Kremsmünster, Austria).
8. BD Vacutainer one-use blood collection tube holder (Becton, Dickinson, and Company, Franklin Lakes, NJ, USA).
9. Disposable sterile syringe (Becton, Dickinson and Company, Franklin Lakes, NJ, USA).
10. 70% isopropyl alcohol (Wal-Mart Stores Inc., Bentonville, AR, USA).
11. Tenderfoot flooring (Nooyen, Mount Sterling, Kentucky, USA).
12. Purina puppy food (Nestle Purina, St. Louis, MO, USA).
13. Grooming supplies: undercoat rake, nail clippers, slicker brush, and comb (PETCO Animal Supplies, San Diego, CA, USA).
14. Animal hair clipper with a #40 blade (Oster Professional Products, Sunbeam Products, Inc., Boca Raton, FL, USA).
15. Non-woven gauze sponges (4 inch × 4 inch) (Medline Industries Inc., Northfield, IL, USA).

2.3 Intramuscular Gene Delivery to Dogs (Fig. 1)

1. Personal protective equipment (PPE), including disposable gowns and/or dedicated scrubs, shoe covers or dedicated shoes, safety glasses or splash goggles, and gloves.

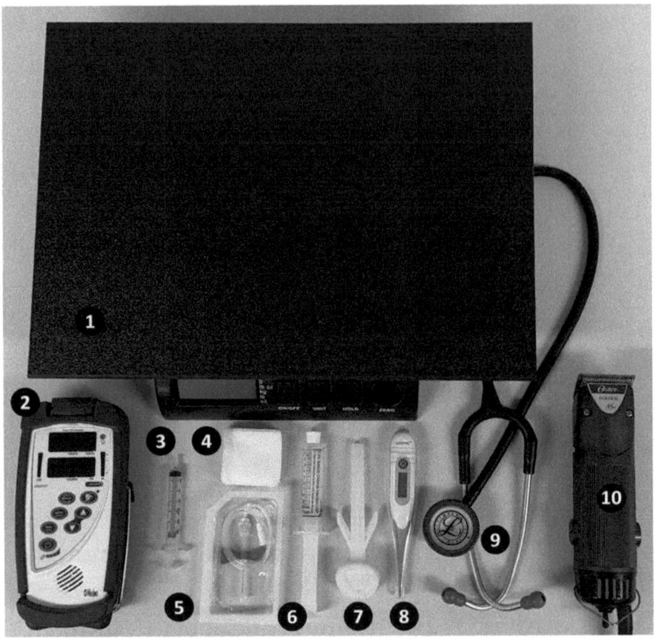

Fig. 2 Materials required for systemic gene delivery to neonatal dogs. (1) table-top digital scale, (2) Handheld pulse oximeter (cables not shown), (3) 3 mL syringe, (4) gauze sponges (2″ × 2″), (5) 23 g × 3/4 butterfly needle infusion set with 7.5″ infusion line, (6) 10 mL pre-filled saline flush, (7) ChloraPrep OneStep applicator, (8) digital thermometer, (9) stethoscope, (10) animal hair clipper. Other materials not shown: personal protective equipment, AAV vector, thermal blanket

7. Animal fine hair clipper (Oster Professional Products, Sunbeam Products, Inc., Boca Raton, FL, USA).
8. Infusion set (23G × 3/4 inch butterfly needle attached to a 7.5″ infusion line with Luer Lock connector) (Greiner Bio-One, Kremsmünster, Austria).
9. Syringe with Luer Lock Tip (3 mL) (Becton Dickinson company, Franklin Lakes, NJ, USA).
10. BD PosiFlush pre-filled saline (0.9% sodium chloride) syringe for intravenous (IV) injection (Becton Dickinson company, Franklin Lakes, NJ, USA).
11. Sterile gauge sponges (2 inch × 2 inch) (Govidien, Mansfield, MA, USA).
12. ChloraPrep one step (2% Chlorhexidine gluconate and 70% Isopropyl alcohol) (CareFusion, Leawood, KS, USA).
13. Recombinant AAV vector (*see* **Note 5**).

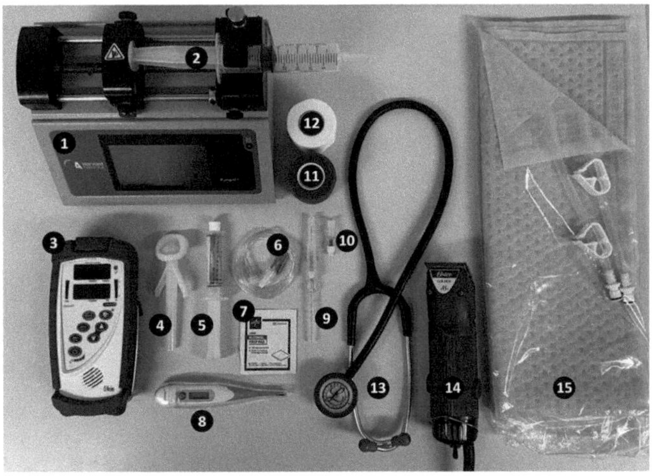

Fig. 3 Materials required for systemic gene delivery to adult dogs. (1) Infusion pump, (2) 30 mL syringe, (3) handheld pulse oximeter (cables not shown), (4) ChloraPrep OneStep applicator, (5) 10 mL pre-filled saline flush, (6) Micro-bore IV extension set (60″), (7) alcohol prep pad, (8) digital thermometer, (9) 22 g × 1 shielded IV catheter, (10) SmartSite Needle-free valve, (11) Vet wrap, (12) Medical tape, (13) stethoscope, (14) animal hair clipper, (15) conductive warming blanket. Other materials not shown: personal protective equipment, AAV vector, hot water bath, digital vet scale

2.5 Systemic Gene Delivery to Young Adult Dogs (Fig. 3)

1. PPE including disposable gowns and/or dedicated scrubs, safety glasses or splash goggles, head cover, shoe covers, surgical mask, and gloves.
2. Digital vet scale (DRE, Louisville, KY, USA).
3. Digital thermometer (ReliOn Inc., Spokane Valley, WA, USA).
4. Stethoscope (3 M Littmann, St. Paul, MN, USA).
5. Handheld pulse oximeter (Masimo Rad-57) (DRE, Louisville, KY, USA).
6. Thermal blanket (100% cotton) (Elite Home Products Inc., Saddle Brook, NJ, USA).
7. Animal worktable.
8. Animal hair clipper equipped with a size 50 blade (Oster professional product, Sunbeam products, Inc., Boca Raton, FL, USA).
9. Shielded I.V. catheter (22G × 1.00 inch needle, or 20G × 1.00 inch needle) (Becton Dickinson company, Franklin Lakes, NJ, USA).
10. SmartSite Needle-free valve (CareFusion, San Diego, CA, USA).

11. Micro-Bore IV Extension set (A 60 inch tube line with 0.3 mL volume capacity, and contains Luer lock connectors as well as a slide clamp) (NeoChild, Oklahoma City, OK, USA).
12. Medical tape (3 M, Maplewood, MN, USA).
13. Vet wrap (3 M, Maplewood, MN, USA).
14. Webcol Alcohol Prep pad (Covidien, Dublin, Republic of Ireland).
15. Syringe with Luer Lock Tip (30 mL) (Becton Dickinson company, Franklin Lakes, NJ, USA).
16. BD PosiFlush pre-filled saline (0.9% sodium chloride) syringe for IV injection (Becton Dickinson company, Franklin Lakes, NJ, USA).
17. Infusion pump (Pump 11 Elite) (Harvard Apparatus, Holliston, MA).
18. Sterile gauge sponges (4 inch × 4 inch) (Govidien, Mansfield, MA, USA).
19. ChloraPrep one step (2% Chlorhexidine gluconate and 70% Isopropyl alcohol) (CareFusion, San Diego, CA, USA).
20. Recombinant AAV vector (*see* **Note 5**).

2.6 Dog Muscle Biopsy (Fig. 4)

1. PPE including disposable gowns and/or dedicated scrubs, safety glasses or splash goggles, head cover, shoe covers, surgical mask, and gloves.
2. Sedation agents: Dexdomitor (Zoetis Inc., Kalamazoo, MI, USA); Nalbuphine (Hospira Inc., Lake Forest, IL, USA).
3. Disposable sterile syringe (Becton, Dickinson and Company, Franklin Lakes, NJ, USA).
4. 20 g × 1 (0.9 mM × 25 mM) PrecisionGlide hypodermic needle (Becton, Dickinson and Company, Franklin Lakes, NJ, USA).
5. Animal hair clipper with a #50 blade (Oster Professional Products, Sunbeam Products, Inc., Boca Raton, FL, USA).
6. Digital thermometer (ReliOn Inc., Spokane Valley, WA, USA).
7. Stethoscope (3 M Littmann, St. Paul, MN, USA).
8. Conductive blankets (Adroit Medical Systems Inc., Loudon, TN, USA) connected to a heated circulating water bath (Thermo Fisher Scientific, Hampton, NH, USA).
9. Medical grade oxygen tank (Airgas, Radnor, PA, USA).
10. Canine anesthesia mask (Shopmedvet Inc., Mettawa, IL, USA).
11. Veterinary vital sign monitor (DRE Veterinary, Louisville, KY, USA).

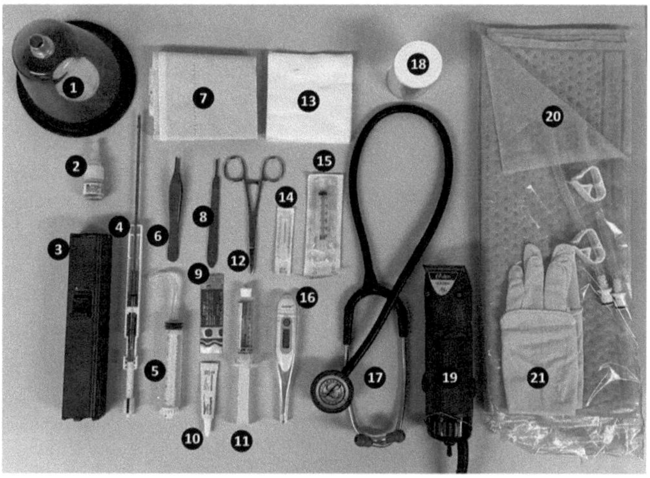

Fig. 4 Materials required for dog muscle biopsy. (1) Canine anesthesia mask, (2) Vetbond tissue adhesive, (3) Vacora® vacuum-assisted biopsy system driver, (4) Vacora® Biopsy probe 10G × 140 mM, (5) vacuum components of the biopsy system, (6) tissue forceps, (7) surgical drape with circular aperture, (8) scalpel holder #3, (9) surgical blade #11, (10) artificial tears eye ointment, (11) 10 mL pre-filled saline flush, (12) needle holder, (13) non-woven gauze sponges (4″ × 4″), (14) 20 g × 1 hypodermic needle, (15) 3 mL syringe, (16) digital thermometer, (17) stethoscope, (18) disposable plastic container with screw-on cap, (19) animal hair clipper, (20) conductive warming blanket, (21) surgical gloves

12. Non-woven gauze sponges (4 inch × 4 inch) (Medline Industries Inc., Northfield, IL, USA).
13. Rugby artificial tears ointment (Wal-Mart Stores Inc., Bentonville, AR, USA).
14. 2% chlorhexidine gluconate (Vedco Inc., Saint Joseph, MO, USA).
15. 70% isopropyl alcohol (Wal-Mart Stores Inc., Bentonville, AR).
16. Surgical gloves (Cardinal Health, Dublin, OH, USA).
17. Surgical drape with circular aperture (24″ × 26″) (Medline, Northfield, IL, USA).
18. Surgical tools: Needle holder, Scalpel handle #3, and tissue forceps (1 × 2 teeth) (Fine Science Tools, Foster City, CA, USA).
19. Surgical blade #11 (Aspen Surgical, Caledonia, MI, USA).
20. Vacora® vacuum-assisted biopsy system driver (Bard Peripheral Vascular Inc., Tempe, Arizona, USA).
21. Vacora® Biopsy probe 10G × 140 mM (Bard Peripheral Vascular Inc., Tempe, Arizona, USA).

22. Vacora® Biopsy probe 14G × 116 mM (Bard Peripheral Vascular Inc., Tempe, Arizona, USA).
23. BD PosiFlush pre-filled saline (0.9% sodium chloride) syringe for intravenous (IV) injection (Becton Dickinson company, Franklin Lakes, NJ, USA).
24. Disposable plastic container with a screw-on cap (Uline, Pleasant Prairie, WI, USA).
25. Pain relief and anti-inflammatory agent: Rimadyl (Zoetis Inc., Kalamazoo, MI, USA).
26. Postanesthetic reversal agent: Antisedan (Zoetis Inc., Kalamazoo, MI, USA).
27. Vetbond tissue adhesive (3 M Animal care product, Saint Paul, MN, USA).

3 Methods

3.1 Generation of Affected Dogs

1. Follow the institutional guidelines for animal care and handling. Obtain approval from the Institutional Animal Care and Use Committee (IACUC). Wear all PPE before working or handling animals.
2. Dog semen selection. PedScope is a software that uses established inbreeding coefficient equations to determine the genetic similarity between the carrier female and the prospective male (semen). Using the PedScope software, semen or a suitable male is selected based on experimental needs and genetic diversity. Breeding is only considered if the inbreeding coefficient is less than 12%. Every newborn puppy or newly acquired semen is entered into the software for future analysis.
3. Heat detection. Carrier females are monitored weekly for signs of heat. Signs include frequent urination and licking at the hind end, vulva swelling, and bloody discharge from the vagina. When suspected that a female may be coming into heat, a blood sample is collected in a red top tube, and the serum progesterone is tested. Progesterone above 2 ng/mL is indicative of impending ovulation. This is the time to determine breeding modality and semen selection (*see* **Note 1**).
4. Dog breeding by artificial transcervical insemination. Place the carrier female on an examination table. Clean the vulva. Restrain the dog's head and hind end without placing pressure on the abdomen. Advance the transcervical insemination endoscope to the opening of the cervix. Advance the transcervical insemination catheter through the endoscope to the body of the uterus. Inseminate the semen through the catheter

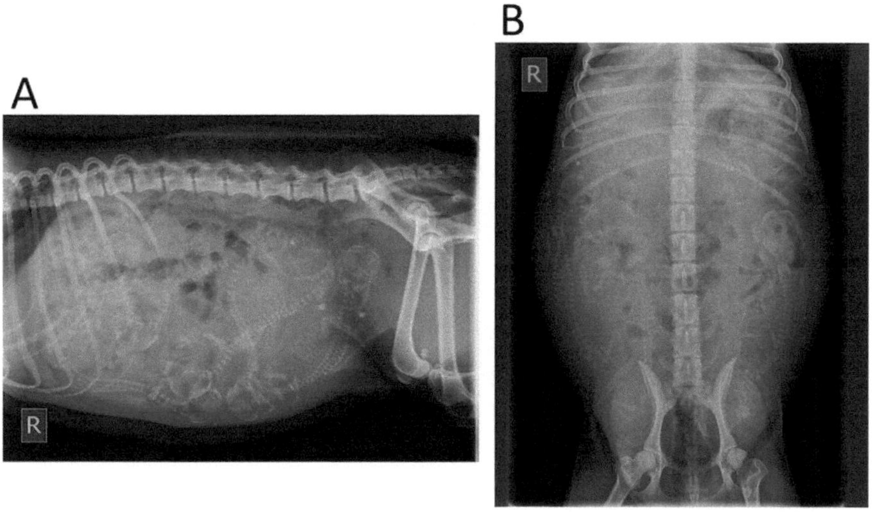

Fig. 5 X-ray determination of the puppy number. Approximately 1 week before the expected whelping date, an abdominal X-ray is performed on the bitch to count the number of puppies. Puppies are readily recognized by their skeleton images. (**a**) Right lateral view. (**b**) Ventrodorsal view

ensuring all of the semen has been cleared from the catheter line (*see* **Notes 1** and **2**).

5. Pregnancy determination. One month post-breeding, confirm pregnancy via ultrasonography checking for fetal heartbeats. Approximately 1 week prior to the expected whelp date, evaluate radiographs to determine an accurate fetal count (Fig. 5).
6. Pre-whelping. Based on the canine gestation length, whelping is expected approximately 63 days after the onset of the estrous cycle (*see* **Note 3**). The mother should be moved into the whelping kennel 1 week prior to the expected due date. Approximately 4 days prior to the expected date, begin taking rectal temperatures daily looking for a drop below 37.2 °C (99 °F) (*see* **Note 4**). A more accurate method to determine whelping date is to measure serum progesterone levels. When the progesterone falls below 2 ng/mL, whelping will occur within 24 h. All whelping supplies should be organized and ready for immediate use upon the initiation of whelping (Fig. 6).
7. Monitor the bitch for any signs of dystocia (difficulty giving birth) such as, prolonged time between contractions and delivery of a puppy, absence of a placenta after a puppy, increased respiratory rate or effort, or cyanosis of the mucous membranes. If any of these signs occur, contact a veterinarian as soon as possible.
8. Allow the bitch to deliver the puppy without interference and with minimal intervention at first. Monitoring using a camera

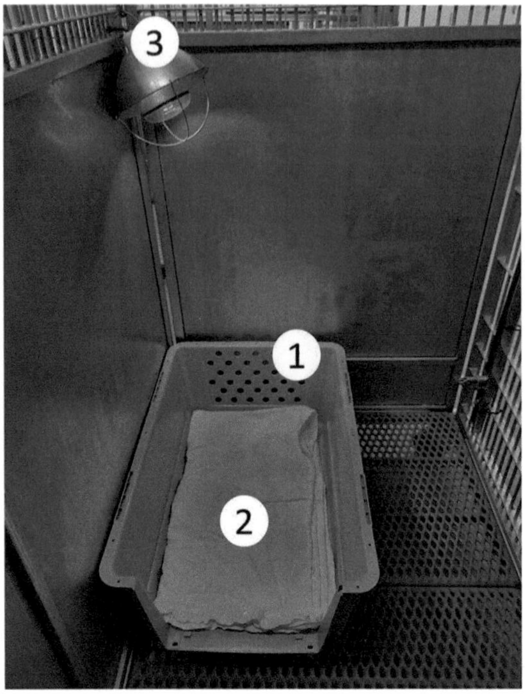

Fig. 6 Whelping kennel. The kennel has (1) a whelping box, (2) white blankets, and (3) an infrared heating lamp

or outside the kennel is best to limit interactions. The bitch should immediately chew off the placenta, clear the amniotic sac, and begin licking the puppy vigorously. This stimulates the puppy to breathe, cry, and pass meconium (first feces). After she has cleaned the puppy, enter the kennel, and confirm it is breathing and crying without abnormal lung sounds such as crackling (*see* **Notes 6, 7** and **8**).

9. Clean and dry the puppy and place it on the gram scale. Record the birth weight and mark the puppy using the nail polish. Shave an area of about 1cm^2, and place a drop of nail polish in the desired location (such as left shoulder, right shoulder, left hip, etc.). The weight sheet should include the puppies in birthing order labeled with their nail polish markings, sex, and birth weight (Fig. 7) (*see* **Note 9**).
10. After marking the puppy, tie off the umbilical cord using the #2 suture, leaving approximately 5 mm remaining. Cut the excess cord with a pair of sterile surgical scissors and place it in a microcentrifuge tube with the puppy birth number and sex written on the cap (*see* **Note 10**). This will serve as the initial identifier for each puppy for genotyping.
11. Repeat the process for all puppies following the same method each time. The puppies may begin nursing immediately after

Name of the bitch: **DOB (Puppies' date of birth):**

Puppy #	#1	#2	#3	#4	#5	#6	#7
Puppy name							
Gender							
Genotype							
Date:		**Time:**			**Caregiver's name:**		
Current Weight (gram)							
Weight after feeding (gram)							
Weight change bewteen shifts (gram)							
Weight change over 24 hours (%)							
Amount of milk drank (mL)							
Nursing/Tube/Bottle							

Fig. 7 An example of the birth and growth record of puppies. The record contains bitch's name, puppies' birth date, puppy name, sex, and genotype. The white dots in puppy cartoons show the location of the markers for each puppy. This record includes the body weight, weight after feeding, weight change between shifts, and the percent of weight change during a 24-h period. Weight changes are automatically calculated by the built-in formulas. This record also includes information on whether a puppy needs nursing, tube feeding, or bottle feeding and how much milk the puppy takes during tube/bottle feeding

cleaning and stimulation or may seek warmth to sleep. Attempts should be made to place the puppies on a teat to nurse if they do not do so themselves (*see* **Note 11**).

12. After all puppies have been born and are marked, weighed, and appear to be doing well, the bitch and puppies should be allowed to rest for the next 12 h. It is best to continue to visually check the litter every 3–4 h to determine if any individual puppy is struggling or is being isolated from the rest of the litter. In the event of a struggling puppy, consult with the veterinarian, and start intervention immediately if needed.
13. Umbilical cord DNA extraction. Mince the cord with a razor blade. Add 750 μL DNA extraction buffer (50 mM Tris-HCl, pH 8.0, 100 mM EDTA, 100 mM NaCl, 1% SDS, 0.5 mg/mL proteinase K). Incubate at 55 °C overnight on a spinning wheel. Add 250 μL 6 M NaCl. Incubate at room temperature for 1 h on a spinning wheel. Centrifuge at 16,000 × g for 30 min. Transfer the supernatant to a new tube. Add 600 μL isopropanol. Mix for 5 min by shaking. Centrifuge at 16,000 × g for 2 min. Wash the pellet with 70% ethanol. Centrifuge at 16,000 × g for 2 min. Air-dry the pellet for 1 h. Resuspend the pellet in 50 μL PCR-quality water as the cord genomic DNA.
14. GRMD dog genotyping [46]. Mix 0.5 μL cord DNA, 4 μL of 5× green GoTaq Flexi buffer, 1.6 μL of 25 mM $MgCl_2$, 0.4 μL of 10 mM dNTP, 0.5 μL of each primer (20 μM stock), 0.1 μL

of GoTaq DNA polymerase and 12.4 μL PCR-quality water. Start PCR with a 2 min initial denaturation at 95 °C for 2 min. Continue PCR with 35 cycles of 95 °C 20 s denaturation, 58 °C 20 s annealing and 72 °C 20 s extension. Digest 10 μL PCR product with two units of Sau96I at 37 °C for 2 h. Examine in 2% agarose gel. Wild-type allele yields a 328 bp band. Affected allele yields two bands at 107 bp and 221 bp. Carrier puppies have three bands at 107 bp, 221 bp, and 328 bp.

15. WCMD dog genotyping [24]. Mix 0.5 μL cord DNA, 4 μL of 5× green GoTaq Flexi buffer, 1.6 μL of 25 mM $MgCl_2$, 0.4 μL of 10 mM dNTP, 0.5 μL of each primer (20 μM stock), 0.1 μL of GoTaq DNA polymerase, and 12.4 μL PCR-quality water. Start PCR with a 2 min initial denaturation at 95 °C for 2 min. Continue PCR with 30 cycles of 95 °C 15 s denaturation, 62 °C 15 s annealing, and 72 °C 15 s extension. At the end of 30 cycles, incubate at 72 °C for 7 min. Examine in 2% agarose gel. Wild-type allele yields a 410 bp band. Affected allele yields a 290 bp band. Carrier puppies have two bands at 290 bp and 410 bp.

3.2 Caring for Affected Dogs

1. Follow the institutional guidelines for animal care and handling. Obtain approval from the Institutional Animal Care and Use Committee (IACUC). Wear all PPE before working or handling animals.
2. Begin weighing all puppies at least three times daily after the first 24 h of life and continue at this frequency for the first 2 weeks (Fig. 7). After the first 2 weeks, the puppies should be weighed at least twice daily until week 4 and then daily until weaning. Record the change in body weight noting the change between each weighing and the 24-h change. Normal puppies should gain between 10% and 15% of their body weight every 24 h; however, affected puppies may gain less. If specific puppies are not gaining any weight over the 24-h period, it may be necessary to supplement feedings with puppy milk replacer or assisted nursing (*see* **Notes 12** and **13**). Puppies will spend much of their time nursing and sleeping. If checking on the puppies using a timed system (every 3–4 h), it may not appear that certain puppies are nursing. It is important to use the weight gain calculations and observation skills to determine a puppy that needs intervention. Sleep is as important as nursing for a young growing puppy.
3. Start establishing interaction with puppies around 2 weeks of age when the eyes and ears open, the puppies will be receptive to human interaction. This is the ideal time to begin establishing a handling routine and conditioning (*see* **Note 14**).

4. Separation of affected puppies. Affected puppies are separated from non-affected litter mates around months 4–5 of life depending on disease status. They may be housed with another affected puppy or singularly if an appropriate pen mate is unavailable. At this time, the food dishes are raised off of the flooring, and a water bowl is added to the cage (*see* **Notes 15** and **16**). Food should be soaked in water to soften the kibble. Affected dogs often have weakened jaw muscles with impaired ability to chew and break up hard foods. Soaking the kibble softens it to facilitate easy mastication and swallowing.
5. Monitoring affected dogs. Affected puppies that have survived to weaning generally remain stable until 5–6 months of age. Six months is usually when model characteristics begin to present (*see* **Note 17**). It is important to carefully monitor the overall appearance of the dog. With the disease model, dogs hypersalivate and may have a thickened tongue causing difficulty swallowing. This may lead to increased mess and dirtiness of the dog. If not cleaned regularly, this may result in superficial skin irritation and/or infection. Affected dogs should be groomed on a regular basis and assessed for any skin issues.
6. Evaluation of affected dogs. It is important to monitor body weight and general health frequently. To do this, a full physical exam (temperature, heart rate and rhythm, and respiratory rate and sound) should be performed in conjunction with the body weight and body condition scoring. Dogs will continue to grow for up to a year depending on breed (*see* **Note 18**).
7. Dogs should also be allowed to move freely around a room or hallway to evaluate limb and gait changes. Observing the dog's natural body movement will enable laboratory personnel to understand changes and potential morphologic complications as they develop.
8. Blood work should be routinely run at least yearly. It may be beneficial to submit blood work more frequently or use a timepoint system for data collection (*see* **Note 19**).
9. Functional assessments and evaluations may be carried out based on desired timepoints for the specific research needs. Periodic assessments at established timepoints allows for accurate disease progression tracking as well as determining normal data for the colony (*see* **Notes 20** and **21**). It is important to note changes and continue to follow disease advancement as every individual may progress differently (*see* **Note 22**).

3.3 Intramuscular Gene Delivery to Dogs

The following protocol describes the procedure for intramuscular injection of gene therapy agents in common, superficial canine skeletal muscles. This protocol describes saturated injection of isolated spots in the biceps femoris (BF), or the whole muscle in the extensor carpi ulnaris (ECU) (*see* **Note 23**).

1. Follow the institutional guidelines for animal care and handling. Obtain approval from the Institutional Animal Care and Use Committee (IACUC). Wear all PPE before working or handling animals. Use aseptic techniques while preparing injections and during the injection procedure (*see* **Note 24**).
2. Pre-fill micro-centrifuge tubes with the desired injection volume of the AAV vector for intramuscular delivery. Outfit syringes with a 25 g × 5/8 hypodermic needle for BF muscle injection. Outfit syringes with a 25 g × 1–1/2 hypodermic needle for ECU muscle injection. Draw up the desired volume of injection solution into the syringe. Draw a small extra volume (approximately 0.1 mL) of air into the syringe to account for syringe and needle dead volume.
3. Fast the dog the night before the procedure if sedation is to be used (*see* **Note 25**).
4. Perform a physical exam, and record the vital signs (heart rate, respiratory rate and effort, and rectal body temperature), capillary refill time (CRT), and mucous membrane color.
5. Gently restrain the animal and shave the body hair superficial to the muscles that will be injected.
6. Prepare a sedative drug cocktail with Dexdomitor (dexmedetomidine HCl) and Torbugesic (butorphanol tartrate) (*see* **Note 26**). Prepare the sedation agents cocktail according to the body weight. First, withdraw the calculated dosage of the Dexdomitor (6–15 μg/kg) into a syringe, and then withdraw the calculated dosage of the Torbugesic (0.2–0.4 mg/kg) into a separate syringe. Next, add the withdrawn volume of the Dexdomitor into the Torbugesic syringe, and gently invert to mix both agents.
7. Inject the sedation agents cocktail intramuscularly in the epaxial muscle at a 90° angle using a 20 g × 1 needle (*see* **Note 27**).
8. Place the animal over the preheated thermal blanket in a left lateral recumbency position. Set the oxygen mask over the dog's muzzle to supply the animal with oxygen throughout the procedure. Start flow to the oxygen mask using medical-grade breathing air supplemented with pure oxygen in a 2:1 ratio (*see* **Note 28**).
9. Monitor the heart rate, electrocardiogram, oxygen saturation (SpO2), blood pressure, and body temperature using the DRE vital sign monitor throughout the entire procedure. Every 15 min, record the vital signs, CRT, mucous membrane color, the palpebral reflex, and the pedal reflex. Place eye lubricant in each eye.
10. Disinfect the injection site using a two-step scrubbing technique. First, place a soaked chlorhexidine gauze at the

proposed incision site, and then move it in a circular motion. Gradually move your way out from the incision site until the whole shaved area is cleaned with chlorhexidine. Discard the chlorhexidine gauze. Next, repeat the same cleaning process using a soaked isopropanol gauze. Repeat the two-step scrubbing technique three·times (*see* **Note 29**).

11. Tattoo the skin of the dog superficial to the planned injection site (*see* **Note 30**). Dip the tip of a 25 g hypodermic needle in a microcentrifuge tube containing green liquid veterinary tattoo dye. Gently insert the tip of the needle in the skin just deep the epidermis. Rotate the needle clockwise and counterclockwise to tattoo the skin (*see* **Note 31**). Next, repeat the two-step scrubbing technique described in **step 10** three times to remove excess tattoo dye and disinfect the injection site.
12. Inject gene delivery agents as planned (*see* **Note 32**). For BF muscle injection, insert the 25 g × 5/8 needle in the muscle body perpendicular to the plane of the muscle. To saturate the injection site, it is recommended to inject 1/5 of the total volume in five directions. In each direction, inject a bolus, and slowly inject additional solution during withdrawal of the needle (*see* **Note 33**). For ECU muscle injection, insert the 25 g × 1–1/2 needle into the center of the muscle body, parallel to the muscle plane (*see* **Note 34**). Position the distal tip of the needle proximal to the insertion of the muscle, with the remainder of the needle in the center of the muscle (*see* **Note 35**) (Fig. 8). Inject a bolus, then slowly withdraw the needle lengthwise, injecting solution throughout the length of the center of the muscle.
13. Clean the injection sites with chlorhexidine-soaked gauze from any bleeding that may have occurred during the injection.
14. Reverse Dexdomitor sedation using Antisedan (atipamezole HCl, 60–200 μg/kg) by intramuscular injection in the contralateral side of the epaxial muscle at a 90° angle using a 20 g × 1 needle (*see* **Note 27**). The administered volume of Antisedan (Stock 5 mg/mL) should be approximately equal to the total volume of Dexdomitor (Stock 0.5 mg/mL) given during the procedure (*see* **Note 36**).
15. Continue monitoring the dog until it can maintain an unassisted sternal position. Then, perform a postoperative assessment for 5 days to confirm no adverse effects from the procedure (*see* **Note 37**).

3.4 Systemic Gene Delivery to Neonatal Dogs

1. Follow the institutional guidelines for animal care and handling. Obtain approval from the Institutional Animal Care and Use Committee (IACUC). Wear all PPE before working with or handling animals. Use aseptic techniques while

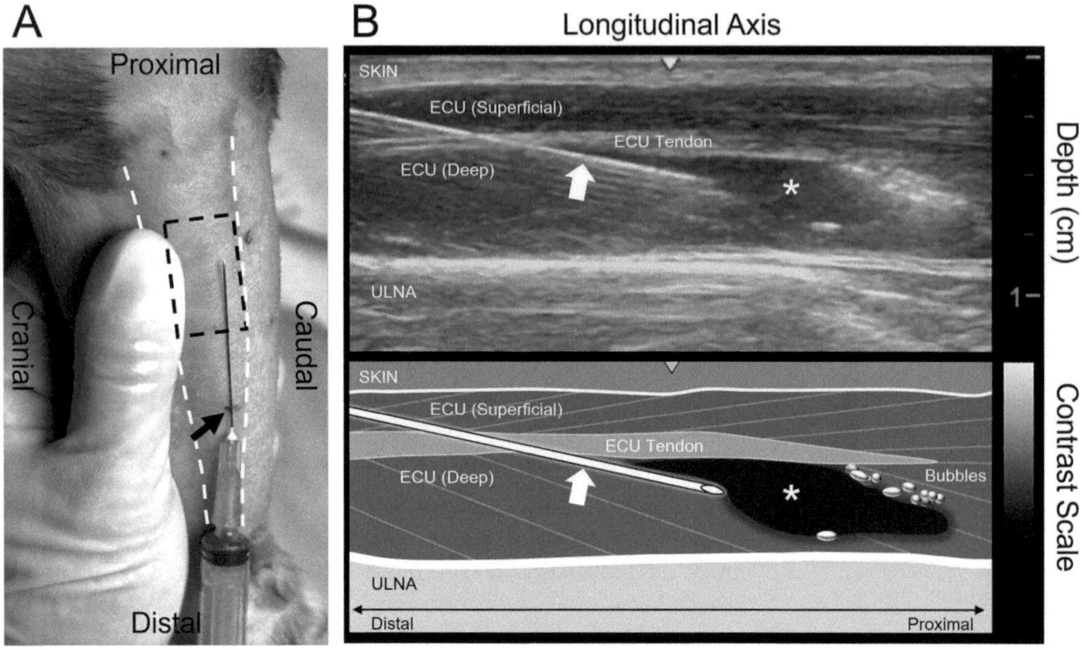

Fig. 8 Ultrasound-guided injection of the extensor carpi ulnaris muscle (ECU). **(a)** A representative image of the canine extensor carpi ulnaris (ECU) muscle. White dashed lines are the relative border of the muscle. Black inset, the region of the muscle viewed under ultrasound during injection. Black arrow, needle insertion site. (**b**) Ultrasound-guided injection of the deep aspect of the ECU muscle. Top panel, freeze-frame longitudinal axis ultrasound image (B-mode, 16 Hz) taken during the delivery of the AAV vector to the deep compartment of the ECU muscle. Bottom panel, a cartoon illustration based on the ultrasound image. White Arrow, 25 g × 1–1/2 hypodermic needle. Asterisk, AAV vector bolus (low contrast/darker area). The Arrowhead at the top of the images represents the center of the ultrasound probe

preparing the AAV vector for injection and during the injection procedure.

2. Monitor the newborn puppy's body weight, physical exam parameters (such as heart rate, respiratory rate, mucous membrane color, and rectal body temperature), and overall mentation frequently (*see* **Note 38**). Consult with the veterinarian immediately for unexpected sudden changes such as a drop in body weight (≥ 10%) or temperature changes (±5 °C).
3. Weight the puppy to determine the AAV injection volume. A normal neonatal puppy can tolerate a maximum volume of 25 μL liquid/g body weight) [47]. (*see* **Note 39**).
4. Measure the vital signs (heart rate, respiration rate, rectal body temperature, and blood oxygen saturation) before injection (*see* **Note 40**).
5. Gently shave the fur on the side of the neck using a #10 clipper blade to better visualize the jugular vein for injection.

6. Connect the prefilled BD PosiFlush saline syringe to the IV extension line and then attach the infusion set. All parts are securely connected using Luer Lock fittings. Prefill the extension line and infusion set with the saline to flush out all air from the lines. Lock the sliding clamp on the extension line. Draw the exact volume of AAV vector (calculated based on body weight and desired dose) into the 3 mL syringe ensuring there are no air bubbles in the Replace the BD PosiFlush syringe with the AAV containing syringe. Check all parts of the infusion system for any signs of remaining air.
7. The injection procedure requires at least three investigators. The first investigator places the preheated thermal blanket on the animal working table and positions the puppy. Gently place the puppy in a supine position on the blanket restraining the forelimbs using both hands. The second investigator restrains the head and scrubs the ventral side of the neck with the ChloraPrep applicator beginning at the injection site and moving outward in a circular motion. The second investigator then gently inserts the needle into the jugular vein when it is safe to do so effectively. The third investigator unlocks the sliding clamp on the extension line and slightly pulls back the plunger to visualize blood in the infusion line. The appearance of blood confirms correct intravenous needle positioning. Inject the AAV vector at a rate of ~2 mL/min, and then flush the infusion line with an appropriate volume of saline to ensure all AAV has been delivered [48].
8. Before exiting the jugular vein, press the area above the needle entry site using two fingers. Retract the butterfly from the vein, and continue applying pressure to the injection site to prevent bleeding.
9. Evaluate the vital signs again understanding the heart rate may be increased due to the stress of restraint.
10. Return the puppy back to the dam and closely monitor the puppy for the following 24 h. It is best to monitor and record vital signs at time points following injection such as: 5 min, 10 min, 30 min, 1 h, 2 h, 4 h, and 8 h. Consult with veterinarian immediately with any concerns.

3.5 Systemic Gene Delivery to Young Adult Dogs

1. Follow the institutional guidelines for animal care and handling. Obtain approval from the Institutional Animal Care and Use Committee (IACUC). Wear all PPE before working with or handling animals. Use aseptic techniques while preparing the AAV vector for injection and during the injection procedure.
2. Weight the dog to determine the AAV injection volume. We found that young adult dystrophic dogs can tolerate a

maximum volume of 6.2 mL/kg body weight (6.24×10^{14} viral genome particles/kg body weight) [38].

3. Measure the vital signs (heart rate, respiration rate, rectal body temperature. and blood oxygen saturation) before injection.
4. Shave the cranial aspect of the lower forelimb using a #40 blade to allow better visualization of the cephalic vein for injection (*see* **Note 41**).
5. Place a thermal blanket on the animal working table. Position the dog on the table and blanket in sternal recumbency with the lower forelimb extended. A second investigator may then occlude the cephalic vein by rolling their thumb at the elbow in a medial to lateral motion. A tourniquet may also be used if only one investigator is performing the procedure.
6. Catheterize the cephalic vein with the help of another investigator. Scrub the cranial aspect of the lower forelimb with ChloraPrep beginning at the injection site and moving outward in a circular motion. Pull the lower forelimb skin distally using one hand to stabilize the vein. Slowly insert the IV catheter needle (bevel up) through the skin and into the vein. A small amount of blood will appear in the catheter hub when the needle punctures the vein. Continue advancing the catheter and needle into the vein as the hub fills with blood. Once confidently in the vein approximately halfway, retract the needle out of the catheter and finish advancing the catheter. At this time, blood will flow out from the catheter. Quickly occlude the vein at the catheter tip using moderate pressure, and attach the SmartSite needle-free valve to the catheter. Secure the catheter in place using tape and then vet wrap to hold it for the procedure. Flush the catheter with saline to ensure there is no leak, it is properly in the vein and flowing, and there is no clot in the catheter.
7. Connect the prefilled BD PosiFlush saline syringe with the IV extension line using the Luer Lock fitting. Remove air from the IV extension line using saline. Lock the sliding clamp on the extension line. Draw the exact volume of AAV vector into the 30 ml syringe. Remove any air bubbles from the syringe. Replace the BD PosiFlush syringe with the AAV containing syringe. Ensure there are no air bubbles in the extension line then secure the syringe to the infusion pump and set the constant infusion rate to 2 mL/min.
8. Connect the IV extension line to the valve and unlock the sliding clamp. Begin the infusion by pressing the start button on the infusion pump. Closely monitor the dog's vital signs during AAV infusion. After completing the infusion, ensure all of the AAV solution was delivered by flushing the line with saline. Disconnect the extension line from the catheter. Before

exiting the cephalic vein, press the area above the needle entry site using two fingers or thumb. Retract the catheter from the vein and continue applying pressure to the injection site to prevent bleeding.

9. Continue monitoring the dog's vital signs for 5-, 10-, 30- and 60-min post-injection. Examine the dog for any adverse reactions such as temperature increase or hives.
10. Return the dog to its housing and continue monitoring the dog for 72 h post-injection. Consult with a veterinarian immediately with any concerns.

3.6 Dog Muscle Biopsy

The following protocol describes a minimally invasive vacuum-assisted muscle biopsy method for sampling the right biceps femoris muscle in the upper hindlimb (*see* **Note 42**).

1. Follow the institutional guidelines for animal care and handling. Obtain approval from the Institutional Animal Care and Use Committee (IACUC). Wear all PPE before working with or handling animals. Use aseptic techniques while performing the needle biopsy procedure.
2. Fast the dog the night before the procedure at least 12 h.
3. Prepare the sedation agents cocktail, Dexdomitor (dexmedetomidine HCl), and Nalbuphine (nalbuphine HCl), according to the fasted body weight. First, withdraw the calculated dosage of Dexdomitor (6–15 μg/kg) into a syringe, and then withdraw the calculated dosage of Nalbuphine (1 mg/kg) into a separate syringe. Add the withdrawn volume of Dexdomitor into the Nalbuphine syringe and gently invert to mix both solutions (*see* **Note 43**).
4. Restrain the dog and shave the fur using a #40 clipper blade on the right lateral hindlimb proximal to the stifle. This will allow visualization of the upper hindlimb musculature (*see* **Note 44**).
5. Perform a physical exam, and record the vital signs (heart rate, respiratory rate and effort, and rectal body temperature), capillary refill time (CRT), and mucous membrane color.
6. Inject the sedation agents cocktail intramuscularly in the epaxial muscle at a 90° angle using a 20 g × 1 needle.
7. Position the sedated dog on the preheated thermal blanket in left lateral recumbency. Apply the oxygen mask over the dog's muzzle to provide supplemental oxygen throughout the procedure. Apply eye lubricant to each eye.
8. Monitor the heart rate, electrocardiogram, oxygen saturation (SpO2), blood pressure, and body temperature using the DRE vital sign monitor throughout the entire experiment. Every

15 min, record the vital signs, CRT, mucous membrane color, the palpebral reflex, and the pedal reflex.

9. Prepare the surgical area by employing a two-step scrubbing method. Begin by applying a chlorhexidine-soaked gauze to the intended incision location, gently rubbing it in a circular motion. Gradually extend the cleansing process outward from the incision site until the entire shaved region is thoroughly disinfected. Dispose of the chlorhexidine gauze appropriately. Subsequently, repeat the cleaning procedure using an isopropanol-soaked gauze. Perform this two-step scrubbing routine three times (*see* **Note 29**).
10. Prepare the surgical site for acquiring the muscle biopsy sample. Replace gloves with appropriately sized sterile gloves, and place a sterile drape over the prepared surgical site. The central opening of the drape should be directly over the intended incision location. Confirm the animal is at the proper sedation depth by squeezing the paw pads and pulling the distal limb simultaneously. There should be no increase in respiratory or heart rate and the dog should not withdraw the paw. Then, using a #11 surgical blade, make a 0.5 cm incision at the proximal end of the biceps femoris muscle.
11. Prepare the Vacora® vacuum-assisted biopsy system by inserting the probe into the driver hub. Once the probe is properly loaded, the drive will automatically prime itself, adjusting the position of the cutting sleeve over the biopsy chamber on the probe.
12. Acquire the muscle biopsy sample by first inserting the probe into the biceps femoris muscle through the incision site parallel with the muscle fibers. Adjust the insertion length and depth of the probe into the muscle as desired to achieve an optimal sample. Press the multi-function button on the driver to acquire a muscle biopsy sample. Pull the probe out of the muscle when the driver's motor stops. Press the multi-function button again to retract the cutting sleeve away from the biopsy chamber. Using sterile tissue forceps, remove the muscle biopsy sample from the probe chamber and place it over a saline wet gauze in a sterile plastic container. Press the multi-function button to reset the probe to acquire additional biopsy samples (*see* **Note 45**).
13. Using chlorhexidine, clean the incision site of any bleeding that may have occurred during the procedure. Dry the incision site with sterile gauze and seal it with one to two drops of the 3 M Vetbond.
14. Administer the calculated dose of Rimadyl (Carprofen) at 4.4 mg/kg subcutaneously along the dorsal portion of the dog's back (*see* **Note 46**).

15. Inject Antisedan (atipamezole) intramuscularly on the contralateral side of the dog in the epaxial muscle at a 90° angle using a 20 g × 1 needle.
16. The volume of atipamezole given is equal to the total volume of dexmedetomidine administered during the procedure.
17. Continue monitoring the dog every 15 min until it can maintain an unassisted sternal position. Continue monitoring the dog postoperatively for 5 days to confirm adequate analgesia and no adverse effects.

4 Notes

1. Female canines generally begin their first heat cycle between 6 and 12 months of age, cycling every 6–8 months after that point. Heat detection is determined using physical changes such as increased urination, vulva swelling, and bloody discharge from the vagina. Heat typically lasts approximately 20 days, with ovulation occurring approximately 14 days after the start of the heat. Optimal breeding is 48–72 h post-ovulation. The best way to determine the optimal breeding window is by monitoring progesterone levels. Ovulation occurs when the progesterone level is above 5 ng/mL. Carriers are not retired from breeding at a specific age. Removal from breeding is determined based on the conception rates, litter size and health, regularity of their estrous cycle, and their overall health. The average age of removed carriers is 6–7 years.
2. We recommend artificial insemination using transcervical insemination by an accredited veterinary theriogenologist rather than natural breeding. Frozen semen can be collected from any affected or normal male after approximately 1 year of age and be frozen in multiple straws for future use. This enables the maintenance of desired characteristics over the years, as well as preserving important genetic lines. Using transcervical insemination, we have an 89% pregnancy success rate. If this is not possible, fresh semen can also be artificially implanted in the vaginal canal using a sterile pipette. If using transvaginal insemination, we recommend delivering the semen to the uterus twice at 48- and 72-h post-ovulation.
3. The gestation length in canines is usually 63–65 days. Pregnancy is confirmed approximately 28 days into the gestation cycle using ultrasound to detect fetal heartbeats. Radiographs may be taken around day 56 of the cycle to estimate the number of puppies to be whelped. The whelping date may be predicted using vaginal cytology or serum progesterone levels.

When the serum progesterone falls below 2 ng/mL, whelping is expected within the following 24 h.

4. The normal canine body temperature ranges from 37.5 to 39°C (99.5 to 102.5 °F). When the temperature drops below 37°C (99 °F) and continues to decrease, labor is likely to begin within 12–24 h. It is important to note that this may not always occur with all females or with every whelping.
5. Many different methods have been developed for the production and purification of recombinant AAV vectors (reviewed in [49, 50]). For the protocols described in this chapter, AAV vectors were generated using the transient plasmid transfection method and purified through two rounds of cesium chloride isopycnic ultracentrifugation [51]. AAV vector titer was determined using a quantitative polymerase chain reaction. AAV vector was dialyzed in HEPES buffer (sodium chloride 8.7 g, HEPES 4.75 g, 10 N sodium hydroxide 1.5 mL, distilled water up to 1 L, pH adjusted to 7.8). AAV quality was confirmed by silver staining. Endotoxin contamination was examined using the Endosafe limulus amebocyte lysate gel clot test kit (Charles Rivers Laboratories, Wilmington, MA).
6. Whelping may take several hours with periods of rest between the delivery of each puppy. Generally, a puppy is expected every 45 min to 1 h followed by the corresponding placenta, but this may vary with the number of puppies to be delivered. It is beneficial to have someone close to the area to continuously monitor the bitch during the entire process. Since whelping is a stressful time for the mother and puppies, it is a good idea to establish a consistent routine of handling and experience with each bitch prior to the whelping date. A bitch may prolong whelping if she is stressed or feels uncomfortable with her surroundings. Generally, more than 3 h between puppies would be considered prolonged and may indicate dystocia.
7. Affected puppies may be born weaker, requiring assistance to clear out mucus from the respiratory tract. Having a trained person enter the room, help clear off the amniotic sac and suction out excess mucus from the nasal passage and throat will increase the survival rate of the puppies. The bulb syringe can be used to suction out excess fluids from the throat and nasal passage. It is critical every puppy is breathing normally and actively making clear vocalizations. A stethoscope may be used to auscultate the lungs for any abnormal sounds.
8. Immediately after birth is a good time to evaluate any unpredictable genetic anomalies. Puppies should urinate and defecate as a result of the initial stimulation from the bitch or the caregiver. Using a warm gauze soaked in saline, rub from the urogenital area to the base of the tail. This should stimulate the

puppy to void urine and feces. If this does not occur, there may be an issue with these body functions. The oral cavity should also be checked for a cleft palate. Cleft palates are genetically linked and will interfere with the puppy's ability to nurse. Puppies with cleft palates should be humanely euthanized and the breeding pair should be re-evaluated.

9. Puppies are weighed immediately after birth to establish a starting weight. It is well-documented that puppies will lose weight in the first 24 h after birth. They should begin gaining weight steadily after this time. Marking the puppies in a consistent manner will help identify individual animals. This will also serve as the birthing order to determine genotyping using the umbilical cords.
10. Once the suture is tied off, the cord may be cut using sterile scissors and monitored for excess bleeding. If bleeding continues, tighten the knot, or tie another knot around the remaining cord.
11. The first puppy's latching will stimulate the let-down reflex via oxytocin release to initiate breast milk flow. Place the clean, dry puppy up to the mammary of the bitch. Guide the teat into the mouth of the puppy gently, and brace the head to prevent the puppy from falling off. Puppies should begin sucking and form a strong suction around the teat. Affected puppies may not physically be strong enough to remain attached or suckle properly; thus assistance may be needed continually. It is best for all puppies to nurse from the bitch at this time to receive colostrum. Colostrum is highly enriched with antibodies, vitamins, electrolytes, and other nutrients. If you have a small litter, you may want to freeze some colostrum for future use.
12. Assist feeding should begin with assisted nursing where the puppy is placed on a teat and supported while it attempts to suckle. If no weight gain is achieved within 30 minutes, this method is considered unsuccessful and tube feeding should be considered. Tube feeding is considered the safest and most efficient method of feeding when properly performed. The 8 French feeding tube is placed along the lateral side of the puppy, measuring from the tip of the nose to the end of the ribs. This is the length of the tube to be inserted into the puppy. Milk replacer can then be made to support the nutritional needs.
13. Puppy milk replacer (Esbilac) will provide the necessary nutrients to the growing puppy but will not supplement the antibodies from the colostrum. The milk replacer should be mixed according to the instructions on the label using clean, warm water. Generally, the formula is made using a 2:1 ratio of water and the powdered replacer. The milk replacer may be heated to

a temperature close to body temperature by placing it in a warm water bath. It should be squirted onto a human wrist and evaluated for temperature before being fed to the puppy. Serum may be separated from the bitch's blood and injected subcutaneously into the scruff of the puppy as an alternative to colostrum. Puppies generally handle assistance well, and if able to survive past the first 72 h, they will have a high chance of surviving to weaning. Utilizing this method, we have achieved an 82% survival rate of affected puppies.

14. Once the eyes and ears are open, the puppies are aware of their surroundings. Establishing a handling routine and physically interacting with the puppies at a young age will ease them into restraint for testing and evaluations. Early conditioning and handling will enable laboratory personnel to efficiently carry out functional assessments with minimal stress to the affected dogs. Stress from restraint can negatively affect the health and safety of affected dogs.
15. Puppies are weaned at 8 weeks and are housed by the litter until approximately 4–5 months at which point they are separated by sex and genotype. The cage flooring should be raised off the ground and have minimal openings where paws may become trapped. Due to joint contracture of the model, it is common for affected dogs to get their paws wedged in large openings in the flooring. We recommend tenderfoot flooring, if possible, to create a surface safe for the dogs' and easy to maintain cleanliness. Dogs are pair-housed, when possible, based on size, age, and disease model status.
16. At approximately 4–5 months, food dishes are placed in a raised feeder to facilitate swallowing. Raising the food dishes helps prevent choking and possible regurgitation in affected dogs. A water bowl is added to the cage to provide a reliable source of water that requires minimal effort to consume, compared to a lixit which requires some effort to obtain the water.
17. Beginning around 5–6 months, DMD characteristics begin to develop at varying rates. Many dogs will begin to progressively show signs of the model; however, it is not uncommon for a percentage of puppies to rapidly demonstrate cardiac, respiratory and/or gastrointestinal illness. It is also possible to observe sudden death in affected dogs.
18. A trained person familiar with normal and affected dogs should perform the physical exam. It is important to understand how medical parameters will change as the model progresses, especially the cardiovascular system. Normal vital signs for a dog are body temperature, 37.5–39°C (99.5–102.5 °F); heart rate, 60–120 beats per minute; and respiration rate, 18–34 breaths per minute. The heart should have the characteristic lub-dub

sound with a synchronous femoral pulse. The lungs should sound clear and free from sounds such as crackling. Body weight and body condition scoring will be based on the size and breed of the dog. A published body condition scoring scale should be used (such as the Purina Body Condition Scale).

19. Minimum bloodwork for an affected DMD dog includes complete blood count, serum biochemistry with CK, ALT, AST, ALP, and GGT, and cardiac troponin. It may also be beneficial to monitor NT-proBNP and GLDH regularly.
20. To monitor disease progression and colony health, period timepoint data may be collected at 3, 6, 9, 12, 24, 36, and 42 months. The level of care required early in puppy development does not predict the degree of disease severity or disease progression.
21. Noninvasive functional assessments may include overnight activity monitoring, bloodwork, electrocardiography, echocardiography, whole limb force analysis, and gait assessment.
22. In our hands, we have occasionally observed affected dogs rapidly decline as early as 6 months, usually due to cardiac or gastrointestinal issues. Affected dogs can be aged to around 40 months with proper care and monitoring for sudden changes. Cardiac and respiratory changes are often noted as the affected dog ages.
23. The protocol can be adapted to deliver AAV vectors to other superficial muscles such as the semitendinosus, semimembranosus, gastrocnemius, vastus lateralis, rectus femoris, cranial tibialis, supraspinatus, infraspinatus, and deltoid muscle. We opt to perform whole muscle-saturated injection in the extensor carpi ulnaris because we have previously developed an in situ protocol to evaluate the contractile property of this muscle [34, 44]. This allows us to study the physiological outcomes of a particular gene therapy strategy with a small amount of the vector. The contralateral extensor carpi ulnaris can be injected with the same volume of excipient and serve as the untreated control for comparison [37].
24. This procedure should be performed by trained investigators familiar with canine musculoskeletal anatomy and have prior experience with intramuscular injection and the handling of sharps such as hypodermic needles used for injection. Investigators should be knowledgeable about the shape, size, muscle fiber orientation, and tendon location of the target muscle.
25. Consult with veterinary staff prior to sedation. Promptly address any medical concerns and review the dog's medical history with anesthesia. Dogs ineligible for sedation may still be used for intramuscular injection; however, restraint is

required. Exceptionally calm or well-behaved dogs may not require sedation.

26. Alternative sedatives may be used at the recommendation of a veterinarian and following the American Animal Hospital Association recommendations for anesthesia [52]. Always follow the recommendations of a veterinarian. Sedative of choice may vary based on prior experience, resource availability, and the conditions or medical history of each individual dog.
27. After insertion in the muscle body, always draw back the plunger of the syringe prior to injection, ensuring that the needle tip is not penetrating a blood vessel. If blood appears in the syringe, readjust the needle position. This applies to both the administration of drugs and gene therapy agents such as AAV.
28. Oxygen supplementation should provide approximately 30–50% fraction of inspired oxygen (FiO2). The flow rate and ratio of breathing air to pure oxygen may vary between animals and gas suppliers. The optimal percentage FiO2 and flow rates for experimental animals should be discussed with a veterinarian or a board-certified veterinary anesthesiologist. Face mask oxygenation generally provides 35–45% FiO2 at a flow rate of 1 to 6 liters per minute [53, 54].
29. Fold the gauze into quarters by bringing the four corners together, and hold by the corners. This generates a smaller contact surface area but limits your finger from touching the skin during scrubbing. Use the scrubbing center-out technique and never return to the starting point (the center point). Discard the scrubbing gauze immediately once it reaches the fur/shaving line. Do not apply excessive pressure on the gauze while scrubbing. This may damage the dermal layer of the skin. Any skin abrasions or inflammation may compromise incision healing.
30. We recommend tattooing the skin superficial to the injection site of large muscles. The tattoo serves as a landmark for future biopsy and dissection of the injected site. Tattooing of smaller, longer, or thinner muscles (e.g., ECU) may not be required due to saturated injection of the whole muscle. Always get approval and recommendations by a veterinarian prior to tattooing. Ensure that the chosen tattoo dye is safe for use in dogs, sterile, and unexpired.
31. Exercise caution and good sharps handling techniques during the tattoo process. We recommend to only tattoo dot-like spots as described in the methods section. We recommend intramuscular injection directly between two tattooed dots, or within the center of a dotted-square tattoo pattern. Keep a detailed record of tattoo locations, tattoo pattern, and

corresponding injection site. Record the angle and distance of tattoo sites to common anatomical landmarks using a goniometer and/or ruler.

32. After insertion in the muscle body, always draw back the plunger of the syringe prior to injection, ensuring that the needle tip is not penetrating a blood vessel. If blood appears in the syringe, readjust the needle position. This applies to both the administration of drugs and gene therapy agents such as AAV.
33. To saturate an area of a larger, thicker muscle body, insert the needle straight into the muscle, perpendicular (90-degree angle) to the plane of the muscle. Slowly inject 1/5 of the volume, withdrawing the needle as the injection is being delivered. Do not fully remove the needle from the skin/muscle. Next, rotate the syringe at approximately a 45-degree angle to the left of the perpendicular angle. Re-insert the needle into the body of the muscle (45 degrees right of the perpendicular injection site). Inject approximately another 1/5 of the injection volume, as previously described. Repeat this step at approximately a 45-degree angle left, cranial, and caudal to the original perpendicular injection site.
34. During ECU muscle injection, it may be helpful to bend the needle using the provided cap or a needle holder tool. Exercise caution and good sharps handling techniques if bending needles.
35. We recommend performing ECU muscle injection under ultrasound guidance (Fig. 8). Ultrasound-guided technique allows for precise injection of deep and superficial aspects of the muscle. Pre-injection ultrasound examination is recommended to evaluate the depth and width of the muscle. Without ultrasound assistance, we do not recommend attempting to inject superficial and deep aspects of the muscle. If using ultrasound-guided injection, take still images during the injection process. These images will serve as references for future dissection/biopsy.
36. If using alternative sedative drugs, proceed with waking the dog according to a veterinarian's recommendation.
37. Pain or discomfort from intramuscular injection of gene therapy agents is uncommon. However, if pain or discomfort is noted following the procedure, inject Rimadyl (carprofen) (4.4 mg/kg) subcutaneously or manage with other suitable nonsteroidal anti-inflammatory drugs at the recommendation of a veterinarian.
38. The body weight should be monitored every 2–4 h as described for puppy rearing. Bowel movements and urination may cause notable transient body weight fluctuations. Puppies

that fail to gain weight should not be used for AAV injection until they are consistently gaining and maintaining weight. The rectal temperature in the first, second, and fourth week of life in a neonatal puppy is in the range of 35–37.2 °C (95–99 °F), 36.1–37.8 °C (97–100 °F), and 37.8–38.9 °C (100–103 °F), respectively [55]. Puppies should maintain body temperature and gradually increase over the weeks. Drops in temperature can result from environmental temperatures being low and may jeopardize puppy health.

39. We have performed neonatal AAV injections in puppies as early as 24 h after birth. Successful systemic AAV injection can also be achieved in puppies up to 1 week of age [47, 56–58]. If there is volume overload, it may lead to pulmonary edema (coughing and increased lung sounds such as crackling on auscultation). If there is an anaphylactic reaction, the puppy may show signs of hives, increased body temperature, or collapse.

40. Ensure the puppy is wrapped in a thermal blanket and kept warm throughout the injection. The thermal blanket should be pre-warmed to 37 °C, so it is prepared if needed during the injection. Body temperature is one of the first signs that will change after injection if an anaphylactic episode occurs.

41. The jugular and/or the saphenous veins may also be used instead of the cephalic vein. Preparation for injection into these veins is similar with the corresponding locations shaved using the #40 clipper blade. For example, for saphenous vein injection, we shave the lateral aspect of the distal tibia above the tarsal joint.

42. Muscle biopsy is a procedure commonly used to obtain skeletal muscle specimens for the diagnosis of neuromuscular disease, evaluation of therapeutic interventions, and collect tissue for research purposes in human patients and donors. Three commonly used approaches are Bergstrom needle biopsy, open biopsy, and vacuum-assisted biopsy system (VAB) [59–61]. We have previously described detailed procedures of these biopsy approaches [41, 48]. The VAB method was first developed as a minimally invasive and cost-effective method for biopsy of breast tissue [62]. VAB was later used to collect muscle biopsies [60, 63]. Of the three approaches, we consider the VAB method to have the lowest risk of complications. Due to the small incision and closed nature of specimen collection, it can be performed outside of the surgery suite. In human patients, the procedure can be performed under local anesthesia, causes minor discomfort, and is generally well tolerated with quick recovery times [60, 63]. VAB has been used to collect biopsies from DMD patients [64]. In this protocol,

we describe the Vacora® biopsy system-based VAB method for collection of biopsies from canine skeletal muscle. The procedure has been published previously [41]. For purposes of simplicity, the protocol described in this chapter presents a method for biopsy of the dog biceps femoris muscle. This procedure can be further adapted to acquire biopsies from other skeletal muscles. We have previously used this technique to biopsy the vastus lateralis, gracilis, cranial tibialis, gastrocnemius, semitendinosus, deltoid, supraspinatus, infraspinatus, and lateral triceps muscles in dogs.

43. We usually start Dexmedetomidine at 6 μg/kg. The dosage may be increased in 3 μg/kg increments up to a maximum of 15 μg/kg. The final dosage is based on the level of sedation as determined by the dog's general mentation and pedal reflex. We have successfully sedated dogs younger than 16 weeks using this cocktail without adverse events.
44. Prior training on canine anatomy and experience with performing dog muscle surgery is recommended before performing the needle biopsy.
45. The person performing the biopsy should have a thorough understanding of the shape, size, and fiber orientation of the muscle to be biopsied. We recommend practicing the procedure on carcasses prior to performing it on a live animal.
46. We recommend an intramuscular injection of 1 mL 2% lidocaine at the biopsy site to relieve pain for dogs that are younger than 16 weeks. If signs of pain are noted during the 5-day post-operation recovery period, a veterinarian should be consulted.

Acknowledgments

The DMD research in the Duan lab is supported by the National Institutes of Health (AI-177600, AR-70517, AR-81544, NS-90634, and NS-131426), Jesse Davidson Foundation-Defeat Duchenne Canada, Hope for Javier, and Jackson Freel DMD Research Fund.

Disclosure RWH is serving on scientific advisory boards for Solid Biosciences and Fractyl Health. RWH received research funding from Hoffman LaRoche. DD is a member of the scientific advisory board for Solid Biosciences and an equity holder of Solid Biosciences. DD is a member of the scientific advisory board for Sardocor Corp. DD is an inventor of several issued and filed patents on AAV vector and DMD gene therapy. The Duan lab has received research support from Elenae Therapeutics and Satellos Bioscience in the last 3 years. All other authors have nothing to disclose.

References

1. Duan D, Goemans N, Takeda S et al (2021) Duchenne muscular dystrophy. Nat Rev Dis Primers 7(1):13. https://doi.org/10.1038/s41572-021-00248-3
2. McGreevy JW, Hakim CH, McIntosh MA, Duan D (2015) Animal models of Duchenne muscular dystrophy: from basic mechanisms to gene therapy. Dis Model Mech 8(3):195–213. https://doi.org/10.1242/dmm.018424
3. Wasala NB, Chen SJ, Duan D (2020) Duchenne muscular dystrophy animal models for high-throughput drug discovery and precision medicine. Expert Opin Drug Discov 15(4): 443–456. https://doi.org/10.1080/17460441.2020.1718100
4. Duan D (2015) Duchenne muscular dystrophy gene therapy in the canine model. Hum Gene Ther Clin Dev 26(1):57–69. https://doi.org/10.1089/humc.2015.006
5. Duan D (2018) CRISPR alleviates muscular dystrophy in dogs. Nat Biomed Eng 2(11): 795–796. https://doi.org/10.1038/s41551-018-0320-0
6. Nghiem PP, Kornegay JN (2019) Gene therapies in canine models for Duchenne muscular dystrophy. Hum Genet 138(5):483–489. https://doi.org/10.1007/s00439-019-01976-z
7. Kornegay JN (2017) The golden retriever model of Duchenne muscular dystrophy. Skelet Muscle 7(1):9. https://doi.org/10.1186/s13395-017-0124-z
8. Cooper BJ, Winand NJ, Stedman H et al (1988) The homologue of the Duchenne locus is defective in X-linked muscular dystrophy of dogs. Nature 334(6178):154–156
9. Shimatsu Y, Katagiri K, Furuta T et al (2003) Canine X-linked muscular dystrophy in Japan (CXMDJ). Exp Anim 52(2):93–97
10. Yokota T, Lu QL, Partridge T et al (2009) Efficacy of systemic morpholino exon-skipping in Duchenne dystrophy dogs. Ann Neurol 65(6):667–676. https://doi.org/10.1002/ana.21627
11. Le Guiner C, Servais L, Montus M et al (2017) Long-term microdystrophin gene therapy is effective in a canine model of Duchenne muscular dystrophy. Nat Commun 8:16105. https://doi.org/10.1038/ncomms16105
12. Birch SM, Lawlor MW, Conlon TJ et al (2023) Assessment of systemic AAV-microdystrophin gene therapy in the GRMD model of Duchenne muscular dystrophy. Sci Transl Med 15(677):eabo1815. https://doi.org/10.1126/scitranslmed.abo1815
13. Le Guiner C, Montus M, Servais L et al (2014) Forelimb treatment in a large cohort of dystrophic dogs supports delivery of a recombinant AAV for exon skipping in Duchenne patients. Mol Ther 22(11):1923–1935. https://doi.org/10.1038/mt.2014.151
14. Vieira NM, Elvers I, Alexander MS et al (2015) Jagged 1 rescues the Duchenne muscular dystrophy phenotype. Cell 163(5):1204–1213. https://doi.org/10.1016/j.cell.2015.10.049
15. Walmsley GL, Arechavala-Gomeza V, Fernandez-Fuente M et al (2010) A Duchenne muscular dystrophy gene hot spot mutation in dystrophin-deficient cavalier king charles spaniels is amenable to exon 51 skipping. PLoS One 5(1):e8647. https://doi.org/10.1371/journal.pone.0008647
16. Piercy RJ, Walmsley G (2009) Muscular dystrophy in Cavalier King Charles spaniels. Vet Rec 165(2):62. 165/2/62-a [pii]
17. Hornby NL, Drees R, Harron R et al (2021) Musculoskeletal magnetic resonance imaging in the DE50-MD dog model of Duchenne muscular dystrophy. Neuromuscul Disord 31(8):736–751. https://doi.org/10.1016/j.nmd.2021.05.010
18. Crawford AH, Hildyard JCW, Rushing SAM et al (2022) Validation of DE50-MD dogs as a model for the brain phenotype of Duchenne muscular dystrophy. Dis Model Mech 15(3). https://doi.org/10.1242/dmm.049291
19. Hildyard JCW, Riddell DO, Harron RCM et al (2022) The skeletal muscle phenotype of the DE50-MD dog model of Duchenne muscular dystrophy. Wellcome Open Res 7:238. https://doi.org/10.12688/wellcomeopenres.18251.1
20. Riddell DO, Hildyard JCW, Harron RCM et al (2022) Serum inflammatory cytokines as disease biomarkers in the DE50-MD dog model of Duchenne muscular dystrophy. Dis Model Mech. https://doi.org/10.1242/dmm.049394
21. Riddell DO, Hildyard JCW, Harron RCM et al (2022) Longitudinal assessment of blood-borne musculoskeletal disease biomarkers in the DE50-MD dog model of Duchenne muscular dystrophy. Wellcome Open Res 6:354. https://doi.org/10.12688/wellcomeopenres.17398.2
22. Riddell DO, Hildyard JCW, Harron RCM et al (2023) Longitudinal assessment of skeletal muscle functional mechanics in the DE50-MD dog model of Duchenne muscular dystrophy. Dis Model Mech 16(12). https://doi.org/10.1242/dmm.050395

23. Amoasii L, Hildyard JCW, Li H et al (2018) Gene editing restores dystrophin expression in a canine model of Duchenne muscular dystrophy. Science 362(6410):86–91. https://doi.org/10.1126/science.aau1549
24. Smith BF, Yue Y, Woods PR et al (2011) An intronic LINE-1 element insertion in the dystrophin gene aborts dystrophin expression and results in Duchenne-like muscular dystrophy in the corgi breed. Lab Investig 91(2):216–231. https://doi.org/10.1038/labinvest.2010.146
25. Smith BF, Kornegay JN, Duan D (2007) Independent canine models of Duchenne muscular dystrophy due to intronic insertions of repetitive DNA. Mol Ther 15(Supplement 1):S51
26. Fine DM, Shin J-H, Yue Y et al (2011) Age-matched comparison reveals early electrocardiography and echocardiography changes in dystrophin-deficient dogs. Neuromuscul Disord 21(7):453–461. https://doi.org/10.1016/j.nmd.2011.03.010
27. Yang HT, Shin J-H, Hakim CH et al (2012) Dystrophin deficiency compromises force production of the extensor carpi ulnaris muscle in the canine model of Duchenne muscular dystrophy. PLoS One 7(9):e44438
28. Shin J-H, Greer B, Hakim CH et al (2013) Quantitative phenotyping of Duchenne muscular dystrophy dogs by comprehensive gait analysis and overnight activity monitoring. PLoS One 8(3):e59875
29. Duan D, Hakim CH, Ambrosio CE et al (2015) Early loss of ambulation is not a representative clinical feature in Duchenne muscular dystrophy dogs. Dis Model Mech 8(3): 193–194. https://doi.org/10.1242/dmm.019216
30. Hakim CH, Peters AA, Feng F et al (2015) Night activity reduction is a signature physiological biomarker for Duchenne muscular dystroophy dogs. J Neuromuscul Dis 2(4): 397–407. https://doi.org/10.3233/JND-150114
31. Hakim CH, Mijailovic A, Lessa TB et al (2017) Non-invasive evaluation of muscle disease in the canine model of Duchenne muscular dystrophy by electrical impedance myography. PLoS One 12(3):e0173557. https://doi.org/10.1371/journal.pone.0173557
32. Kodippili K, Hakim CH, Yang HT et al (2018) Nitric oxide dependent attenuation of norepinephrine-induced vasoconstriction is impaired in the canine model of Duchenne muscular dystrophy. J Physiol 596(21): 5199–5266. https://doi.org/10.1113/JP275672
33. White Z, Hakim CH, Theret M et al (2020) High prevalence of plasma lipid abnormalities in human and canine Duchenne and Becker muscular dystrophies depicts a new type of primary genetic dyslipidemia. J Clin Lipidol 14(4):459–469. https://doi.org/10.1016/j.jacl.2020.05.098
34. Hakim CH, Yang HT, Burke MJ et al (2021) Extensor carpi ulnaris muscle shows unexpected slow-to-fast fiber type switch in Duchenne muscular dystrophy dogs. Dis Model Mech 14(12):dmm049006. https://doi.org/10.1242/dmm.049006
35. Kodippili K, Thorne PK, Laughlin MH, Duan D (2021) Dystrophin deficiency impairs vascular structure and function in the canine model of Duchenne muscular dystrophy. J Pathol 254(5):589–605. https://doi.org/10.1002/path.5704
36. Shin J-H, Pan X, Hakim CH et al (2013) Microdystrophin ameliorates muscular dystrophy in the canine model of Duchenne muscular dystrophy. Mol Ther 21(4):750–757. https://doi.org/10.1038/mt.2012.283
37. Kodippili K, Hakim CH, Pan X et al (2018) Dual AAV gene therapy for Duchenne muscular dystrophy with a 7-kb mini-dystrophin gene in the canine model. Hum Gene Ther 29(3): 299–311. https://doi.org/10.1089/hum.2017.095
38. Yue Y, Pan X, Hakim CH et al (2015) Safe and bodywide muscle transduction in young adult Duchenne muscular dystrophy dogs with adeno-associated virus. Hum Mol Genet 24(20):5880–5890
39. Duan D (2018) Micro-dystrophin gene therapy goes systemic in Duchenne muscular dystrophy patients. Hum Gene Ther 29(7): 733–736. https://doi.org/10.1089/hum.2018.012
40. Hakim CH, Kumar SRP, Perez-Lopez DO et al (2021) Cas9-specific immune responses compromise local and systemic AAV CRISPR therapy in multiple dystrophic canine models. Nat Commun 12(1):6769. https://doi.org/10.1038/s41467-021-26830-7
41. Hakim CH, Burke MJ, Teixeira J, Duan D (2023) Histological assessment of gene therapy in the canine DMD model. Methods Mol Biol 2587:303–338. https://doi.org/10.1007/978-1-0716-2772-3_16
42. Hakim CH, Kumar SRP, Perez-Lopez D et al (2023) Assessment of the gene therapy immune response in the canine muscular dystrophy model. Methods Mol Biol 2587:353–375. https://doi.org/10.1007/978-1-0716-2772-3_18
43. Hakim CH, Perez-Lopez D, Burke MJ et al (2023) Molecular and biochemical assessment of gene therapy in the canine model of Duchenne muscular dystrophy. Methods Mol Biol

2587:255–301. https://doi.org/10.1007/978-1-0716-2772-3_15
44. Hakim CH, Teixeira J, Leach SB, Duan D (2023) Physiological assessment of muscle, heart, and whole body function in the canine model of Duchenne muscular dystrophy. Methods Mol Biol 2587:67–103. https://doi.org/10.1007/978-1-0716-2772-3_5
45. Zalcman AR, Hakim CH, Lattimer JC et al (2023) MRI evaluation of gene therapy in the canine model of Duchenne muscular dystrophy. Methods Mol Biol 2587:339–352. https://doi.org/10.1007/978-1-0716-2772-3_17
46. Bartlett RJ, Winand NJ, Secore SL et al (1996) Mutation segregation and rapid carrier detection of X-linked muscular dystrophy in dogs. Am J Vet Res 57(5):650–654
47. Yue Y, Ghosh A, Long C et al (2008) A single intravenous injection of adeno-associated virus serotype-9 leads to whole body skeletal muscle transduction in dogs. Mol Ther 16(12): 1944–1952. https://doi.org/10.1038/mt.2008.207
48. Yue Y, Shin J-H, Duan D (2011) Whole body skeletal muscle transduction in neonatal dogs with AAV-9. Methods Mol Biol 709:313–329. https://doi.org/10.1007/978-1-61737-982-6_21
49. Kotin RM, Snyder RO (2017) Manufacturing clinical grade recombinant adeno-associated virus using invertebrate cell lines. Hum Gene Ther 28(4):350–360. https://doi.org/10.1089/hum.2017.042
50. Clement N, Grieger JC (2016) Manufacturing of recombinant adeno-associated viral vectors for clinical trials. Mol Ther Methods Clin Dev 3:16002. https://doi.org/10.1038/mtm.2016.2
51. Shin J-H, Yue Y, Duan D (2012) Recombinant adeno-associated viral vector production and purification. Methods Mol Biol 798:267–284. https://doi.org/10.1007/978-1-61779-343-1_15
52. Grubb T, Sager J, Gaynor JS et al (2020) 2020 AAHA Anesthesia and monitoring guidelines for dogs and cats. J Am Anim Hosp Assoc 56(2):59–82. https://doi.org/10.5326/JAAHA-MS-7055
53. Manning AM (2002) Oxygen therapy and toxicity. Vet Clin North Am Small Anim Pract 32(5):1005–1020., v. https://doi.org/10.1016/s0195-5616(02)00043-8
54. Ramesh M, Thomovsky E, Johnson P (2021) Conventional versus high-flow oxygen therapy in dogs with lower airway injury. Can J Vet Res 85(4):241–250
55. Davidson AP (2023) Management of the neonate in dogs and cats. Merck veterinary manual. Merck & Co, Rahway
56. Pan X, Yue Y, Zhang K et al (2013) Long-term robust myocardial transduction of the dog heart from a peripheral vein by adeno-associated virus serotype-8. Hum Gene Ther 24(6):584–594. https://doi.org/10.1089/hum.2013.044
57. Hakim CH, Yue Y, Shin J-H et al (2014) Systemic gene transfer reveals distinctive muscle transduction profile of tyrosine mutant AAV-1, -6, and -9 in neonatal dogs. Mol Ther Methods Clin Dev 1:14002. https://doi.org/10.1038/mtm.2014.2
58. Pan X, Yue Y, Zhang K et al (2015) AAV-8 is more efficient than AAV-9 in transducing neonatal dog heart. Human Gene Ther Methods 26(4):54–61
59. Heckmatt JZ, Moosa A, Hutson C et al (1984) Diagnostic needle muscle biopsy. A practical and reliable alternative to open biopsy. Arch Dis Child 59(6):528–532. https://doi.org/10.1136/adc.59.6.528
60. Akarolo-Anthony SN, Ogundiran TO, Nkwodimmah C et al (2012) Office based muscle biopsy using Vacora vacuum assisted biopsy system. Afr J Med Med Sci 41(3):313–316
61. Mikell CB, Chan AK, Stein GE et al (2013) Muscle and nerve biopsies: techniques for the neurologist and neurosurgeon. Clin Neurol Neurosurg 115(8):1206–1214. https://doi.org/10.1016/j.clineuro.2013.05.008
62. Pistolese CA, Ciarrapico AM, Della Gatta F et al (2009) Cost-effectiveness analysis of two vacuum-assisted breast biopsy systems: Mammotome and Vacora. Radiol Med 114(5): 743–756. https://doi.org/10.1007/s11547-009-0404-8
63. Gallo A, Abraham A, Katzberg HD et al (2018) Muscle biopsy technical safety and quality using a self-contained, vacuum-assisted biopsy technique. Neuromuscul Disord 28(5): 450–453. https://doi.org/10.1016/j.nmd.2018.02.006
64. Barthelemy F, Woods JD, Nieves-Rodriguez S et al (2020) A well-tolerated core needle muscle biopsy process suitable for children and adults. Muscle Nerve 62(6):688–698. https://doi.org/10.1002/mus.27041

Chapter 15

In Vivo Electroporation of Plasmid DNA into the Skeletal Muscle of Dystrophic Mouse Models

Miguel A. Lopez Perez, Gianni N. Giarrano, and Noah L. Weisleder

Abstract

Gene transfer into living, adult skeletal muscle in experimental animals is a powerful tool for investigating the fundamental molecular mechanisms of neuromuscular disease progression. This methodology has been leveraged in the development of potential therapeutics for Duchenne muscular dystrophy (DMD), limb girdle muscular dystrophies (LGMD), and other skeletal muscle maladies. There are several modalities for gene transfer into skeletal muscle, including the use of viral vectors, nucleic acid encapsulation, and the development of transgenic mouse models. However, there are distinct advantages to the use of electroporation to deliver plasmid DNA in the preclinical setting. Electroporation applies an electric current to biological tissues that produce small, transient pores in the plasma membrane of cells. At high concentrations, plasmid DNA or other nucleic acid constructs can enter through these membrane disruptions and become enclosed within the cell after membrane resealing. This allows for rapid modification of gene expression or gene editing in target cells. Thus, electroporation represents a tractable and efficient method for tissue-specific gene delivery into the muscles of many mouse models, including those for DMD and other muscular dystrophies. In the specific approach detailed here, transdermal electric pulses applied to the hindlimb permeabilize the sarcolemmal membrane in vivo, allowing for plasmid DNA transfection of adult myofibers in dystrophic mice. We delineate the methodology for in vivo electroporation of the flexor digitorum brevis muscle for efficient plasmid gene transfer in dystrophic and wild-type mouse models.

Key words Electroporation, Transdermal, Gene transfer, Plasmid DNA, Electrotransfer, In vivo transfection, Gene delivery, Muscular dystrophy

1 Introduction

Tissue-specific gene expression in differentiated mouse skeletal muscle is a dynamic methodology to understand the molecular basis of Duchenne muscular dystrophy (DMD) and limb-girdle muscular dystrophies (LGMD). It is also an effective tool to screen for efficacious gene therapies. Historically, the generation of transgenic mouse models has been widely used to investigate the

Authors Miguel A. Lopez Perez and Gianni N. Giarrano have equally contributed to this chapter.

Kevin Flanigan and Stefan Nicolau (eds.), *Duchenne Muscular Dystrophy: Methods and Protocols*, Methods in Molecular Biology, vol. 2975, https://doi.org/10.1007/978-1-0716-4811-7_15,

functional role of single proteins and protein complexes in the pathogenesis of DMD [1, 2]. While polymer encapsulation and lipid nanoparticles (LNPs) have also been used to deliver mRNA, DNA, and CRISPR-based therapies to dystrophic muscle, these approaches have been hindered by suboptimal manufacturing throughput, inefficient targeting to desired muscle tissue, and slow development of transgene expression [3]. Alternatively, viral-mediated delivery of genes using muscle-specific promoters in adeno-associated viral vectors (AAV), or other viral vectors (retroviral or lentiviral), has been extensively leveraged to express, screen, and develop therapies for DMD that have led to FDA approval for use in humans [4, 5]. Although AAV-based therapies have emerged as the only approved gene replacement therapy for DMD, the utility of virus-encapsulated DNA sequences in the research and development of therapies is significantly limited by the inability of genes larger than 4.7 kilobases to be packaged into AAV particles [3]. The methods of genetic manipulation and gene transfer can be costly, time-consuming, and necessitate specialized expertise and production capabilities. With the success of gene therapy for the treatment of DMD, an impetus exists to continue to develop more efficacious treatments. Thus, to conduct medium- to high-throughput screening of potential therapeutics requires methodology that is rapid, highly efficient, and accessible across diverse expertise levels and laboratory resources.

To that end, fundamental biophysical studies in the mid-twentieth century demonstrated increased permeabilization of the plasma membrane of cells upon exposure to electrical stimulus [6–9]. Subsequent studies elucidated a mechanism by which sustained electric pulses generate pores within the plasma membrane and contribute to an increased membrane potential, allowing for the transfer of negatively charged nucleic acids across the plasma membrane [6, 9, 10]. The application of an electric field increases the mammalian cell transmembrane potential (TMP) from −0.07 V to 0.2–0.5 V, generating random hydrophobic pores. If TMP remains elevated, water enters the pores, forcing lipid reorientation and the formation of DNA-accessible hydrophilic pores [11]. This method has been optimized to directly transfer genes into skeletal muscle cells in vivo [12–16]. The adaptability of this approach is powerful and can be applied to various experimental designs, including either DNA or RNA genetic material. Plasmid DNA constructs can be engineered to contain open reading frames that overexpress a gene of interest (GOI) through a constitutive promoter (e.g., CMV) or a skeletal muscle specific promoter (e.g., MHCK7). Similarly, mutant versions of GOIs can be encoded to evaluate reported pathological variants. Alternatively, shRNA containing plasmids can be designed to elicit RNA interference (RNAi) and knockdown one or several genes of interest, allowing for rigorous interrogation of a GOI or pathway [17]. Recent advances in genome editing utilizing CRISPR/Cas9 systems also present the ability to evaluate the capacity of gene correction in dystrophic mouse models. Here, a dual plasmid system encoding the CRISPR/Cas9 nuclease and guide RNA, respectively, can be directly electroporated into dystrophic mouse muscle with the aim of correcting aberrant gene expression [18, 19].

Electroporation (EP), or the use of electric pulses to transfer plasmid-encoded DNA sequences into adult myofibers, has thus become a robust tool for the study of DMD and other muscle diseases [6, 20–23]. EP has been used to screen therapeutics, investigate protein function, and uncover new insights into the pathophysiology of dystrophic muscle [24–27]. Furthermore, EP overcomes the manufacturing hurdles and gene size limitations of AAV and other viral and non-viral-based approaches. Large quantities of DNA are readily manufacturable by standard molecular biology procedures, and dystrophic muscle can be efficiently transfected with full-length dystrophin encoding plasmids [23]. Many previous studies using EP in dystrophic mouse muscle perform the procedure in the tibialis anterior (TA) muscle [24, 28]. Whole muscle evaluation of the TA is difficult due to the large size of the muscle. Additionally, applying uniform electric pulses across the entirety of the TA is technically challenging [24]. We have found that injection of plasmid DNA and transdermal electrical stimulation of the hindlimb footpad allows for high transfection efficiency of the entire flexor digitorum brevis (FDB) muscle. Assessment of gene expression can be performed by traditional biochemical techniques and downstream analysis of the muscle can be conducted through contractility measurements (*see* Subheading 3.3). For experiments investigating physiology at the single myofiber level, a fluorescent reporter (e.g., eGFP, mRuby) can be fused at the amino- or carboxy-terminus of the GOI coding sequence to select only positively electroporated myofibers. Similarly, a control plasmid containing a fluorescent tag without the gene of interest can be co-injected during electroporation. Previous work has described the EP of the FDB muscle in wild-type mice [30–33]. Here, we describe an optimized EP protocol for gene transfer into the FDB muscle of the dystrophic mouse hindlimb.

2 Materials

Prepare all solutions using filtered distilled water. All solutions and buffers introduced into experimental animals should be sterilized by filtration using a 0.2 micron cellulose acetate filter or appropriate methods as mandated by local regulations.

2.1 Anesthesia and Injection Equipment

1. Isoflurane vaporizer (SomnoFlo® Low-Flow Electronic Vaporizer or a VetEquip anesthesia vaporizer and compressed oxygen).
2. Anesthesia chamber and nose cone.
3. 0.2 μm cellulose acetate syringe filters.
4. 28G insulin syringes.
5. Electroporator: e.g., ECM 830 Square Wave Pulse Generator by BTX.
6. BTX 7 mM external paddle electrodes.

2.2 Reagents

1. Phosphate buffered saline: 0.1 M sodium phosphate, 0.15 M NaCl, pH 7.4.
2. Saline: 0.9% NaCl.
3. 70% ethanol.
4. 1× hyaluronidase solution (20 μg/μL): 400 μg (20 μL) of hyaluronidase is utilized per each FDB electroporated. Measure 400 μg of lyophilized hyaluronidase per FDB planned for electroporation. Dissolve measured hyaluronidase in sterile phosphate buffered saline (PBS) pH 7.4 to achieve a final concentration of 20 μg/μL. Sterilize the solution using a 0.2 μm cellulose acetate filter. Store at room temperature (*see* **Note 1**).
5. 25× NaCl stock solution (22.5% NaCl): Dissolve 4.5 g of NaCl in distilled water and adjust final volume to 20 mL. Sterilize the solution using a 0.2 μm cellulose acetate filter. Store at room temperature.
6. Plasmid DNA solution for injection (1 μg/μL in 0.9% saline): 20 μL (20 ug) of plasmid DNA solution is used per FDB being electroporated. Prepare the plasmid DNA solution by diluting endotoxin-free plasmid DNA containing your gene of interest to a final concentration of 1 μg/μL in 1× NaCl using 25× NaCl stock solution (*see* **Note 2**). Repeat the process to prepare your control plasmid injection (*see* **Note 3**).

3 Methods

All procedures can be completed at room temperature. Mice should be carefully monitored throughout the experimental procedure and during the subsequent recovery period.

3.1 FDB Preparation

1. Anesthetize the mouse in concordance with local regulations, preferably by isoflurane inhalation or intraperitoneal (IP) injection of ketamine and xylazine. We utilize isoflurane

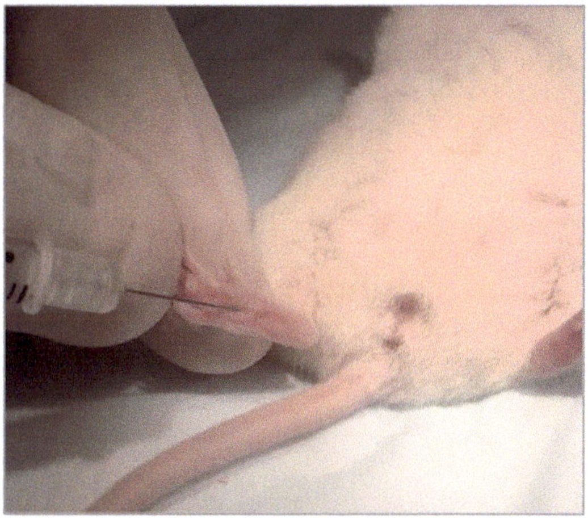

Fig. 1 Subcutaneous injection of the hindlimb footpad. The footpad cavity is prepared for electroporation by injecting hyaluronidase to break down the extracellular matrix, followed by injection of plasmid DNA. Both solutions are injected at a 20–30° degree angle with the needle bevel face up. A small injection bubble will form in the middle of the footpad and disappear after the solution has been absorbed by the surrounding tissue

inhalation using a SomnoFlo® Low-Flow Electronic Vaporizer or a VetEquip anesthesia vaporizer and compressed oxygen.

2. Place the mouse in the induction chamber, and anesthetize the animal using the following settings:
   (a) VetEquip vaporizer:
      (i) Isofluorane: 1.5–2%.
      (ii) Oxygen: 1 L/min.
   (b) SomnoFlo® Low-Flow Electronic Vaporizer:
      (i) Isofluorane: 4%.
      (ii) Oxygen: 500 mL/min.
3. Remove the anesthetized mouse from the induction chamber, and place supine on the procedure platform. Secure anesthesia nose cone on mouse, and ensure proper sedation by toe pinch.
4. Prepare a sterile insulin syringe with 20 μL of 1× hyaluronidase solution.
5. Wipe the footpad of the mouse with 70% ethanol (or an alcohol prep pad) to remove debris prior to injection.
6. Inject 15–20 μL of hyaluronidase solution subcutaneously (SQ) into the footpad of the mouse, right below the skin (Fig. 1). This should be performed using a shallow angle to ensure the underlying FDB muscle is not damaged. If completed correctly, a small bubble underneath the skin should be transiently visible (*see* **Note 4**).

7. After injection, gently massage the footpad to assist in the efficient diffusion of hyaluronidase solution (*see* **Note 5**).
8. Return mouse to cage and allow to recover for 60 min (*see* **Note 6**). Monitor recovery every 30 min.

3.2 Electroporation of Plasmid DNA

1. After the initial recovery period, place the mouse in the SomnoFlo® Low-Flow Electronic Vaporizer or a VetEquip vaporizer induction chamber and anesthetize the animal using the following settings:
   (a) VetEquip vaporizer:
      (i) Isofluorane: 1.5–2%.
      (ii) Oxygen: 1 L/min.
   (b) SomnoFlo® Low-Flow Electronic Vaporizer:
      (i) Isofluorane: 4%.
      (ii) Oxygen: 500 mL/min.
2. Prepare a sterile insulin syringe with 20 μL of plasmid DNA solution.
3. Remove the anesthetized mouse from the induction chamber and place supine on the procedure platform. Secure anesthesia nose cone on mouse and ensure proper sedation by toe pinch.
4. Wipe the foot pad of the mouse with 70% ethanol (or an alcohol prep pad) to remove debris prior to injection.
5. Inject 20 μL plasmid DNA solution (20 μg) into the footpad of the mouse, right below the skin (Fig. **1**). This should be completed using a shallow angle to ensure the underlying FDB muscle is not damaged. A small bubble underneath the skin should be transiently visible if completed correctly.
6. After injection, gently massage the footpad to assist in efficient diffusion of plasmid DNA (*see* **Note 5**).
7. Clamp the middle of the footpad using BTX 7 mM calipers (or equivalent) as depicted in (Fig. **2**). Electroporate the footpad using the following settings (*see* **Note** 7).
   (a) Mode: LV.
   (b) Voltage: 100 V.
   (c) Pulse length: 2 ms.
   (d) Pulses: ~15.
   (e) Intervals: 267 ms.
8. Repeat electroporation process with the contralateral foot and/or fluorescent control plasmid.
9. Return mouse to cage and allow it to recover for 60 min. Monitor recovery every 30 min (*see* **Note 8**).

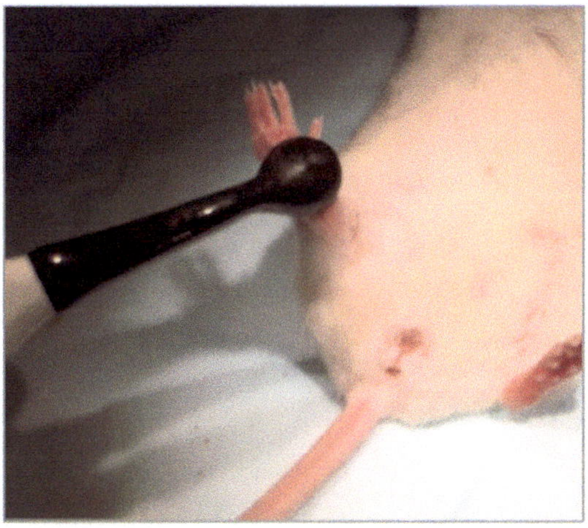

Fig. 2 Caliper electrode placement for the transdermal electroporation of the flexor digitorum brevis (FDB). Electrodes are placed in the middle of the hindlimb footpad with enough pressure to secure the foot but limit unnecessary injury to the muscle. Correct placement will be indicated by involuntary twitching of the hindlimb digits

10. Mice subjected to electroporation are allowed 2 weeks of recovery for adequate muscle repair (*see* **Note 9**). Autoclaved cage bedding should be used while the mice are recovering to minimize the chance of infection (*see* **Note 10**). Downstream analysis should not be conducted during the recovery period. Gene expression can be easily observed at the 14-day time point. The time that expression will continue can vary depending on the plasmid construct; however, it can be expected that expression will be maintained for weeks to months from a single round of electroporation [12, 16].

3.3 Downstream Analysis

Electroporated muscle tissue allows the end user to perform a variety of analytical assays to interrogate their respective GOI. Electroporated plasmids can be designed to elicit RNAi-mediated knockdown, overexpression of target genes, or introduce gene mutants in a naive background. Delivery of DNA constructs to utilize CRISPR/Cas9 or other gene editing technologies are also possible. Below is a non-comprehensive list of analyses that can be performed on successfully electroporated muscle. When conducting downstream analysis of electroporated muscle tissue, it is important to consider the transfection efficiency of muscle fibers. While these approaches should deliver substantial amounts of DNA into numerous muscle fibers in the FDB, there should not be an expectation that all the muscle fibers will take up DNA. As with all transfection methodology, it should not be assumed that all the muscle fibers will have exactly equivalent amounts of plasmid DNA. The most effective way to assess the number of muscle fibers with

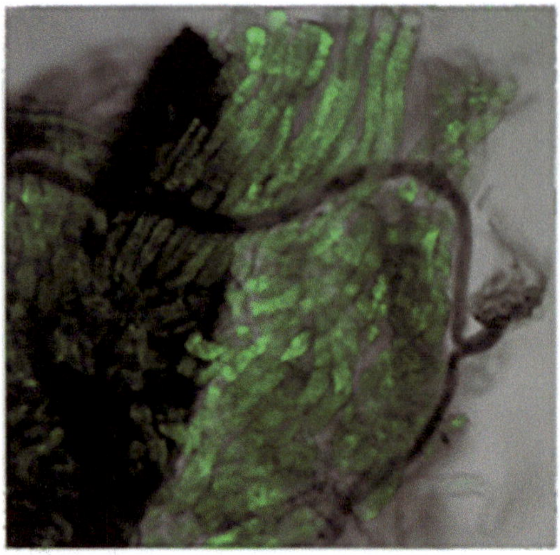

Fig. 3 Fluorescent imaging of isolated flexor digitorum brevis (FDB) muscle bundle. Isolated FDB muscle 14 days post-electroporation with eGFP control plasmid. Green pseudocoloring indicates positively transfected muscle fibers. A high percentage of muscle fibers are expected to be transfected through this approach

DNA delivery is to include a fluorescent marker protein (e.g., eGFP, mRuby) in the designed plasmid construct (s) electroporated into the muscle and then image dissected muscles using epifluorescence microscopy (Fig. 3).

3.3.1 Biochemical

Electroporation efficiency can be measured by epifluorescence microscopy if a fluorescent protein tag is included in the experimental design or by standard biochemical approaches. Electroporated muscle can be isolated, and total protein content can be measured via Bradford or other comparable colorimetric assays. Further downstream analyses include western blotting, qPCR, and co-immunoprecipitation (co-IP).

3.3.2 Functional/ Biophysical

Functional assays can also be utilized to evaluate how the gene of interest influences overall muscle function. These analyses can be performed ex vivo, in vivo, or in situ. For example, electroporated flexor digitorum brevis (FDB) muscle bundles can be carefully isolated and used ex vivo in a laser-induced injury assay to measure sarcolemma resealing kinetics [29]. Many other assessments can be made on living muscle fibers using this approach, bearing in mind experimental transfection efficiency. Inclusion of a fluorescent protein marker allows for the selection of only positively transfected muscle fibers [29, 30]. This allows for single-molecule analysis of individual muscle fibers in contrast to ensemble measurements of whole muscle bundles. Moreover, force production can be measured in vivo from an electroporated gastrocnemius muscle using a contractility apparatus, such as the Aurora Scientific 3-in-1 whole animal system [34]. In situ experiments can also be conducted to measure the contractility and force output of the FDB [31]. These analytical methods present a robust way of interrogating functional output by the gene of interest.

3.3.3 Histological

Electroporated muscle can be processed for cryosectioning and/or paraffin embedding for histological analysis. Histology stains, such as Hematoxylin and Eosin (H&E) and Trichrome can be used to evaluate critical cellular structures. For specific biomarkers, immunohistochemistry (IHC) can be used. These analyses can evaluate muscle parameters such as fiber size, centralized nuclei, fiber regeneration, and fibrosis. IHC can also be useful for identifying positively transfected muscle fibers.

4 Notes

1. It is recommended to make extra hyaluronidase solution to account for any pipette error or syringe filter retention loss. We have found that making this solution fresh delivers the best results.
2. We suggest preparing 25 μL of plasmid injection per FDB to account for any pipette error.
3. Control plasmids normally contain a fluorescent tag (e.g., eGFP, mRuby) without the gene of interest. If the GOI containing plasmid does not include a fluorescent tag, co-electroporation with the control plasmid can be used to identify electroporated fibers. In this case, generate a 1 μg/μL dilution in 1× NaCl of each plasmid. Add the diluted plasmids together in a 1:1 ratio to generate a total DNA concentration of 1 μg/μL comprised of 500 ng/μL each of the GOI plasmid and control plasmid. This method can result in ~75% or more of the positively electroporated fibers expressing both genes [33].
4. This step will require practice. We have found the most success with a 20–30° injection angle with the needle bevel face up.
5. Use mild pressure when massaging the footpad to avoid unnecessary injury to the underlying muscle. The injection bubble will quickly resolve as the injected fluid diffuses throughout the cavity.
6. The recovery period should be done for the entire 60 min to allow for effective hyaluronidase-mediated enzymatic degradation of the extracellular matrix. This is critical for successful and efficient electroporation of plasmid DNA into myofibers [15].

7. Caliper positioning and pressure are crucial throughout the electroporation step. Successful positioning and pressure will elicit involuntary twitching of the hindlimb digits with each pulse due to stimulation of the muscles. If twitching is not observed, abort electroporation and reposition the calipers until desired twitching is observed.
8. Mice exhibiting visible inflammation of the footpad at any point following electroporation should be excluded from the study. The parameters used here should not cause damage to the muscle or surrounding tissue as plasmid introduction is performed subcutaneously rather than intramuscularly. Any inflammation is likely due to infection at the injection site.
9. We conduct regular monitoring of the experimental mice during the recovery period. Mice showing notable behavioral changes or inflammation of the footpad should be removed from the study.
10. Autoclaved cage bedding is absolutely required for mice following injection into their foot pads; otherwise significant infection can occur in the injected feet. With the use of autoclaved bedding this should be an extremely rare complication.

References

1. van Putten M, Lloyd EM, de Greef JC, Raz V, Willmann R, Grounds MD (2020) Mouse models for muscular dystrophies: an overview. Dis Model Mech 13(2):dmm043562. https://doi.org/10.1242/dmm.043562
2. Bez Batti Angulski A, Hosny N, Cohen H et al (2023) Duchenne muscular dystrophy: disease mechanism and therapeutic strategies. Front Physiol 14:1183101. https://doi.org/10.3389/fphys.2023.1183101
3. Nance ME, Hakim CH, Yang NN, Duan D (2018) Nanotherapy for Duchenne muscular dystrophy. Wiley Interdiscip Rev Nanomed Nanobiotechnol 10(2):e1472. https://doi.org/10.1002/wnan.1472
4. Heydemann A, Siemionow M (2023) A brief review of duchenne muscular dystrophy treatment options, with an emphasis on two novel strategies. Biomedicine 11(3):830. https://doi.org/10.3390/biomedicines11030830
5. Sarepta Therapeutics Announces FDA Approval of ELEVIDYS, the First Gene Therapy to Treat Duchenne Muscular Dystrophy (2023) Press release. Accessed 30 May 2024
6. Hughes DC, Hardee JP, Waddell DS, Goodman CA (1985) CORP: Gene delivery into murine skeletal muscle using in vivo electroporation. J Appl Physiol 133(1):41–59. https://doi.org/10.1152/japplphysiol.00088.2022
7. Neumann E, Kakorin S, Toensing K (1999) Fundamentals of electroporative delivery of drugs and genes. Bioelectrochem Bioenerg 48(1):3–16. https://doi.org/10.1016/s0302-4598(99)00008-2
8. Coster HG (1965) A quantitative analysis of the voltage-current relationships of fixed charge membranes and the associated property of "punch-through". Biophys J 5(5):669–686. https://doi.org/10.1016/S0006-3495(65)86745-5
9. Coster HG, George EP, Simons R (1969) The electrical characteristics of fixed charge membranes: solution of the field equations. Biophys J 9(5):666–684. https://doi.org/10.1016/S0006-3495(69)86411-8
10. Golzio M, Teissie J, Rols MP (2002) Direct visualization at the single-cell level of electrically mediated gene delivery. Proc Natl Acad Sci USA 99(3):1292–1297. https://doi.org/10.1073/pnas.022646499
11. Glaser RW et al (1988) Reversible electrical breakdown of lipid bilayers: formation and evolution of pores. Biochim Biophys Acta 940(2):275–287. https://doi.org/10.1016/0005-2736(88)90202-7
12. Wolff JA, Malone RW, Williams P et al (1990) Direct gene transfer into mouse muscle in vivo. Science 247(4949 Pt 1):1465–1468. https://doi.org/10.1126/science.1690918

13. Gehl J, Sorensen TH, Nielsen K et al (1999) In vivo electroporation of skeletal muscle: threshold, efficacy and relation to electric field distribution. Biochim Biophys Acta 1428(2–3): 233–240. https://doi.org/10.1016/s0304-4165(99)00094-x
14. Mir LM, Bureau MF, Gehl J et al (1999) High-efficiency gene transfer into skeletal muscle mediated by electric pulses. Proc Natl Acad Sci USA 96(8):4262–4267. https://doi.org/10.1073/pnas.96.8.4262
15. McMahon JM, Signori E, Wells KE, Fazio VM, Wells DJ (2001) Optimisation of electrotransfer of plasmid into skeletal muscle by pretreatment with hyaluronidase – increased expression with reduced muscle damage. Gene Ther 8(16):1264–1270. https://doi.org/10.1038/sj.gt.3301522
16. Molnar MJ, Gilbert R, Lu Y et al (2004) Factors influencing the efficacy, longevity, and safety of electroporation-assisted plasmid-based gene transfer into mouse muscles. Mol Ther 10(3):447–455. https://doi.org/10.1016/j.ymthe.2004.06.642
17. Isenmann M et al (2023) Basic principles of RNA interference: nucleic acid types and in vitro intracellular delivery methods. Micromachines (Basel) 14(7). https://doi.org/10.3390/mi14071321
18. Ousterout DG et al (2015) Multiplex CRISPR/Cas9-based genome editing for correction of dystrophin mutations that cause Duchenne muscular dystrophy. Nat Commun 6:6244. https://doi.org/10.1038/ncomms7244
19. Xu L et al (2016) CRISPR-mediated genome editing restores dystrophin expression and function in mdx mice. Mol Ther 24(3): 564–569. https://doi.org/10.1038/mt.2015.192
20. Gollins H, McMahon J, Wells KE, Wells DJ (2003) High-efficiency plasmid gene transfer into dystrophic muscle. Gene Ther 10(6): 504–512. https://doi.org/10.1038/sj.gt.3301927
21. Jørgensen LH, Larochelle N, Orlopp K et al (2009) Efficient and fast functional screening of microdystrophin constructs in vivo and in vitro for therapy of duchenne muscular dystrophy. Hum Gene Ther 20(6):641–650. https://doi.org/10.1089/hum.2008.162
22. Wong SH, Lowes KN, Quigley AF et al (2005) DNA electroporation in vivo targets mature fibres in dystrophic mdx muscle. Neuromuscul Disord 15(9–10):630–641. https://doi.org/10.1016/j.nmd.2005.04.008
23. Vilquin JT, Kennel PF, Paturneau-Jouas M et al (2001) Electrotransfer of naked DNA in the skeletal muscles of animal models of muscular dystrophies. Gene Ther 8(14): 1097–1107. https://doi.org/10.1038/sj.gt.3301484
24. Sokołowska E, Błachnio-Zabielska AU (2019) A critical review of electroporation as a plasmid delivery system in mouse skeletal muscle. Int J Mol Sci 20(11):2776. https://doi.org/10.3390/ijms20112776
25. Choi E, Koo T (2021) CRISPR technologies for the treatment of Duchenne muscular dystrophy. Mol Ther 29(11):3179–3191. https://doi.org/10.1016/j.ymthe.2021.04.002
26. Brolin C, Shiraishi T, Hojman P, Krag TO, Nielsen PE, Gehl J (2015) Electroporation enhanced effect of dystrophin splice switching PNA oligomers in normal and dystrophic muscle. Mol Ther Nucleic Acids 4(12):e267. https://doi.org/10.1038/mtna.2015.41
27. Derenne A, Tassin A, Nguyen TH et al (2020) Induction of a local muscular dystrophy using electroporation in vivo: an easy tool for screening therapeutics. Sci Rep 10(1):11301. https://doi.org/10.1038/s41598-020-68135-7
28. Schertzer JD, Plant DR, Lynch GS (2006) Optimizing plasmid-based gene transfer for investigating skeletal muscle structure and function. Mol Ther 13(4):795–803. https://doi.org/10.1016/j.ymthe.2005.09.019
29. Demonbreun AR, McNally EM (2015) DNA electroporation, isolation and imaging of myofibers. J Vis Exp 106:e53551. https://doi.org/10.3791/53551
30. DiFranco M, Quinonez M, Capote J, Vergara J (2009) DNA transfection of mammalian skeletal muscles using in vivo electroporation. J Vis Exp 32:1520. https://doi.org/10.3791/1520
31. Hain BA, Waning DL (2022) Electroporation of plasmid DNA into mouse skeletal muscle. J Vis Exp 182. https://doi.org/10.3791/63916
32. Foltz SJ, Hartzell CH, Choo HJ (2023) In vivo electroporation of skeletal muscle fibers in mice. Bio Protoc 13(13):e4759. https://doi.org/10.21769/BioProtoc.4759
33. Rana ZA, Ekmark M, Gundersen K (2004) Coexpression after electroporation of plasmid mixtures into muscle in vivo. Acta Physiol Scand 181(2):233–238. https://doi.org/10.1111/j.1365-201X.2004.01282.x
34. Gallot YS et al (2019) PERK regulates skeletal muscle mass and contractile function in adult mice. FASEB J 33(2):1946–1962. https://doi.org/10.1096/fj.201800683RR

Chapter 16

Intramuscular and Intravenous AAV-Mediated Gene Delivery in Mouse Models

Liubov V. Gushchina

Abstract

Recombinant adeno-associated virus (rAAV)-mediated gene therapy is a powerful approach for precise delivery of desired eukaryotic genes to a wide range of target cells and tissues to treat various types of muscular and neuromuscular diseases, including spinal muscular atrophy, Duchenne/Becker muscular dystrophy (DMD/BMD), and limb-girdle muscular dystrophy.

Proof-of-concept, preclinical, and clinical studies in these and other fields have demonstrated high transduction efficiency of naturally occurring and engineered rAAV serotypes across multiple mammalian species, from rodents and dogs to nonhuman primates and humans, to target skeletal muscles, heart, diaphragm, and central nervous system via various routes of administration. Among them, intramuscular and intravenous injections are the most widely used routes for administering therapeutic molecules such as rAAV vector particles in dystrophic animals.

The protocols compiled in this chapter describe a simple, reliable, and reproducible method for injecting rAAV vectors using both administration techniques to deliver genes of interest into neonates and adult mice to study muscular and neuromuscular diseases and develop therapeutic options for their treatment.

Key words Neuromuscular disorder, Adeno-associated virus, Intramuscular injections, Tail vein and temporal face vein injections, Skeletal muscle, Tissue dissection

1 Introduction

The era of adeno-associated virus (AAV)-mediated gene therapy began 60 years ago with the discovery of "virus-like" particles in a sample of simian adenovirus type 15 [1]. Since then, numerous recombinant AAV (rAAV)-mediated approaches have been developed to treat monogenic disorders, some of which have already been approved by the Food and Drug Administration (FDA), while many others are in preclinical and clinical studies (ClinicalTrials.gov) [2].

In proof-of-concept and preclinical studies, the selection of animal model and route of the delivery of the material are very important elements of experimental design, which cannot be

Kevin Flanigan and Stefan Nicolau (eds.), *Duchenne Muscular Dystrophy: Methods and Protocols*, Methods in Molecular Biology, vol. 2975, https://doi.org/10.1007/978-1-0716-4811-7_16,

Table 1
Delivery methods and injection volumes

Injection method	Age	Volume
Face vein (FV)	P0-P2 (1- to 3-day-old)	Up to 50 μL
Intravenous, tail vein (IV)	P21 or older	Up to 300 μL (*see* **Note 2**)
Intramuscular (IM)	P21 or older	Up to 50 μL in each muscle

overestimated, as they allow to mimic human conditions, accurately assess safety and efficacy profiles, and estimate the dosage required for successful transduction of target cells in vivo using rAAV vectors [3–6]. Although the main advantage of rAAV-mediated gene therapy is its ability to provide a long-term therapeutic effect after a single administration, it also induces an immune response against the AAV capsid proteins or the transgene, or sometimes both, precluding any further rAAV infusions [7, 8].

Rodents are the most widely used animals for a range of experiments, including biological, medical, chemical, toxicological, pharmacological, and genetic. Among the many routes to deliver rAAV to mice, the most common are intraperitoneal, intramuscular, intracerebral, intrathecal, and intravenous injections. Each delivery method has its own advantages and limitations that should be taken into account when selecting an injection method for specific studies.

In this chapter, we describe the intramuscular and intravenous types of injection, as they are still considered the most common routes rAAV vectors administration (Table 1). We also provide the pros (for intravenous administration via the tail vein, at least two independent injections are allowed if the rAAV vector titer is too low to perform a single injection; for intramuscular administration, simultaneous injection into several skeletal muscles is also allowed) and cons (for most techniques, this is the maximum volume of rAAV vector that is allowed to be injected) for each gene delivery method and guidelines for preparing rAAV vector doses and their storing conditions in the case if injections cannot be performed on the day of dose preparation as recommended.

For additional information on tissue collection and analysis post-rAAV administration, see Chapter 8, "Preparation, Sectioning, and Histochemical Staining of Muscle Tissue."

2 Materials

For animal injections, all aqueous solutions should be sterilized using a 0.22-micron polyethersulfone (PES) or cellulose acetate membrane filters or alternative methods as mandated by local regulations.

2.1 General Materials and Reagents

1. Class II A2 Biological Safety Cabinet (BSC) or equivalent.
2. Calibrated L10, L20, L200, and L1000 manual micropipettes or equivalent electronic pipettes.
3. Sterile low-retention filter micropipette tips for manual and/or electronic pipettes.
4. Sterile DNase- and RNase-free 1.7 mL DNA LoBind microcentrifuge tubes.
5. NeedleSafe II device or equivalent.
6. 0.22 micron PES or cellulose acetate filters.
7. Sterile insulin syringe for FV injection: Exel International Syringe & Needle Combo, 0.3 mL, permanent needle, 30G × 5/16″, or equivalent.
8. Sterile insulin syringe for IM and IV injections: Exel International insulin syringe, 0.5 mL, permanent needle, 29G × 1/2″, or equivalent.
9. Isoflurane vaporizer: SomnoSuite Low-Flow Anesthesia System with integrated digital vaporizer and compressed oxygen or alternative systems.
10. Isoflurane anesthesia chamber and mouse nose cone.
11. Forane or isoflurane, liquid for inhalation.
12. Mouse restrainer.
13. Sterile 6 or 10 cm tissue culture-treated plastic dishes.
14. Sterile cotton swabs.
15. 70% ethanol or isopropanol prep pads.

2.2 Test Articles, Diluents, and Mice

1. Test article/rAAV vector containing a transgene of interest.
2. Sterile TMNP200P: 20 mM Tris-HCl, pH 8.0, 1 mM $MgCl_2$, 200 mM NaCl, 0.001% (w/w) Poloxamer 188 (Diluent).
3. Sterile saline: 0.9% sodium chloride (Diluent).
4. Sterile phosphate buffered saline (PBS): 137 mM NaCl, 2.7 mM KCl, 10 mM Na_2HPO_4, and 1.8 mM KH_2PO_4, pH 7.4 (Diluent).
5. Adult (P21 or older) and neonatal (P0-P2) Dup2 [3] or C57Bl/6 J mice. C57Bl/6 J mice can be purchased from The Jackson Laboratory (stock no. 000664).

3 Methods

All study procedures must be performed by trained personnel with the advance approval of the site's Institutional Animal Care and Use Committee (IACUC).

All described procedures can be completed at room temperature. Animals must be constantly monitored throughout the test article injection procedure and recovery period. Any abnormal observations should be documented and appropriately reported.

3.1 Test Article/rAAV Vector Preparation for Nonclinical Studies

Perform all test article preparation within a biosafety cabinet (BSC).

1. Calculate the total volumes of test article and diluent that are needed to prepare an adequate volume of working solution (*see* **Note 1**). Calculate the appropriate dose for each animal, based on body weight or number of muscles to inject (*see* **Note 2**).
2. Clean and disinfect the test article preparation area per institutional procedures, as categorized in the institutional biosafety committee (IBC) protocol (i.e., BSL-1 or BSL-2 conditions).
3. Remove the correct number of test article and diluent (TMNP200P, saline, PBS or alternative solutions) vials from the freezer (−80 °C), and allow them to thaw and reach room temperature. Remove enough extra diluent to prime each syringe/needle (*see* **Note 3**).
4. Check contents for ice crystals every 15 min, and invert gently at least three times to homogenize the vial content until the last visible ice crystals have disappeared in all vials. Do not shake the vials.
5. After all vials are thawed, perform a visual inspection of the vials for the presence of particulate matter or discoloration. The clear to slightly opaque solution should be free of visible particles.
6. While the product is reaching room temperature, disinfect the rest of the materials needed for the preparation and transfer them to the test article preparation area.
7. Disinfect the vials with a sterilizing agent and tap the vials gently on top (5×) to allow the vial content to settle down.
8. Gently remove the cap from the vial(s), making sure that aseptic conditions are maintained throughout the process.
9. Use pipettes to dispense the appropriate volumes of rAAV test article and diluent into a 1.7 mL LoBind microcentrifuge tube.
10. Use the pipette to very gently triturate the working test article five to ten times.
11. Prepare the syringes by removing the wrappers and pre-labeling them with the animal IDs.

12. Prime each syringe/needle by drawing diluent into the syringe, through the needle, until the syringe is full. Then, dispense and dispose of the diluent.
13. Fill the first syringe with the appropriate volume of diluted test article and inspect for air bubbles. Gently tap the barrel of the syringe to move all air bubbles trapped in the diluted test article toward the needle. Once the air bubbles are at the top of the syringe, gently push the plunger until the air bubbles are removed. A small amount of diluted test article ($\leq$ 1%) may squirt out during this step. Once all air bubbles are removed, inspect the amount of diluted rAAV test article remaining in the syringe, making sure that it corresponds to the correct dose.
14. Repeat the above step for each syringe (*see* **Note 4**).
15. If the prepared syringe must be transported, place a safety cap over the needle using the NeedleSafe II device.
16. Confirm that the correct animal ID is on the syringe label, and then pack the syringe in an appropriately labeled plastic bag.
17. The prepared syringe(s) may be stored in the refrigerator (4 °C) prior to transferring to the dosing room (*see* **Note 5**).
18. Transport prepared syringes to the dosing room per institutional procedures. Avoid any shaking or excessive agitation during transport. If there will be more than 30 min between transport to the dosing room and dose administration, transport the syringes on ice.
19. Test articles should be brought to room temperature prior to injection (*see* **Note 6**).

3.2 Face or Temporal Vein (FV) Injections

1. For FV injection, only newborn pups at P0-P2 (1- to 3-day--old) can be used. Two types of anesthesia are appropriate for FV injection: cryoanesthesia and isoflurane inhalation. For cryoanesthesia, place the pup on a Petri dish, and then place the Petri dish on wet ice for 30–60 s. This time is sufficient to slow the pup's movements to allow for injection. To avoid hypothermia and other complications, do not leave pups on ice for more than 60 s. Alternatively, if deeper anesthesia is required, immobilize pups using inhalants such as 1.5–2.5% isoflurane in 80–100% oxygen for 5–10 min.
2. Prepare individual insulin syringes with 50 μL of the rAAV test article working solution or diluent.
3. When the animal is anesthetized, which can be confirmed by the lack of movement while still breathing, move the animal onto the procedure platform under the light source to locate the temporal vein. To identify the temporal vein and control

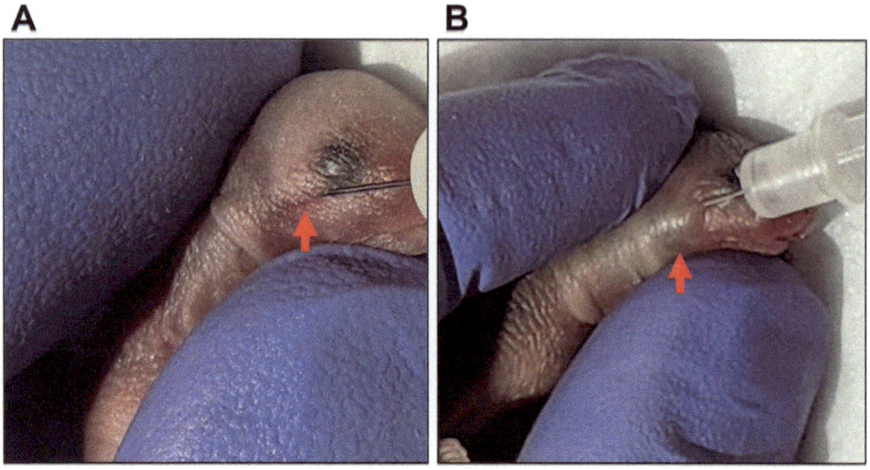

Fig. 1 Administration of rAAV test article or diluent into the temporal vein of neonatal mice. (**a**) The right temporal vein of a newborn Dup2 mouse. The pup is held between fingers for better vein visualization and localization, which appears as a thin dark vessel under the skin (red arrow). (**b**) A bolus formation at the injection site during temporal vein injection in a neonatal C57Bl/6 mouse. This indicates that the injection material did not enter the vein, and the pup should be excluded from the study

the entire injection procedure, each pup can also be placed under the surgical microscope.

4. Clean the injection site with disinfectant using 70% ethyl alcohol or 70% isopropyl prep pad, and let it dry for about 1 min.
5. For a right-handed injection, face the animal's muzzle to the right. Place the left index finger on the muzzle and the left middle finger caudal to the ear bud so that the ear bud is between the index and middle fingers.
6. Locate the superficial capillary that moves when skin is manipulated. This is NOT the target. However, identification is important for the temporal vein identification. Now locate the temporal vein, which appears dark and shadowy and runs dorsal to ventral. It feeds into the jugular vein.
7. Insert the needle into the temporal vein with the bevel facing up. If correctly inserted, it is possible to view the needle bevel fill with blood through the skin (Fig. 1a).
8. Depress the plunger slowly and note blanching of the vein down the side of the face. Do not remove the needle from the vein for 5–15 s to prevent backflow.
9. A lack of bolus formation at the injection site signals a successful injection. A bolus formation at the injection site is a clear indicator that at least some of the injection material was not successfully injected into the vein (Fig. 1b).
10. Remove the needle and use a cotton swab to gently apply pressure to the injection site until the blood clots.

11. Monitor the pup for signs of distress. Allow the pup to recover and rewarm for 2–3 min by carefully holding it in gloved hands or placing it in a cage with a heating pad underneath.
12. Make sure the pup is conscious, upright, and moving before returning it to its cage.
13. Cover the pup with bedding and/or nestlets to ensure reacceptance by the dam.
14. Monitor the pup daily for at least the next 2 days for any abnormalities. Document all observations and abnormalities.

3.3 Tail Vein (IV) Injections

1. For IV via tail vein injection, place the animal in a cylindrical restraint tube with an opening for the tail. This procedure can be done with or without anesthesia. If anesthesia is chosen, place the animal in the isoflurane inhalation chamber, close the lid tightly, and administer *isoflurane at* 1.5–2.5% in 80–100% oxygen for 5–10 min.
2. Warm the entire body of the animal by indirect application of a heating pad for about 2–5 min or placement in an incubator at 32 °C for about 30–60 min. Continue to warm the animal, until the blood vessels become dilated, that is noticeably larger. Remember, the lateral caudal tail veins are located on the sides of the tail. The artery that is in the ventral midline is not suitable for injection purposes. Rotate the restraint device, and position the tail such that the lateral tail veins are facing up and held under tension. Do not apply excessive tension or the vessel can be stretched and the blood flow diminished.
3. Prepare individual insulin syringes with 300 μL of the rAAV test article working solution or diluent (TMNP200P, saline, PBS or alternative solutions).
4. Clean the tail with disinfectant using 70% ethyl alcohol or 70% isopropyl prep pad, and let it dry for about 1 min.
5. Position the needle with the bevel up, directly above and parallel to the blood vessel, as distally as possible, and apply slight pressure to slide the needle into the tail vein.
6. Inject the material in a slow, fluid motion and notice that the vessel blanches as blood is pushed out by the injection material (Fig. 2a). If the needle is not positioned correctly in the vessel, there will be strong resistance during injection. If it is injected with force, the material will fill the subcutaneous space which may be evidenced by swelling of the tail (Fig. 2b). If this occurs, stop immediately as the material—that is designed to be injected intravenously—may be caustic to the surrounding tissues. Attempt another injection more proximally on the tail (*see* **Note** 7).

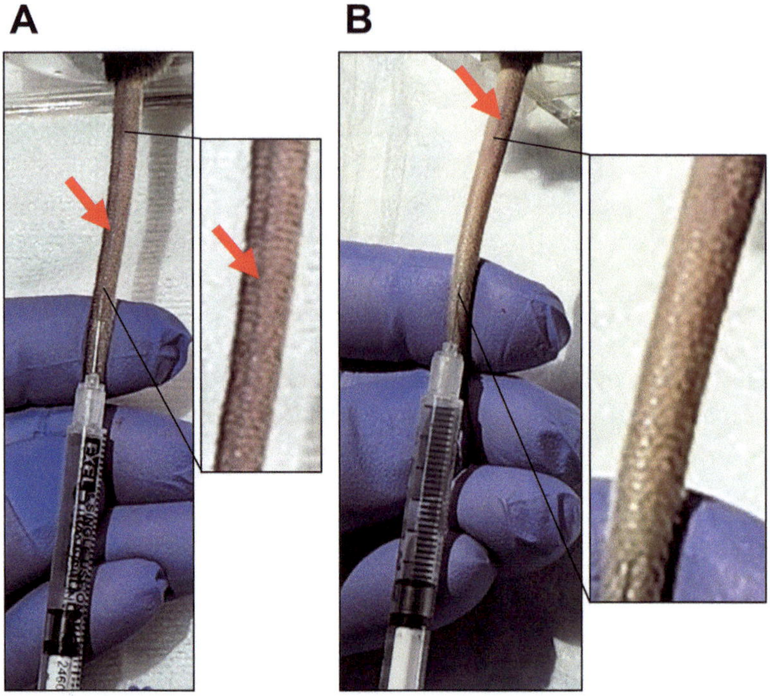

Fig. 2 Administration of rAAV test article or diluent into the lateral tail veins of mice. (**a**) The right lateral tail vein of a Dup2 mouse, which can be clearly identified as a thin dark vessel under the skin (red arrows). (**b**) The left lateral tail vein of a C57Bl/6 J mouse. The material was injected with force, resulting in subcutaneous space injection and consequent tail swelling

7. After successful injection, withdraw the needle, and apply pressure to the injection site to ensure good hemostasis before releasing the animal from the restrainer.
8. After injection, return the mouse to the cage and monitor for approximately 5–10 min. If any abnormalities are observed, document them on the animal injection form.

3.4 Intramuscular (IM) Injections into the Tibialis Anterior and Gastrocnemius Muscles

1. For IM injections into the tibialis anterior (TA) and gastrocnemius (Gast) muscles, place the animal in the isoflurane inhalation chamber, close lid tightly, and administer *isoflurane at* 1.5–2.5% in 80–100% oxygen for 5–10 min. Lack of pedal reflex (reaction to a toe pinch) can be used to confirm an adequate plane of anesthesia. If the mouse still responds to the toe pinch, return the animal to the anesthesia chamber for another 1–2 min, and try again until adequate anesthesia is reached.
2. Prepare individual insulin syringes with 50 μL of the rAAV test article working solution or diluent (TMNP200P, saline, PBS or alternative solutions) (*see* **Note 8**).

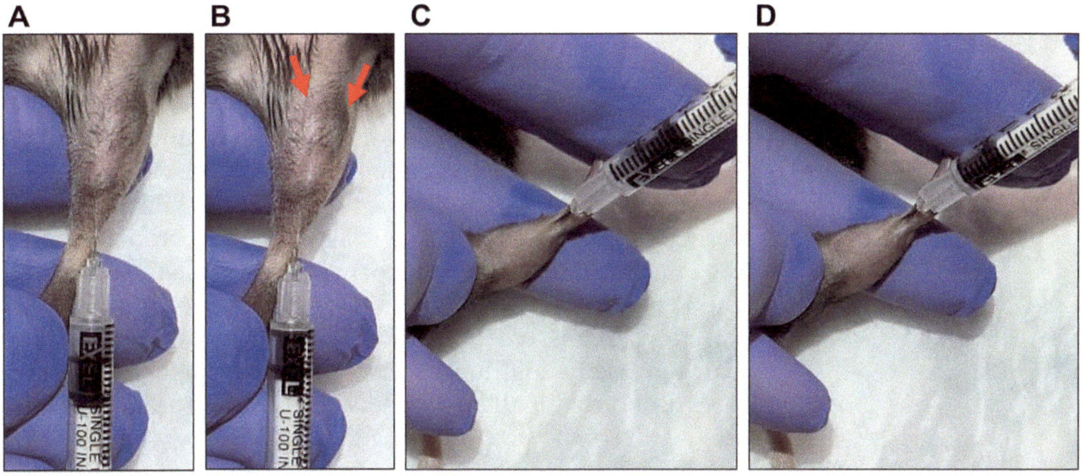

Fig. 3 Intramuscular administration of rAAV test article or diluent into the shaved left hind limbs of a Dup2 mouse. The tibialis anterior muscle (**a**) before and (**b**) after injection. If the injection is performed correctly, a small bubble (red arrows) underneath the skin over the TA will be visible. The gastrocnemius muscle (**c**) before and (**d**) after injection

3. Transfer the animal from the induction chamber and place supine on the procedure platform.
4. Place the nose mask over the mouse's muzzle to maintain the isoflurane delivery.
5. Identify the TA and Gast muscles that will be injected, and then remove fur from both TA and Gast muscles using a trimmer or hair removal cream (Fig. 3).
6. Clean the hind limbs with disinfectant using 70% ethyl alcohol or 70% isopropyl prep pad, and let it dry for about 1 min.
7. Insert the needle tip through the skin at a 10–20° angle into the desired muscle, and inject the material slowly and steadily. Do not remove the needle from the muscle for 5–10 s to prevent backflow (Fig. 3).
8. Remove the needle and gently apply pressure with a cotton swab to the injection site until the blood clots.
9. Inject other muscles by repeating **steps 6–8**.
10. If the TA muscle injection is completed correctly, a small bubble underneath the skin over the TA should be visible (Fig. 3b) (*see* **Note 9**).
11. Return the mouse to the cage and monitor for approximately 5–10 min. If any abnormalities are observed, document them on the animal injection form.

4 Notes

1. The rAAV test article working solution is prepared by diluting endotoxin-free rAAV vector stock to a final volume of 50 μL or 300 μL using one of the diluents: saline, PBS, or TMN200P vector diluent. As an injection control, use the same volume of desired diluent: saline, PBS, or TMN200P vector diluent. If the same dose needs to be prepared for several animals (for FV and IV injections) or several muscles of the same mouse (for IM injections), it is recommended to make extra 1× rAAV test article working solution to account for any pipette error. We have found that making 1× rAAV test article working solution fresh delivers the best results.
2. For face vein (FV) injections, the dose volume depends on the body weight of the newborn pups, which can range from 0.7 g to 1.5 g, with affected pups (with neuromuscular disorders) noticeable smaller than their C57Bl/6 J littermates. The dose volume is calculated using the following formula: 0.01 mL of the rAAV test article working solution or diluent per 1 g of body weight. However, if the higher volume is required, a maximum volume of 50 μL of the rAAV test article or diluent can be administered with a maximal dose of 1E15 rAAV vector genomes per kilogram (vg/kg) [5]. For intravenous (IV) injections via tail vein, it is recommended to administer a maximum volume of 300 μL (for a 30 g mouse) of the rAAV test article or diluent with a maximal dose of 1E15 rAAV vector genomes per kilogram (vg/kg). For this type of injection, the volume is also calculated based on the mouse body weight using the following formula: 0.01 mL of the rAAV test article working solution or diluent per 1 g of body weight. If the desired intravenous dose exceeds the maximum allowable injection volume, two separate injections may be given with a minimum 30 min recovery period between injections. For intramuscular (IM) injections, it is recommended to administer a maximum volume of 50 μL of the rAAV test article working solution or diluent with a maximal dose of 1E15 rAAV vector genomes per muscle.
3. If saline or PBS solutions are used as a diluent, they can be stored at room temperature.
4. Be extremely cautious not to touch the needle at any point during the preparation process. If you touch the needle by mistake, dispose of the syringe and start over with a new one.
5. If animals are not injected on the day of dose preparation, the prepared syringes can be maintained at 4 °C. However, the total time from complete thawing of test article to fully administering it should not exceed 12 h.

6. Since the recommended temperature of the injected material is in the range of 22–25 °C or higher (close to the animal's body temperature), syringes should be removed from the refrigerator at least 30 min prior to injections to allow the temperature of vector solution to rise to the recommended level.
7. It is important to start from the end of the tail (farthest from the body) and ascend toward the base of the tail (closest to the body) with each injection attempt. This allows to avoid any leakage due to tissue damage caused by the previous injection.
8. If left and right TA and Gast muscles of each mouse are to be injected, prepare four sterile insulin syringes with 50 μL of the rAAV test article working solution or diluent in each syringe for each individual muscle injection.
9. Due to different sizes of the TA and Gast muscles, a small bubble is usually visible only after TA muscle injection (Fig. 3c, d).

Acknowledgments

This work is supported by a grant from the American Neuromuscular Foundation (LVG).

References

1. Atchison RW, Casto BC, Hammon WM (1965) Adenovirus-associated defective virus particles. Science 149(3685):754–756
2. Zwi-Dantsis L et al (2025) Adeno-associated virus vectors: principles, practices, and prospects in gene therapy. Viruses 17(2)
3. Vulin A et al (2015) The first exon duplication mouse model of Duchenne muscular dystrophy: a tool for therapeutic development. Neuromuscul Disord 25(11):827–834
4. Simmons TR et al (2021) Pre-clinical dose-escalation studies establish a therapeutic range for U7snRNA-mediated DMD exon 2 skipping. Mol Ther Methods Clin Dev 21:325–340
5. Wein N et al (2022) Systemic delivery of an AAV9 exon-skipping vector significantly improves or prevents features of Duchenne muscular dystrophy in the Dup2 mouse. Mol Ther Methods Clin Dev 26:279–293
6. Gushchina LV et al (2023) Persistence of exon 2 skipping and dystrophin expression at 18 months after U7snRNA-mediated therapy in the Dup2 mouse model. Mol Ther Methods Clin Dev 31:101144
7. Mingozzi F, High KA (2013) Immune responses to AAV vectors: overcoming barriers to successful gene therapy. Blood 122(1):23–36
8. Ertl HCJ (2022) Immunogenicity and toxicity of AAV gene therapy. Front Immunol 13: 975803

Chapter 17

Assessment of Contractile and Kinetic Properties of Skeletal and Cardiac Multicellular Preparations in Mouse Models: A Comprehensive Methodological Guide

Mohammed A. Mashali, Nancy S. Saad, Neha Rastogi, and Paul M. L. Janssen

Abstract

The mechanical functions of skeletal and cardiac muscles are fundamentally defined by their contractile and passive properties, reflecting their unique physiological roles and operational dynamics. Contractile properties, which include the generation of force during muscle contraction, the duration required to reach peak force, and the subsequent relaxation period, are pivotal in assessing muscle performance, strength, injury, fatigue resistance, and overall health. These properties are crucial for understanding muscle function across various physiological and pathological conditions, including muscular dystrophies, myopathies, and during recovery post-injury or therapy. Muscular disorders, such as Duchenne muscular dystrophy (DMD), significantly impair several of these mechanical properties, causing a progressive decline in muscle strength, loss of mobility, respiratory muscle weakness, and early mortality most commonly due to respiratory failure. DMD also adversely affects cardiac muscle, leading to progressive deterioration and complications like dilated cardiomyopathy with congestive heart failure or arrhythmias, significant contributors to mortality in affected individuals.

Here we detail the methodologies and experimental techniques used for ex vivo assessment of contractile and kinetic properties in multicellular preparations from mouse models, specifically focusing on the extensor digitorum longus (EDL) muscle, the diaphragm, and cardiac trabeculae and papillary muscles. We describe protocols for measuring EDL muscle force output and twitch characteristics, tetanic contractions and eccentric contractions, and the force-frequency and length-tension relationships, as well as temperature effects. We elaborate on protocols assessing diaphragm muscle strength, including fatigue resistance evaluations. Additionally, we detail the experimental techniques used to investigate the three main mechanisms regulating cardiac contractility: the Frank-Starling mechanism, the force-frequency relationship, and ß-adrenergic stimulation through the use of isolated trabeculae and papillary muscles from mouse hearts. These techniques enable researchers and clinicians to quantify the effects of diseases, drugs, training regimens, and genetic modifications on muscle function, contributing to the development of treatments and interventions for muscle-related conditions, offering hope for improved patient outcomes in conditions like DMD.

Authors Mohammed A. Mashali and Nancy S. Saad contributed equally as first authors to this chapter.

Kevin Flanigan and Stefan Nicolau (eds.), *Duchenne Muscular Dystrophy: Methods and Protocols*, Methods in Molecular Biology, vol. 2975, https://doi.org/10.1007/978-1-0716-4811-7_17,

Key words Cardiac trabeculae and papillary muscles, Contractile force, Diaphragm, Duchenne muscular dystrophy (DMD), Extensor digitorum longus (EDL), Fatigue resistance, Force-frequency relationship, Mouse models, Skeletal muscles, Tetanic and eccentric contraction

1 Introduction

1.1 Mechanical Properties of Skeletal and Cardiac Muscles

The mechanical function of both skeletal and cardiac muscles is critically defined by their contractile and passive properties, which play a central role in the assessment of muscle performance, strength, injury recovery, fatigue resistance, and overall health [1, 2]. The contractile properties encompass the force generated during muscle contraction, the duration required to achieve this force, and the period of muscle relaxation. Critical factors influencing these properties include the interaction between actin and myosin filaments, the maintenance of calcium homeostasis, and the ATP/ADP ratio, all of which play a central role in the efficiency of muscle contraction and relaxation [3]. In the heart muscle, the regulation of contractility is mediated through three primary mechanisms: The Frank-Starling mechanism, which adjusts the force of contraction in response to changes in ventricular filling volume, serves as a fundamental principle in cardiac function [4, 5]. The Bowditch effect, or the force-frequency relationship, enhances myocardial contractility as the interval between beats shortens [6–9]. Additionally, the response to β-adrenergic stimulation enhances contractility and accelerates contraction/relaxation kinetics [6, 10], with both the force of contraction and the speed of contraction and relaxation being critical determinants of cardiac performance. The passive properties, which relate to the muscle's elasticity and viscosity when not contracting, are influenced by structural components like titin and connective tissues, mainly collagen [11–13]. Both sets of properties are vital for maintaining the structural integrity and functional efficiency of muscle tissues across various physiological and pathological states, including muscular dystrophies, myopathies, and during the phases of recovery following injury or therapeutic interventions.

1.2 Impact of Duchenne Muscular Dystrophy on Muscle Function

Muscle diseases such as Duchenne muscular dystrophy (DMD) significantly impair several mechanical properties of muscles. DMD, a severe muscle-wasting disease caused by dystrophin deficiency [14], leads to damage of the sarcolemma during muscle contraction [15]. This damage manifests as muscle cell death, loss of contractile components, reduced muscle force, and increased stiffness due to the replacement of healthy muscle fibers with fibrotic tissue [16, 17], leading to the eventual loss of mobility, loss of respiratory muscle strength, and early mortality in the early to mid-20 s due to respiratory failure [18]. In the disease's early stages, the diaphragm, the principal muscle responsible for inspiration [19], suffers more significant impairment than other skeletal muscles, likely because of its constant activation as the main facilitator of ventilation while at rest [20]. As DMD progresses, diaphragm weakness worsens with age, significantly contributing to respiratory insufficiency [21, 22]. Although DMD is primarily recognized for its profound impact on skeletal muscles, it also significantly affects cardiac function. In DMD patients, the heart muscle experiences a progressive deterioration, similar to that observed in skeletal and diaphragm muscles, primarily due to the absence of dystrophin. This condition leads to cardiomyopathy, characterized by an enlarged and weakened heart muscle that impairs the efficiency of blood pumping [23]. Notably, cardiac complications, including dilated cardiomyopathy with congestive heart failure or ventricular arrhythmias, have emerged as leading causes of mortality in DMD, accounting for 40% to 50% of patient deaths [24, 25]. Despite medical advancements that have extended the life expectancy of DMD patients, effectively managing cardiac complications is essential in their overall treatment plan.

1.3 The mdx Mouse Model for Duchenne Muscular Dystrophy Research

Given the profound impact of DMD on skeletal, diaphragm, and cardiac muscle functions, the precise measurement of changes in contractile and kinetic properties of the muscle is vital for tracking disease progression and assessing the efficacy of new treatments. In this context, the *mdx* mouse model, which features a mutation in the dystrophin gene leading to a loss of dystrophin protein expression in most skeletal myofibers [26, 27], is particularly insightful. The *mdx* mouse mirrors some of the muscle pathology found in DMD patients, making it a preferred and widely-used model for testing therapeutic approaches to this disease [28].

1.4 Eccentric Contraction Protocols and Muscle Testing

One of the most robust tests of skeletal muscle damage is the response to eccentric contraction protocols. During forced lengthening under stimulation, *mdx* muscle fibers typically exhibit greater damage and force loss in comparison to wild-type muscles (WT) [29, 30]. However, it is noteworthy that WT muscles also experience force reduction in such protocols [29, 30]. Thus, measurements of force drop during eccentric contractions provide insights into the damage process leading to muscle strength loss in DMD and the damage occurring in non-dystrophic muscles. Additionally, the general contractile properties of skeletal muscles, including twitch force, tetanic force, time to peak tension (TTP), and time to 50% twitch relaxation (RT50), alongside comparisons of force-frequency response curves, are crucial for evaluating muscle functionality. Focusing on specific muscle groups, like the extensor digitorum longus (EDL), enriches our understanding due to its unique physiological and pharmacological characteristics [31]. Located in the anterior compartment of the lower leg, the EDL muscle plays a key role in digit extension and locomotion in mice. Its distinct fiber composition, characterized by a higher proportion of fast-twitch (type II) fibers compared to slow-twitch (type I) fibers, makes the EDL muscle particularly useful for studies on muscle contraction, fatigue, regeneration, and pharmacological effects [32–34]. Fast-twitch fibers generate greater force and contract more rapidly but fatigue quickly, making them ideal for experiments on high-intensity, quick-response actions, or fatigue mechanisms, especially in conditions like DMD. In DMD, the absence of dystrophin accelerates muscle damage and weakening, with fast-twitch fibers being primarily affected. The straightforward isolation of the EDL muscle also facilitates precise experimental manipulation, providing valuable insights into muscle biology. Researchers use the EDL to study phenomena such as muscle hypertrophy, atrophy, neuromuscular diseases, and the testing of new therapeutic interventions [35].

1.5 Diaphragm and Cardiac Muscle Assessment in *mdx* Mice

Expanding the use of the *mdx* mouse model beyond skeletal muscle assessment, it is equally instrumental in evaluating diaphragm muscle strength, a critical aspect of DMD research for understanding respiratory insufficiency. The strength of the diaphragm, an essential physiological indicator, is determined by its force output in response to increased stimulation frequencies and its resistance to fatigue protocols. Therefore, analyzing the decrease in force during fatigue provides valuable insights into the mechanisms behind muscle strength deterioration in DMD and the damage experienced by non-dystrophic muscles [36]. Rodent diaphragm force assessments are typically conducted on a muscle strip isolated from the costal hemidiaphragm, the part of the diaphragm connected to the lower ribs on one side of the body, either left or right. This technique, originally developed by Ritchie for rat diaphragm studies [37], is crucial for accurately evaluating diaphragm force, thereby enhancing the utility of these models in preclinical research.

Moreover, rodent models, including the *mdx* mouse, have become an indispensable tool in cardiovascular disease research due to their anatomical similarity to humans, ease of genetic manipulation, and relative cost-effectiveness [38]. They play a pivotal role in advancing the understanding of the pathogenesis and progression of human cardiovascular diseases [39, 40]. Specifically, small, linear trabeculae or papillary muscles dissected from the ventricle are exemplary models for examining cardiac contractile and relaxation properties. These intact preparations retain their calcium-handling properties, allowing for electrical stimulation, drug treatments, and measurements of calcium transients [41]. Trabeculae and papillary muscles comprise all major cell types found in the ventricle, such as cardiomyocytes, endothelial cells, and fibroblasts [42, 43]. They provide a comprehensive platform for studying various aspects of cardiac contractile regulation, including ejection properties [43, 44], owing to their linear structure, which is ideal for mechanical studies of sarcomeric myofilaments. Thus, this model allows for detailed studies on cardiac mechanics, advancing our understanding of DMD's cardiac implications.

1.6 Overview of Experimental Techniques and Methodologies

This chapter provides an overview of the methodologies and experimental techniques employed for the ex vivo assessment of contractile and kinetic properties in multicellular preparations from mouse models. It focuses specifically on the EDL, the diaphragm, and cardiac trabeculae and papillary muscles. The chapter outlines the procedures for isolating the EDL muscle, along with the appropriate techniques for handling and conducting experimental protocols crucial for evaluating muscle damage and functionality. It also covers the protocols essential for assessing the strength of diaphragm muscles, including evaluations of fatigue resistance. Furthermore, it details the experimental techniques used to investigate the three primary mechanisms regulating cardiac contractility: the Frank-Starling mechanism, the force-frequency relationship, and ß-adrenergic stimulation, utilizing isolated trabeculae and papillary muscles from mouse hearts.

For over a decade, these techniques and protocols have been used, adapted, and developed in our laboratory and consistently applied by our research group and collaborators. This collaborative effort has yielded a substantial body of research findings, as documented in numerous publications focused on the EDL [45–52] and diaphragm [47, 49–56] muscles, as well as cardiac muscles [49, 52, 57–61] derived from mice as well as various other animal species, or humans. In addition, these protocols have been designed to be used collectively on the tissues of a single animal. This approach allows for the correlation of deficits in these distinctly different muscle groups within the same subject. In several of our past studies, we concurrently assessed the EDL, diaphragm, and cardiac muscles [49, 52], which not only provided potential correlation within an individual animal's tissues but also helped reduce the overall number of animals required for a comprehensive study.

In summary, the techniques described in this chapter serve as critical tools for researchers and clinicians to quantify the effects of diseases, drugs, training regimens, and genetic modifications on muscle function. By enabling precise measurement and analysis, these methods contribute significantly to the development of targeted treatments and interventions for muscle-related conditions and offer the potential for substantially improved patient outcomes, particularly in debilitating conditions like DMD.

2 Materials

2.1 Solutions

1. *Krebs-Henseleit (K-H) Solution*: The K-H solution is designed to closely mimic the ionic composition of the extracellular fluid. It contains 120 mm NaCl, 5 mm KCl, 1.2 mm NaH_2PO_4, 2 mm $MgSO_4$, 20 mm $NaHCO_3$, 10 mm D-glucose (dextrose), and 0.25 mm $CaCl_2$. To prepare 1 L of K-H solution, start by adding approximately 800 mL of distilled water into a 1 L graduated cylinder. Next, add the appropriate amounts of each component to the cylinder to achieve the final desired molar concentrations listed above. Then, top up the solution with distilled water to reach a final volume of 1 L. Mix the solution thoroughly with a stir bar until all the components are completely dissolved. After mixing, adjust the pH of the solution to approximately 7.4. Transfer the prepared solution to a glass container and label it with the date of preparation. It is recommended to use the solution on the same day for best results. Prior to and during experimental use, gently bubble the solution with a gas mixture of 95% O_2 and 5% CO_2 to ensure proper oxygenation (*see* **Note 1**).
2. *Modified K-H Solution (Dissection Buffer)*: This variant of the K-H solution includes the addition of 20 mm 2,3-butanedione monoxime (BDM) to adapt it for use in dissection procedures (*see* **Note 2**). To prepare 1 L of Modified K-H Solution, start by preparing the K-H solution as per the above instructions, and then add BDM to achieve a final concentration of 20 mm. Stir the solution with a stir bar to ensure the BDM is fully dissolved. Cover the cylinder, for example, with parafilm, and continue mixing for approximately 5 min. It is advisable to prepare this modified solution fresh on the day of the experiment for optimal effectiveness.
3. *Isoproterenol Stock Solution A (10 mm)*: Weigh out 0.260 g of Isoproterenol (molecular weight (MW): 260 g/moL) and 0.176 g of ascorbic acid (MW: 176 g/moL). Begin by adding distilled water to a 100 mL graduated cylinder until it is almost full. Then, add the weighed Isoproterenol and ascorbic acid into the cylinder. Stir the solution until both components are completely dissolved. Finally, adjust the volume to the 100 mL mark with distilled water and stir thoroughly to ensure a homogenous mixture. Aliquot the solution into 2 mL centrifuge tubes and store at −20 °C (*see* **Note 3**).
4. *Isoproterenol Stock Solution B (1 mm)*: Add 100 μL of isoproterenol stock solution A into a 2 mL centrifuge tube containing 900 μL of distilled water, effectively creating a 1 mm solution (*see* **Note 3**).

5. *Isoproterenol Stock Solution C (10 μM)*: Add 10 μL of isoproterenol stock solution B to a 2 mL centrifuge tube containing 990 μL of distilled water to achieve a 10 μM concentration (*see* **Note 3**).

2.2 Drugs and Chemicals

1. Heparin Sodium Injection, USP—10,000 USP Units/mL, 1 mL single-dose vial (*see* **Note 4**).
2. Calcium Chloride Solution, 1.0 M $CaCl_2$, ≥98%—Fluka™ volumetric solution.

2.3 Experimental Equipment and Software

1. Circulating water bath (e.g., Isotemp 3016H, or similar; *see* **Note 5**).
2. Digital variable-speed pump (perfusion pump) (*see* **Note 6**).
3. O_2/CO_2 gas tank (95% O_2, 5% CO_2) (*see* **Note 7**).
4. Jacketed beaker (*see* **Note 8**).
5. Muscle bath (*see* **Note 9**).
6. Electrical stimulation electrodes (*see* **Note 10**).
7. Force transducers (range 0–2 N and 0–5 mN, e.g., SI-KG2 and SI-KG7, World Precision Instruments, or similar; *see* **Note 11**).
8. Micro displacement device (*see* **Note 12**).
9. Stereo microscope.
10. Optical microscope illuminator light source.
11. Pulse generator (e.g., A310 Accupulser; World Precision Instruments, or similar; *see* **Note 13**).
12. Stimulus isolator (e.g., A365; World Precision Instruments, or similar; *see* **Note 14**).
13. Optical force transducer amplifier (e.g., SI-BAM21-LC; World Precision Instruments, or similar; *see* **Note 15**).
14. High-speed linear servo-controlled motor (e.g., MOT1; Scientific Instruments, or similar; *see* **Note 16**).
15. Data acquisition card (e.g., NI USB-6009 DAQ; National Instruments, or similar; *see* **Note 17**).
16. Data acquisition PC and screen.
17. Waste beaker.
18. LabVIEW Software (National Instruments; *see* **Note 18**).

2.4 Additional Equipment

1. Scissors and forceps.
2. Scalpel.
3. pH meter.
4. Liquid nitrogen tank.

5. 6-0 braided silk sutures.
6. Dissection Board.

3 Methods

3.1 Experimental Procedures for Assessing the Contractile and Kinetic Properties of the Mouse Extensor Digitorum Longus Muscle

3.1.1 Dissection of the EDL Muscle in Mice

1. Begin by administering ten units of heparin to the experimental mouse via intraperitoneal injection (*see* **Note 4**). Wait approximately 5 min after the heparin injection, and then proceed to euthanize the mouse using cervical dislocation.
2. Position the mouse in a supine posture on the dissection board. Secure the mouse to the board using four dressmaker pins: two pins to stabilize the front legs and two to secure the hind legs, ensuring the mouse is appropriately fixed for the dissection procedure.
3. Using scissors and forceps, carefully peel off the skin from the leg to expose the underlying hind limb muscles. Ensure to continuously superfuse all exposed muscles with oxygenated modified K-H solution (K-H containing BDM).
4. Using forceps, excise the fascia surrounding the lower leg muscles with utmost care to ensure minimal damage to the muscle tissue.
5. Expose the tibialis anterior (TA) muscle's tendon and the extensor ligament under a stereomicroscope by dissecting the skin toward the foot. Cut the extensor ligament to release the TA tendon.
6. Cut the TA tendon and use it to peel off the TA muscle (*see* Fig. 1a). Carefully remove the TA muscle at its proximal attachment from both legs. Throughout this process, continuously superfuse the EDL with oxygenated modified K-H solution (*see* **Note 19**).
7. Using forceps, thread 6-0 braided silk suture under the lower tendons of the EDL (*see* **Note 20**).
8. Carefully slide the suture upward until it is under the top of the EDL.
9. At the juncture where tendon meets muscle, tie a standard knot, and tighten securely.
10. Continue by tying two square knots (*see* Fig. 1b), making sure to tie around the tendon both times (*see* **Note 21**).
11. After tying the two square knots, create a braid used to mount the suture to the setup by tying five normal knots (*see* **Note 22**).

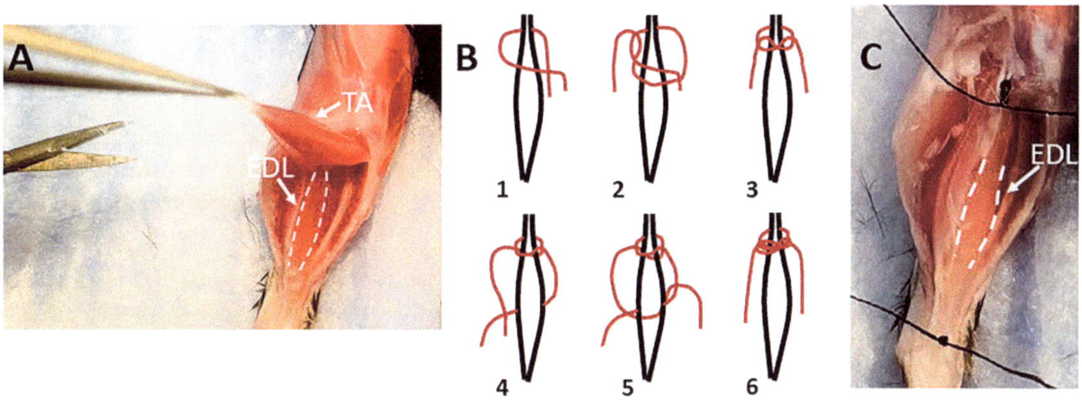

Fig. 1 Illustration of the dissection and suturing techniques for the EDL muscle. (**a**) Cutting the TA tendon to peel off the muscle, thereby exposing the underlying EDL muscle. (**b**) Steps of securing the EDL muscle tendon with a double square knot, followed by a loop knot for stabilization. (**c**) Shows two sets of braids, each consisting of five standard knots (circles), tied at the junction where the tendon meets the EDL muscle. These braids are prepared for mounting the suture within the experimental setup. *EDL* extensor digitorum longus muscle, *TA* tibialis anterior muscle

12. Pass another suture under the muscle's lower tendons, and complete the tying process as described in **steps 9–11** (*see* Fig. 1c).
13. Remove the EDL muscle, being sure to cut as much tendons as possible along with the muscle. Place in a weighing dish with oxygenated modified K-H solution.
14. Repeat the suture threading and knot-tying process for the EDL muscle of the opposite leg, and then place it into the oxygenated modified K-H solution.
15. Prior to placing the muscles in the experimental setup, trim the sutures leaving approximately ¼ inch remaining.

3.1.2 Finding Optimal Length and Stabilization of the EDL Muscle

In the EDL muscle experimental setup, the jacketed beaker is set to 30 °C. It uses two parallel platinum electrical stimulation electrodes. The force transducer, with a range of 0–2 N, is connected to a pointed mounting hook. The setup also includes a high-speed linear servo-controlled motor, which is connected to a pointed mounting hook, and a micro displacement device attached to the servo-controlled motor.

In the experimental procedures, a standard stimulation frequency of 150 Hz is the frequency to be applied, except when particular protocols specify the use of alternative frequencies. The force and kinetics of the EDL muscle are recorded and analyzed using LabVIEW Software (*see* **Note 18**). Key parameters such as active developed twitch tension, TTP, and RT50 are measured from the single twitch contraction at the optimal length of each individual EDL muscle. The optimal length is defined as the length at which the twitch developed force is maximal. Stabilization of the EDL muscles at their optimal length is crucial to ensure accurate experimental outcomes.

1. Start the circulating water bath and set the temperature to ensure that the muscle bath maintains a constant 30 °C. Allow adequate time for the temperature to stabilize before starting the experiment.
2. Mix 87.5 μL of 1.0 M $CaCl_2$ with 50 mL of K-H solution. Transfer this mixture into the jacketed beaker and start oxygenation with a gas mixture of 95% O_2 and 5% CO_2 (*see* **Note 23**).
3. Turn on the perfusion pump to fill the muscle bath with the warm, oxygenated K-H solution. Once the bath is filled, switch off the pump to create a stable environment conducive to muscle transfer.
4. Ensure that the electrical stimulator is in the "off" position and configured with the appropriate settings, typically including "bipolar" and "single" options for this type of procedure. For the force transducer amplifier, adjust to the sensitivity level and voltage amplitude suitable for your setup; typically, a voltage greater than ~20% of the threshold voltage is sufficient to stimulate the entire muscle.
5. With great care, transfer the muscles into the experimental setup. Handle them delicately by the suture to minimize any disturbance.
6. Set the micro displacement device, so that the distance between the hooks of the force transducer and the servo-controlled motor is approximately 8 mm (*see* **Note 24**).
7. Using the dissection microscope and two pairs of forceps, pierce the sutures (between the braid create from 5 normal knots) over the needles of both hooks.
8. Adjust the height of the muscle so that it is parallel to the base of the bath (*see* **Note 25**).
9. Reactivate the pump, setting it to maintain a consistent flow rate that ensures uniform perfusion and adequate refreshing of the muscle bath solution (*see* **Note 26**).
10. Using the micro displacement device, adjust the muscle's length to a relaxed state.
11. Launch LabVIEW Software to start recording EDL muscle data (*see* **Note 18**).
12. Activate the stimulator and gently stretch the muscle until it is no longer slack, and then run the program to stimulate the muscle twitch contraction (evoked by a single stimulation

pulse). Record the first contraction after this adjustment, labeling it, for example, as "T1."

13. Following the initial contraction, incrementally stretch the muscle using small, consistent adjustments appropriate to your micro displacement device. Allow a brief interval, typically around 20 s, before each new stimulation to ensure muscle recovery. Continue this process, recording each subsequent contraction sequentially, and label the responses in an organized manner for analysis (e.g., "T2" for the second contraction, "T3" for the third, and so on).
14. If the total contraction size (from baseline to peak) has increased since the last measurement, continue to incrementally adjust the muscle length, as described above, until no further increase in force response is observed, indicating that the muscle has reached its optimal length for maximal force generation.
15. Allow the muscle to stabilize for 10 min before starting the protocols.
16. During the stabilization period, measure the muscle dimensions using a calibration reticule in the ocular of the dissection microscope, as practiced in the instant protocol, or employ a calibrated measurement tool that is appropriate for your experimental setup (*see* **Note 27**). Record the length (from suture to suture) and the width (at the widest portion of the muscle) in mm (*see* **Note 28**).
17. Record the measured length, which is necessary to adjust length changes needed during the eccentric contraction experiments.
18. Proceed to one of the protocols outlined below.

3.1.3 EDL Muscle Tetanic and Eccentric Contraction Protocol

This protocol assesses the muscle's ability to generate force when stimulated rapidly (tetanic contractions) and its resistance to stretch under load (eccentric contractions). It measures the muscle's maximal force production and its structural integrity under conditions that simulate muscle overstretching during activity, which can provide insights into muscle health, functionality, and potential damage mechanisms. In this protocol, a tetanus contraction is performed by applying a 250 ms tetanus at 150 Hz, with each pulse lasting 1 ms. Following a 5-min rest period, muscles are subjected to six eccentric contractions. These are characterized by a 450 ms period of stimulation at 150 Hz, with a 3% muscle stretch applied during the final 200 ms of contraction. Intervals of 5 min (300 s) separate the initial five contractions, with a longer 15-min interval before the sixth, allowing for recovery.

1. Start by stabilizing the muscle at its optimal length, as described above (*see* Subheading 3.1.2).
2. After the 10-min stabilization period, initiate a tetanic contraction by setting the "total time" on the program interface to 250 ms and the stimulus delay to 100 ms. Ensure the amplifier sensitivity remains set at 1. Label and save the resulting contraction, for example, as "tet."
3. Allow a 5-min rest period post-tetanic contraction to avoid muscle fatigue.
4. Initiate the eccentric contraction phase by adjusting the total time to 450 ms and setting the motor amplitude to 3%. This adjustment triggers the motor to automatically stretch the muscle by 3% of its optimal length during the last 200 ms of the contraction phase. Program the repeat counter for five runs with 300-s intervals between each. Label and save the initial run, for example, as "ecc."
5. The program will automatically stimulate the muscle at 5-min intervals, executing a total of 20 min of eccentric contractions. It will save each contraction sequentially as "ecc2," "ecc3," "ecc4," and "ecc5."
6. After the completion of the fifth contraction, allow a 15-min rest period for the muscle's recovery.
7. Conduct a final stimulation and save the result as "ecc6."
8. Turn off the pump and gas supply. Carefully remove the muscle from the setup.
9. Detach the suture from the muscle. To measure the muscle weight accurately, wrap the muscle in a Kimwipe, and briefly place a 10 g weight atop for 10 s to absorb excess water (*see* **Note 29**). This weight will be integral in subsequent analyses (*see* **Note 30**). Post-weighing, the tissue can be discarded.
10. Repeat the entire protocol with the other muscle to ensure comparative data analysis.

3.1.4 EDL Muscle Length-Tension Protocol

This protocol investigates how varying muscle lengths of the EDL affect its twitch contraction force production capabilities: a fundamental aspect of muscle physiology known as the length-tension relationship. This protocol involves systematically stretching the muscle from a slackened position to its optimal length using twitch contractions. At each length, the muscle is allowed to stabilize before its length is measured, and the resulting contraction force is recorded.

1. Follow **steps 1–11** as outlined in "Finding Optimal Length and Stabilization of the EDL Muscle" (*See* Subheading 3.1.2) to prepare the muscle for length measurements.

2. Activate the stimulator and gently stretch the muscle until it is no longer slack. Use a calibration reticule in the ocular of the dissection microscope to measure the muscle length. Enter this measurement into the data acquisition system and then run the program. Record the resulting contraction, labeling it for example, as "L1."
3. Incrementally stretch the muscle using the micro displacement device, applying small and consistent increases in length. After each incremental adjustment, allow a brief period, typically around 20 s, for the muscle to stabilize. Measure the new muscle length, enter it into the data acquisition system, and stimulate the muscle. Record the contraction force at each new length, continuing with the established labeling sequence such as "L2," "L3," and so forth.
4. Continue the process of stretching, measuring, and stimulating the muscle until the force response no longer increases, indicating that the optimal muscle length has been reached.
5. Before exiting the program, measure and record the width of the muscle.
6. If no further protocols are to be run, follow **steps 8–10** as described in "EDL Muscle Tetanic and Eccentric Contraction Protocol" (*See* Subheading 3.1.3) for the proper shutdown of the experiment.

3.1.5 EDL Muscle Force-Frequency Protocol

This protocol evaluates the muscle's response to a range of stimulation frequencies, creating a force-frequency response curve to determine its impact on muscle twitch force. The curve is generated by progressively increasing the stimulation frequency to induce isometric contractions over a broad spectrum from 1 Hz up to 180 Hz.

1. Start by stabilizing the muscle at its optimal length, as described above (*see* Subheading 3.1.2).
2. After the 10-min stabilization period, begin a series of muscle stimulations at 14 different frequencies: 1, 3, 5, 10, 15, 20, 30, 40, 50, 60, 75, 100, 150, and 180 Hz. After stimulating at each frequency and recording the contraction force, allow a 90-s recovery interval before proceeding to the next frequency.
3. Label and save the recorded contractions sequentially from the lowest frequency to the highest (e.g., "F1" for 1 Hz, "F2" for 3 Hz, ..., "F14" for 180 Hz), ensuring consistent documentation for analysis.
4. In adapting this protocol, adjust settings such as stimulation duration and delay according to the capabilities and requirements of your specific equipment. The objective is to maintain

consistency in stimulus application and force recording across all frequencies.

5. Upon completing all force-frequency response curve, if there are no additional protocols to be run, proceed with **steps 8–10** from the "EDL Muscle Tetanic and Eccentric Contraction Protocol" (*See* Subheading 3.1.3) for the proper shutdown of the experiment.

3.1.6 EDL Muscle Temperature Response Protocol

This protocol evaluates the effect of temperature changes on twitch force in two phases. In the first phase, temperature is systematically decreased from the physiological baseline of 30 °C down to 22.5 °C. Twitch force measurements are recorded at steady-state conditions at intervals of 2.5 °C during this cooling phase. The temperature adjustments are made gradually, at less than 0.5 °C per min, to ensure steady-state conditions are maintained. In the second phase, the temperature is gradually increased from 22.5 °C back to 30 °C and then further to an elevated temperature of 37.5 °C. The purpose of the second phase is to validate the reproducibility of twitch force measurements by ensuring that steady-state conditions are achieved consistently as temperature returns to and exceeds the physiological baseline. This validation step is crucial to confirm that the muscle's response to temperature variation is reliable and repeatable in both the cooling and warming phases of the experiment.

1. Start by stabilizing the muscle at its optimal length, as described above (*see* Subheading 3.1.2).
2. After the 10-min stabilization period, insert the tip of a wired thermometer into the muscle bath and record the initial temperature, which should be around 30 °C. At this baseline temperature, stimulate the muscle and record the force generated by the muscle contractions.
3. Then, decrease the temperature of the circulating water bath by increments of 2.5 °C. After each decrement, as indicated by the thermometer in the muscle bath, stimulate the muscle and record the force at the new temperature.
4. Continue this process, stimulating the muscle and recording the force after each 2.5 °C temperature drop in the muscle bath, until it reaches 22.5 °C.
5. Once the minimum temperature is reached, gradually increase the temperature of the circulating water bath to raise the muscle bath temperature in increments of 2.5 °C. At each incremental temperature rise as measured in the muscle bath, stimulate the muscle and record its response. Continue this process until the muscle bath temperature returns to and surpasses the initial baseline, reaching 37.5 °C.

6. If there are no additional protocols to be run, proceed with **steps 8–10** from the "EDL Muscle Tetanic and Eccentric Contraction Protocol" (*See* Subheading 3.1.3) for the proper shutdown of the experiment.

3.1.7 Cleaning Procedure of the Experimental Setup

The cleaning process for the experimental setup at the end of the experiment is critical for maintaining equipment functionality and ensuring accurate results in future experiments.

1. Use a large syringe to carefully remove the K-H solution from the jacketed beaker.
2. Add approximately 50 mL of distilled water to the beaker to begin the rinsing process.
3. Disconnect the drainage tubing from the jacketed beaker, and place it into an alternative waste collection container.
4. Allow the distilled water to flow completely through the set-up's tubing to ensure that all internal components are adequately rinsed.
5. Once all the water has passed through the system, proceed to turn off the pump and all related equipment (*see* **Note 31**).

3.2 Force Output and Fatigue Resistance as Key Indicators of Diaphragm Muscle Health in Mice

3.2.1 Diaphragm Muscle Preparation

1. Begin the experiment by administering an intraperitoneal injection of ten units of heparin to the mouse (*see* **Note 4**). Allow a 5-min waiting period to confirm effective anticoagulation before proceeding to euthanize the mouse through cervical dislocation.
2. Place the euthanized mouse on its back on a dissection board, carefully remove the skin from the torso, and perform a bilateral thoracotomy to open the chest cavity, and then quickly excise the heart.
3. Make complete cuts through the body just below where the heart was located and another below the xiphoid process (cutting through the liver and above the intestines) (*see* **Note 32**).
4. Immerse the body section containing the diaphragm in a small beaker (~50 mL) filled with oxygenated modified K-H solution (K-H containing BDM) to cleanse the tissue of any hair or blood.
5. Transfer the section to a small dissection plate filled with the same oxygenated modified K-H solution to trim the tissue.
6. Carefully remove the liver, lungs, and surrounding tissues from atop and below the diaphragm. Trim the rib tissue from around the section, leaving a thin margin (~3 mm) of rib tissue around the diaphragm.
7. Pin this section in a dissection plate, inserting pins into the peripheral rib tissue as illustrated in Fig. 2a (*see* **Note 33**).

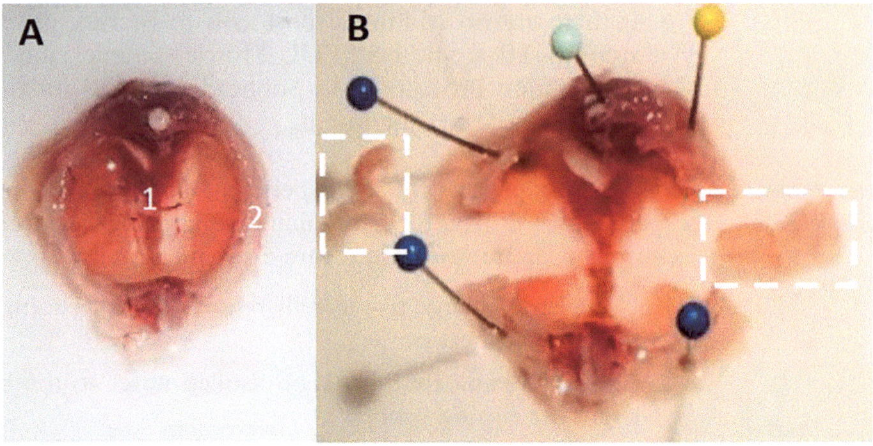

Fig. 2 Mouse diaphragm muscle anatomy and preparation. (**a**) A representative image of a fully dissected mouse diaphragm muscle, pinned in a dissection plate. This image highlights the central connective tissue tendon (1) along a thin rib tissue margin around the diaphragm (2), illustrating the diaphragm's anatomical features. (**b**) The white dotted rectangles highlight the muscle strips, approximately 2.5–3.5 mm in width, which have been cut for force measurement

8. Using a scalpel, cut two thin strips of diaphragm tissue (~2.5–3.5 mm wide each) by rolling the blade from the central white connective tissue across the diaphragm and into the rib tissue (*see* **Note 34**).
9. Hold the rib end of each muscle strip with forceps, and carefully trim the opposite end, ensuring sufficient white tissue remains for mounting the strip in the experimental setup as shown in Fig. 2b.

3.2.2 Finding Optimal Length and Stabilization of the Diaphragm Muscle

In the diaphragm muscle experimental setup, the jacketed beaker is set to 37 °C instead of 30 °C. It also uses two parallel platinum electrical stimulation electrodes. The force transducer (range ~ 0–2-N) is connected to a platinum-iridium basket-shaped mounting device instead of a pointed mounting hook, while the micro displacement device remains connected to a pointed mounting hook.

The experimental protocols start with the electrical stimulation of the diaphragm muscle using single electrical pulses delivered through two parallel platinum-iridium electrodes positioned around the muscle to determine the optimal length. The optimal length is defined as the length at which the maximum twitch force is recorded. Ensuring the diaphragm muscles are stabilized at this optimal length is crucial for the accuracy of the experimental results.

Force measurements of the diaphragm muscle are recorded and analyzed using LabVIEW Software (*see* **Note 18**). Force measurements are normalized per unit cross-sectional area (CSA) to express isometric force or tension (mN/mm^2). CSA is calculated using the following equation: muscle mass in mg/[(optimal length in mm) × (muscle density in mg/mm^3)], where muscle density is taken to be 1.056 mg/mm^3.

1. Start by warming up the circulating water bath, allowing sufficient time for the temperature to stabilize before starting the experiment.
2. Measure 50 mL of K-H solution using a graduated cylinder. Transfer this solution to the jacketed beaker and initiate oxygenation with a gas mixture of 95% O_2 and 5% CO_2 (*see* **Note 23**).
3. Turn on the perfusion pump to fill the muscle bath with the warmed, oxygenated K-H solution. Once the bath is filled, switch off the pump to ensure a stable environment for transferring the muscle.
4. Ensure that the stimulator is in the "off" position and configured with the appropriate settings, typically including "bipolar" and "single" options for this type of procedure.
5. Transfer the diaphragm strip to the muscle bath carefully by holding the rib tissue with forceps.
6. Set the micro displacement device so that the distance between the basket and the hook within the bath is approximately 2 mm (*see* **Note 35**).
7. Under the dissection microscope, use two pairs of forceps to feed the strip's white end (the central tendon on the interior end of the diaphragm strip) through the basket connected to the force transducer, piercing the white tissue over the hook needle at the bath's opposite end. (*see* **Note 36**).
8. Reactivate the pump, setting it to maintain a consistent flow rate that ensures uniform perfusion and adequate refreshing of the muscle bath solution (*see* **Note 26**).
9. Using the micro displacement device, adjust the muscle's length to a relaxed state.
10. Initiate data acquisition by starting the software you are using for the experiment. Configure the settings to suit the experimental parameters needed for accurate muscle stimulation and force recording.
11. Activate the stimulator and then run the program to stimulate the muscle twitch contraction via a single pulse.
12. Prior to muscle stretching, utilize the zero-adjustment feature on your force transducer amplifier to calibrate the baseline force to zero (*see* **Note 37**).
13. Gradually and incrementally stretch the muscle using the micro displacement device, applying consistent adjustments to its length. After each adjustment, allow a 25-s period for the

muscle to stabilize before stimulation and recording the force of contraction. If the measured contraction size shows an increase from the previous measurement, continue with further incremental stretches until no additional increase in force response is observed, indicating that the muscle has reached its optimal length for maximal force generation capability.

14. Allow the muscle to stabilize for 10 min before starting the protocol.
15. During the stabilization period, measure the muscle strip's dimensions using a calibration reticule in the ocular of the dissection microscope, as practiced in the instant protocol, or employ a calibrated measurement tool that is appropriate for your experimental setup (*see* **Note 27**). Record the length by measuring the distance from the basket to the white connective tissue and the width at the middle of the muscle in mm (*see* **Note 28**).

3.2.3 Diaphragm Muscle Tetanic Contractions and Fatigue Protocol

The purpose of this protocol is to evaluate the effects of increasing stimulation frequencies on muscle twitch force production during rapid stimulation (tetanic contractions) and to assess the muscle's resistance to fatigue, evidenced by its capacity to maintain contraction under load (eccentric contractions).

Initially, the muscle is subjected to a series of six tetanic contractions at ascending frequencies of 20, 50, 80, 120, 150, and 180 Hz, each lasting 250 ms, with a 2-min interval between consecutive stimulations. Subsequently, the protocol transitions to evaluate the muscle's resistance to fatigue through an eccentric contraction regimen. Starting 5 min after the final tetanic contraction, this phase involves stimuli at a constant frequency of 100 Hz, each pulse lasting 250 ms, delivered consecutively every second over a span of 65 s. Following a rest period of 15 min, the fatigue assessment is repeated to gauge the muscle's response consistency.

1. Start by stabilizing the muscle at its optimal length, as previously described (*see* Subheading 3.2.2).
2. Following the 10-min stabilization period, turn the stimulator off and prepare the system for the tetanic contractions by setting the "duration" to 250 ms on the program interface.
3. Start the tetanic contraction series by setting the stimulator to an initial frequency of 20 Hz. Activate the stimulator to deliver a pulse lasting 250 ms, and record the resulting muscle contraction, labeling the output for example, as "20 Hz," and then allow the muscle to rest for a period of 2 min before proceeding to the next frequency.
4. Proceed with the sequence, increasing the frequency to 50, 80, 120, 150, and finally 180 Hz. After each 250 ms pulse, ensure a

rest period of 2 min and record each contraction, labeling sequentially (e.g., "50 Hz," "80 Hz," and so on).

5. After the last contraction at 180 Hz, allow the muscle a 5-min recovery phase before initiating the fatigue protocol.
6. For the fatigue protocol, set the "duration" to 250 ms and the "frequency" to 100 Hz on the program interface, timing the session to last 65 s.
7. Following a 5-min rest after the tetanic series, begin the fatigue protocol: administer consecutive pulses, each lasting 250 ms, delivered every second for a total duration of 65 s, and document the muscle's response.
8. Conclude with a 15-min rest period, then repeat the fatigue protocol.
9. Turn off both the pump and gas supply and carefully remove the muscle from the setup by lifting the white tissue from the hook and pulling the strip back through the basket.
10. Cut the chunk of rib tissue from the strip. Accurately measure muscle weight by wrapping the tissue in a Kimwipe, placing a 10 g weight on top for 10 s to remove excess water (*see* **Note 29**). This weight will be crucial for subsequent analysis (*see* **Note 30**). Post-weighing, the tissue can be discarded.
11. Repeat the entire protocol with the second strip of diaphragm tissue from the same mouse.

3.2.4 Cleaning Procedure of the Experimental Setup

For detailed instructions on the cleaning procedure, please refer to the protocol provided above (*see* Subheading 3.1.7).

3.3 Cardiac Muscle Contractile and Kinetic Properties Assessment in Mouse Models: Specialized Protocols for Trabeculae and Papillary Multicellular Preparations

3.3.1 Cardiac Muscle Preparation

1. Begin the experiment by administering an intraperitoneal injection of ten units heparin to the mouse (*see* **Note 4**). This is followed by a waiting period of approximately 5 min to ensure effective anticoagulation. Subsequently, euthanize the mouse using the cervical dislocation method.
2. Place the euthanized mouse on its back on a dissection board, securing it with four dressmaker pins: two for the forelimbs and two for the hind limbs. Perform a bilateral thoracotomy to open the chest cavity, and then rapidly excise the heart, transferring it immediately into a small dissection plate filled with oxygenated modified K-H solution (K-H containing BDM) (*see* **Note 38**).
3. Cannulate the excised heart through the ascending aorta, followed by retrograde perfusion with the same modified K-H solution for several min (*see* **Note 39**).
4. From the right ventricle, delicately dissect thin, uniform, non-branched trabeculae (*see* **Note 40**) or small papillary-like muscles (three muscles typically extend from the septal wall, near

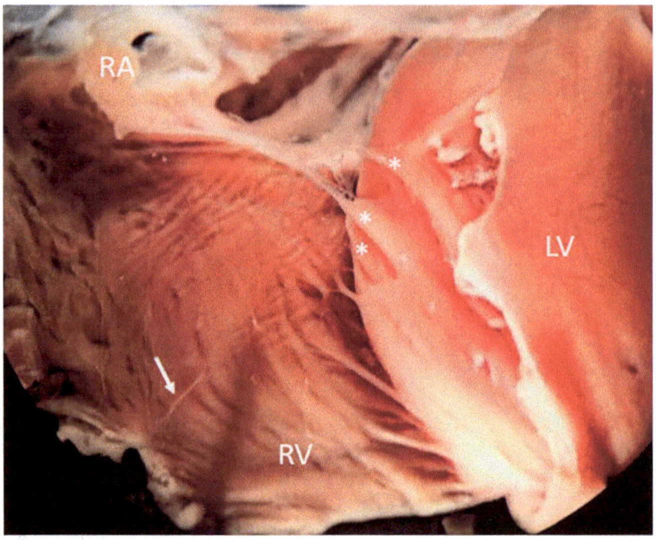

Fig. 3 Mouse trabeculae and papillary muscles. This image highlights the mouse heart's interior, showing trabeculae muscle (arrow) along the RV wall and three papillary muscles (asterisks) connected to the tricuspid valve. *RA* right atrium, *RV* right ventricle, *LV* left ventricle

the right ventricular free wall, to the tricuspid valve) (*see* Fig. 3). During dissection, ensure to leave a small block of tissue at one end, attached to the right ventricular free wall or septum, and a portion of the valve tissue at the other end to facilitate the effective mounting.

5. Papillary muscles from the left ventricle are not selected for use in these experiments, as their dimensions typically exceed the critical cutoff size of approximately 200 μm (*see* **Note 41**).

3.3.2 Finding Optimal Length and Stabilization of the Cardiac Muscle

In the cardiac muscle experimental setup (*see* Fig. 4), the jacketed beaker is set to 37 °C, similar to the diaphragm setup. It uses a single electrical stimulation electrode. The force transducer, instead of the 0–2 N transducers used in the EDL and diaphragm setups, has a range of 0–5 mN and is connected to a pointed mounting hook. Additionally, the micro displacement device is connected to the single electrical stimulation electrode and a pointed mounting hook.

The experimental protocols begin by electrically stimulating the trabeculae or papillary muscles at a frequency of 4 Hz, allowing them to stabilize at their optimal length before initiating any experimental protocol. The optimal length is defined as the length at which a small increase in length results in nearly equal increases in resting tension and active developed tension. This length is selected to be comparable to the maximum length attained in vivo at the end of diastole (around 2.2 μm sarcomere length). Stabilizing the

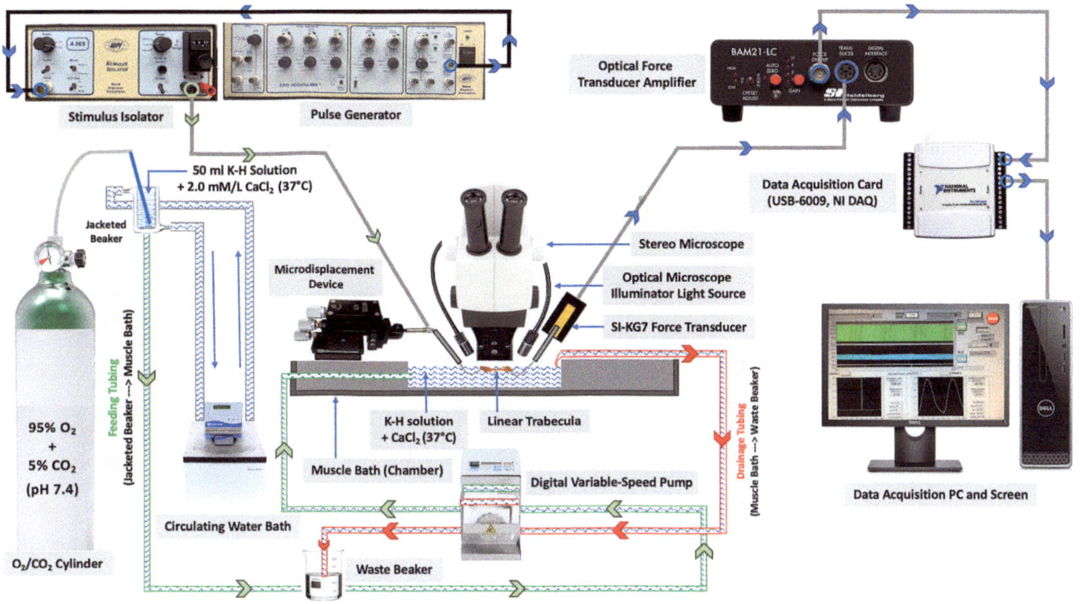

Fig. 4 Schematic diagram of the experimental custom-designed setup for cardiac muscles. This diagram presents the components of the custom-designed setup for cardiac muscle experiments. In this setup, muscles are mounted between two pointed hook-like extensions; one is attached to a sensitive force transducer, and the other is connected to a micro displacement device for adjusting muscle length. The muscles are bathed in a continuous flow of BDM-free Krebs-Henseleit (K-H) solution with 2 mM $CaCl_2$, maintained at 37 °C. The solution is continuously oxygenated with a gas mixture of 95% O_2 and 5% CO_2, ensuring a stable pH of 7.4

cardiac muscles at their optimal length is essential for ensuring accurate experimental outcomes.

1. Begin by warming up the circulating water bath, allowing enough time for the temperature to stabilize before starting the experiment.
2. Accurately measure 50 mL of K-H solution using a graduated cylinder. Transfer this solution to the jacketed beaker and initiate oxygenation with a gas mixture of 95% O_2 and 5% CO_2 (*see* **Note 23**).
3. Set the stimulation frequency to 4 Hz (*see* **Note 42**). Make sure the stimulation is turned off before mounting the muscle.
4. Launch LabVIEW Software to start recording cardiac muscle data (*see* **Note 18**).
5. Turn on the perfusion pump to fill the muscle bath with the warm, oxygenated K-H solution. Once filled, turn off the pump to stabilize the environment for muscle transfer.
6. Under the dissection microscope, carefully transfer the muscle into the muscle bath without directly handling the muscle tissue. Connect one end of the muscle chunk to the force

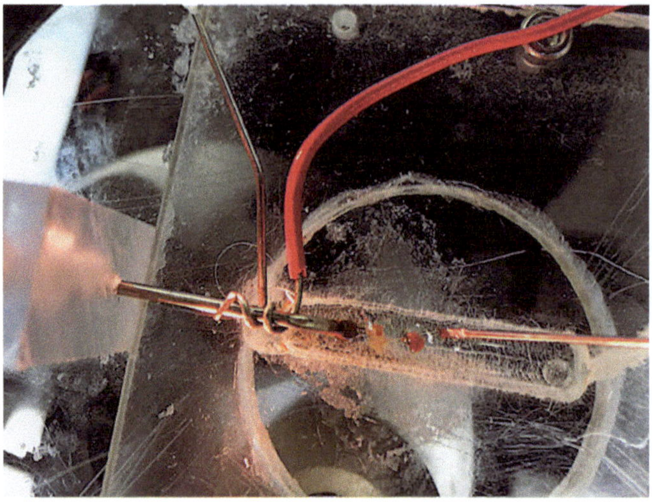

Fig. 5 Mounting papillary muscle. A papillary muscle positioned within the muscle bath. The muscle's one end is connected to a force transducer on the right, while its valve end is connected to the stimulator on the left while it is continuously perfused with Krebs-Henseleit (K-H) solution

transducer hook and the other end, including the valve tissue, to the stimulator-connected hook (*see* Fig. 5 and *see* **Note 43**).

7. Reactivate the pump, setting it to maintain a consistent flow rate that ensures uniform perfusion and adequate refreshing of the muscle bath solution (*see* **Note 26**).
8. Using the micro displacement device, adjust the muscle's length to a relaxed state.
9. Activate the stimulator and add 87.5 μL of 1.0 M $CaCl_2$ to the K-H solution in the jacketed beaker, adjusting the extracellular calcium concentration to 2 mM.
10. Allow the muscle to stabilize for at least 30 min to achieve a stable state.
11. Utilize the zero-adjustment feature on your force transducer amplifier to calibrate the baseline force to zero prior to initiating muscle stimulation (*see* **Note 37**).
12. Carefully stretch the muscle using the micro displacement device or a similar instrument, applying small and consistent increases in length. Allow the muscle sufficient time to adapt after each adjustment before proceeding with further incremental stretches.
13. Continue the incremental stretching until you observe proportional increases in both resting tension and active developed tension with each small increase in length. This indicates the optimal muscle length (L100). Record the setting on the displacement device at this point, as it will be important for

the length-tension protocol. This recorded measurement will be used to precisely adjust the muscle length to specified percentages as required by the protocol.

14. Measure the muscle dimensions using a calibration reticule in the ocular of the dissection microscope, as practiced in the instant protocol, or employ a calibrated measurement tool that is appropriate for your experimental setup. Record the muscle length by measuring the distance between the two hooks and the width at the muscle's midpoint in mm (*see* **Note 27**). In the current protocol, where the muscle is assumed to have an ellipsoidal shape, thickness is estimated by multiplying the muscle width by two and dividing by three. Apply the standard or experiment-specific formula that is most appropriate for the muscle shape you are studying to accurately calculate the CSA (*see* **Note 44**).
15. Allow 15–20 min for the muscle to stabilize at this length. Only proceed with further protocols if both resting and developed tensions are stable.
16. Proceed to one of the protocols outlined below.

3.3.3 Cardiac Muscle Length-Tension Protocol

This protocol assesses the impact of changing muscle length on force production. Initially, the muscle's slack length (L85)—the length without preload—and its optimal length (L100) are determined. The process involves slackening the muscle from L100 to L85, followed by dividing the difference between these lengths into three equal increments. The muscle is then gradually stretched through these increments until it returns to the baseline length (L100). Force production and kinetic parameters are recorded after the muscle stabilizes at each specified length.

1. Start by stabilizing the muscle at its optimal length, as described above (*see* Subheading 3.3.2).
2. Once the muscle is at L100 (optimal length), initiate the data recording using your chosen data acquisition software to capture the force and kinetics generated by the muscle at this length. The muscle is stimulated at a constant frequency of 4 Hz for the duration of the length-tension protocol.
3. Adjusting to L85: Calculate 15% of the muscle length. Use the micro displacement device to decrease the muscle length by this value. For example, if L100 is 2 mm with a starting reading of 400 μm on the micro displacement device, to achieve L85 (85% of L100), the length should be adjusted to 1.7 mm, which is a reduction of 0.3 mm (300 μm) from L100. Using the micro displacement device, this adjustment translates to a change in the device's reading from 400 μm to 100 μm. Allow the muscle to stabilize for 1 min then record.

4. The muscle is gradually stretched back to L100 through three increments, L90, L95, and L100, each involving a 5% increase in muscle length. Following the provided example, this increment corresponds to an 0.1 mm (or 100 μm) increase per step.
5. Adjusting to L90: Based on the example, increase the muscle length by 5%, corresponding to a 100 μm increase, and adjust the micro displacement device to 200 μm. Allow the muscle to stabilize for 1 min before recording.
6. Adjusting to L95: Apply another 5% increase, adjusting the muscle length by an additional 100 μm, and set the micro displacement device to 300 μm. Stabilize for 1 min before recording.
7. Adjusting back to L100: Increase the muscle length by a final 5%, returning the setting to 400 μm on the micro displacement device. After 1 min of stabilization, record the response.
8. Upon completion of this protocol, proceed with the "Cardiac Muscle Freezing Protocol" (*see* Subheading 3.3.6) if no further protocols are to be run. Otherwise, proceed to the next protocol.

3.3.4 Cardiac Muscle Force-Frequency Protocol

This protocol assesses the effect of increasing stimulation frequencies from 4 to 14 Hz, covering the entire in vivo range of the mouse heart rate. At each frequency, forces are allowed to reach a steady state before data are recorded.

1. Start by stabilizing the muscle at its optimal length, as described above (*see* Subheading 3.3.2). If this protocol follows the "Cardiac Muscle Length-Tension Protocol" (*see* Subheading 3.3.3), proceed to increase the frequency of stimulation accordingly.
2. Adjust the stimulator frequency to 6 Hz and wait for 3 min or longer, as needed, to achieve a stable response. Once stabilization is reached, initiate data recording using your chosen data acquisition software to capture the force and kinetics generated by the muscle at this frequency.
3. Sequentially increase the stimulation frequency, first to 8 Hz and then to 10 Hz, allowing a 3-min stabilization period at each frequency. After stabilization at each increment, proceed to record the force and kinetics generated by the muscle.
4. Increase the stimulation to 12 Hz and stabilize for 2 min then record. If you notice a decline in developed force prior to the completion of 2 min, record immediately without delay, and then proceed to the next step.
5. Increase the stimulation to 14 Hz and allow 1 min stabilization then record. If you notice a decline in developed force before

the end of 1 min, record immediately, and then proceed to the next step.

6. Upon completion of this protocol, proceed with the "Cardiac Muscle Freezing Protocol" (*see* Subheading 3.3.6) if no further protocols are to be run. Otherwise, reset the stimulator to its initial setting of 4 Hz to proceed to the next protocol.

3.3.5 Cardiac Muscle Isoproterenol Response Protocol

This protocol evaluates the response to ß-adrenergic activation by establishing a concentration-response curve for isoproterenol. The cardiac muscle is exposed to increasing concentrations of isoproterenol in semilogarithmic increments, ranging from 10^{-9} mol/L (1 nM) to 10^{-6} mol/L (1 μM), at optimal length and a baseline stimulation frequency of 4 Hz.

1. Following the completion of the previous protocol, allow the muscle to stabilize for 15 min. During this stabilization period, begin thawing the isoproterenol stock solution A and freshly prepare stock solutions B and C.
2. Once the stabilization period has ended, start recording the force and kinetics produced by the muscle, using your chosen data acquisition system, prior to the application of any isoproterenol.
3. Begin with isoproterenol stock solution C, pipetting 5 μL (corresponding to a concentration of 10^{-9} mol/L or 1 nM) into the K-H solution in the jacketed beaker. After mixing gently using the pipette tip, wait for the muscle to fully stabilize for at least 3 min before recording the muscle's response.
4. Continue with isoproterenol stock solution C, incrementally adding 10 μL for 3×10^{-9} mol/L (3 nM), followed by 35 μL for 10^{-8} mol/L (10 nM), and then 100 μL for 3×10^{-8} mol/L (30 nM), allowing 3 min for the muscle to fully stabilize before recording the response and proceeding to the next concentration.
5. Switch to isoproterenol stock solution B for higher concentrations, adding 3.5 μL for 10^{-7} mol/L (100 nM), followed by 10 μL for 3×10^{-7} mol/L (300 nM), and finally 35 μL for 10^{-6} mol/L (1 μM). Similar to the previous steps, after each dose, allow the muscle to stabilize completely for at least 3 min before recording its response and proceeding to the next dose.
6. Upon completion of this protocol, proceed with the "Cardiac Muscle Freezing Protocol" (*see* Subheading 3.3.6).

3.3.6 Cardiac Muscle Freezing Protocol

1. After completing all experimental protocols, begin preparations for freezing of the muscle without turning off any equipment. Collect the necessary tools, including micro scissors, tweezers, and a small tube accurately labeled with the mouse

number and specific protocol details. Pre-cool these items by placing them in liquid nitrogen.

2. Switch off the perfusion pump while keeping the stimulation active. Carefully remove as much liquid as possible from the muscle bath. Then, apply liquid nitrogen directly onto the muscle. Using the pre-cooled tweezers, quickly cut and transfer the muscle into the previously prepared and labeled tube, and then immediately immerse it back into the liquid nitrogen to ensure rapid and efficient freezing.
3. Stop both the stimulation and the data acquisition program to allow the setup to gradually return to room temperature. Safely dispose any residual muscle tissue.

3.3.7 Cleaning Procedure of the Experimental Setup

For detailed instructions on the cleaning procedure, please refer to the protocol provided above (*see* Subheading 3.1.7).

3.3.8 Adaptation of Cardiac Contractility Research Protocols Across Species

While this section focuses on the use of isolated trabeculae and papillary muscles from mice for assessing cardiac contractile and kinetic properties, it is important to note that the fundamental principles and experimental protocols can be readily adapted for application in other species, including humans, canines, rabbits, and rats as previously documented [38, 62–67].

To accommodate the specific physiological and anatomical characteristics of each species, adjustments primarily involve modifying the stimulation frequency to align with the species-specific cardiac physiology and more closely reflect the natural heart rate. For instance, the baseline stimulation frequency used for mouse models, which have a naturally higher heart rate, may need to be reduced when adapting these protocols for species with slower heart rates, such as humans or larger mammals. Specifically, when adapting the protocols for human subjects, the stimulation frequencies are generally set to a range between 0.5 and 3 Hz [38, 62–64]. In studies involving canines, the frequencies used are 1, 2, 3, 4, and 5 Hz [65]. Rabbit experiments typically utilize frequencies of 1, 2, 3, and 4 Hz [66]. For rat models, a wider frequency spectrum from 0.1 to 12 Hz is used to account for their varied heart rate conditions [67]. Such adaptations allow for a broader application of these methodologies, facilitating comparative studies across species and enhancing understanding of cardiac function and dysfunction in a variety of mammalian models.

4 Notes

1. This step is crucial for oxygenating the solution and maintaining the correct pH balance, which should be around 7.4. The bubbling should be performed gently to ensure adequate gas

exchange without causing excessive turbulence in the solution. If necessary, particularly due to the altitude of the laboratory which can affect pH levels, the amount of $NaHCO_3$ may need to be adjusted to achieve the correct pH.

2. The addition of BDM prevents cutting injury during dissection by helping to relax muscle tissue.
3. Thaw stock A and prepare stocks B and C fresh prior to running the experiment.
4. Heparin is an anticoagulant administered intraperitoneally to prevent blood clotting during initial muscle preparations. This preparation is critical for smooth dissection and extraction, particularly in protocols involving delicate muscle preparations such as excising the EDL muscle, performing thoracotomies for diaphragm extraction, and obtaining cardiac tissue samples. Ensuring anticoagulation simplifies the procedures by minimizing blood clot formation that could obstruct tissue separation or compromise the muscle sample quality.
5. The circulating water bath maintains a controlled temperature environment essential for preserving the physiological conditions of the muscle tissue.
6. The digital variable-speed pump (perfusion pump; e.g., Ismatec, or similar) regulates the flow rate of the K-H solution through the muscle bath, which can be adjusted to match physiological conditions or experimental needs. This pump provides precise flow control, vital for maintaining a stable environment and ensuring uniform oxygenation and waste removal in the muscle bath.
7. The O_2/CO_2 gas tank (95% O_2, 5% CO_2) supplies a gas mixture for continuous oxygenation of the K-H solution, which is essential for preserving cellular function and pH balance. The gas mixture is delivered through a finely controlled regulator to ensure consistent bubbling within the solution.
8. The jacketed beaker holds the K-H solution and is equipped to consistently warm the solution to the required temperature (30 °C or 37 °C, depending on the muscle type). For EDL muscle experiments, the beaker is set to 30 °C, while for cardiac and diaphragm muscle experiments, it is set to 37 °C to maintain the K-H solution at their respective physiological temperatures. An oxygenation rod inside the beaker continuously bubbles the solution with a gas mixture of 95% O_2 and 5% CO_2, maintaining an optimal pH of 7.4 for muscle function.
9. The muscle bath is a chamber that holds the muscle. It is superfused with warm, oxygenated K-H solution circulated using a perfusion pump. It is custom-designed to allow easy access to the muscle, precise positioning, and quick

adjustments during experiments. The pump's flow rate is typically adjusted to equal 5 times the bath's volume per min to ensure efficient oxygenation and removal of metabolic waste, crucial for maintaining the muscle's functional state.

10. The electrical stimulation electrodes provide electrical stimulation to the muscle, ensuring consistent and precise activation of muscle fibers. The EDL and diaphragm setups use two parallel platinum electrodes positioned longitudinally 2–3 mm from the muscle, designed for synchronous stimulation. The cardiac setup, however, uses a single electrode that allows for targeted, localized stimulation suited to cardiac tissue characteristics.
11. The force transducer measures the force generated by the muscle with high accuracy. In the EDL and diaphragm setups, the transducer has a range of ~0–2 N, suitable for detecting higher force outputs typical of skeletal muscle contractions. In the cardiac setup, a more sensitive transducer with a range of ~0–5 mN is used to accurately measure the lower force levels characteristic of cardiac muscle contractions.
12. The micro displacement device enables precise, incremental adjustments of muscle position and length, either directly or through fine control of attached components such as the servo-controlled motor. This device is crucial for ensuring accurate alignment and fine-tuning of muscle length during experiments, allowing for consistent force measurements and minimizing positional errors that could affect experimental outcomes.
13. The pulse generator generates controlled electrical pulses to stimulate the muscle during experiments, with adjustable pulse frequency, duration, and amplitude to meet specific experimental requirements.
14. The stimulus isolator ensures precise delivery of electrical pulses to the muscle, eliminating interference from external noise or variability and providing reliable control over pulse parameters for accurate stimulation protocols.
15. The optical force transducer amplifier amplifies signals from the force transducer for precise measurement of muscle contraction force, with adjustable sensitivity to capture minute force changes accurately.
16. The servo-controlled motor provides precise real-time control over muscle length and tension, essential for experiments involving the EDL muscle where fine length regulation is necessary to assess contractile properties and muscle responses under varying tension.
17. The data acquisition card digitally records experimental data for analysis, interfacing with the data acquisition PC to capture

high-resolution data from the force transducer and other sensors, enabling real-time monitoring and post-experiment analysis of muscle responses.

18. LabVIEW Software is utilized for precise data acquisition and analysis in the EDL, diaphragm, and cardiac muscle protocols. Equivalent software with similar data recording and analysis capabilities can be substituted based on availability and experimental setup compatibility.
19. Throughout the dissection, ensure that the muscles are consistently moistened with oxygenated modified K-H solution to prevent desiccation and maintain tissue health. All steps should be performed with precision and care to ensure the structural integrity of the muscle is maintained.
20. Make sure to include all EDL tendons—there is a bundle of 3 above the suture.
21. The two square knots ensure that the suture is firmly attached to the muscle-tendon junction. This stability is vital for experiments involving muscle contraction, stretching, or any mechanical testing, as it allows for accurate transfer of force without slippage.
22. This braid is essential for effectively mounting the muscle in the experimental setup. The series of knots creates a robust and stable structure which can be easily manipulated and attached to the setup.
23. Cover the beaker with parafilm to prevent any solution from splashing during the oxygenation process and to minimize evaporation effectively.
24. This distance ensures that the EDL muscle is mounted without being overstretched to avoid any damage to the muscle fibers.
25. Aligning the muscle parallel to the bath base aims to mimic the muscle's natural orientation as it would be in the body. Muscles in living organisms typically align with the direction of force they exert or resist, and this natural alignment is crucial for their proper functioning.
26. It is imperative to maintain the pump's speed consistently throughout the entire experiment to ensure uniform perfusion speed. Consistent flow is vital for maintaining a constant temperature, ensuring efficient oxygenation, and facilitating the removal of metabolic waste, all of which are necessary to preserve the muscle's functional state. The pump's flow rate is carefully adjusted to equal approximately five times the bath's volume per min, a setting that contributes to the stability of experimental conditions. Variations in flow rate could significantly impact the experimental results by altering the physiological environment of the muscle tissue.

27. It is important to adjust any measured dimension by dividing it by the magnification factor indicated on the microscope dial to accurately determine the muscle's actual dimensions.
28. The length and width measurements of the muscle are integral for calculating the muscle's cross-sectional area (CSA), a key step in the data analysis process. This calculation facilitates the normalization of force measurements, allowing for a comparison of muscle strengths that is independent of size differences.
29. The procedure employs a standardized technique using a Kimwipe and a specified, known standard weight to uniformly remove excess water from all muscle samples. This consistent application of pressure is crucial, as it guarantees that each sample undergoes the same water removal process. By applying a uniform method, the data obtained from different samples can be accurately compared, ensuring that any weight measurements reflect the true muscle mass, rather than discrepancies due to varying water content, which is a significant consideration in biological tissues.
30. Muscle weight is a key variable that needs to be controlled or accounted for normalizing data obtained during the experiment.
31. To maintain optimal functioning of the experimental setup, it is recommended to perform a thorough cleaning of the tubing and muscle bath once a month by circulating a solution composed of one part vinegar to two parts distilled water through the system. After the vinegar solution has been run through, flush the system several times with water to ensure all traces of vinegar are removed. Additionally, every 2–3 months, replace all tubing and tubing connectors to prevent deterioration and potential contamination. This routine maintenance is essential for preventing the accumulation of deposits and ensuring the long-term functionality of the setup.
32. It is important to avoid touching the diaphragm tissue to prevent any damage or alteration to its structure, which is essential for the accuracy of subsequent experiments.
33. Apply a slight stretch when pinning the diaphragm to ensure it lays flat, thereby enhancing the precision and consistency of cutting diaphragm strips required for the experiment.
34. This width has been identified as optimal for diaphragm muscle studies, as documented in previous research [53].
35. This will ensure not stretching the diaphragm strip during the mounting as it could damage the muscle tissue, affecting its ability to contract and produce reliable data.
36. Ensure the diaphragm strip is horizontally aligned and untwisted in the bath to allow for even electrical stimulation

across the entire muscle. A properly aligned and untwisted strip ensures that the force measured is directly due to muscle contractions and not influenced by bending artifacts.

37. Adjusting the baseline force reading to zero before starting muscle stimulation is a calibration step. It ensures that any force measured during the experiment is due to muscle contraction in response to stimulation, not any pre-existing tension or artifact in the setup. This calibration helps in obtaining accurate and consistent measurements of the muscle's contractile properties.
38. During the excision of the heart, carefully hold the aorta with forceps, and avoid direct contact with the heart tissue itself to minimize mechanical damage.
39. The primary purpose of this perfusion technique, conducted over a period of several min, is to comprehensively remove all blood from the heart's vascular system. This thorough cleansing is crucial for enhancing the clarity of the tissue, thereby aiding in the accurate identification of anatomical structures and improving the precision of the dissection technique.
40. During dissection, ensure that all trabeculae muscles are derived from a small, well-defined region of the ventricle. This targeted approach significantly minimizes variations that could arise due to differences in location within the ventricle, thereby enhancing the consistency and reliability of the experimental data.
41. When muscle samples exceed 200 μm in size, core hypoxia tends to develop under the experimental baseline conditions of the protocol, which includes a temperature of 37 °C and stimulation at 4 Hz [68]. Consequently, we chose muscles from the right ventricle rather than the left, primarily because of their generally smaller size. This careful consideration of muscle size is crucial for ensuring the consistency and reliability of the experimental results, avoiding complications like core hypoxia.
42. The baseline of 4 Hz is selected instead of the more physiological 12 Hz because isometric contractions at 12 Hz induce super-physiological stress on the muscle and demand an energetic output that likely surpasses the muscle's capacity to generate ATP. While isometric contractions at 12 Hz can only be maintained for 1–2 min with minimal loss of force, contractions at 4 Hz can be sustained for several hours. However, to study contractions under more physiological conditions, 12 Hz contractions are assessed, but only for brief periods.
43. This mounting method minimizes handling of the muscle tissue by attaching only the ends of the muscle chunk to the force transducer and stimulator-connected hooks. By avoiding

direct contact with the muscle, this approach significantly reduces damage at the attachment points, preventing excessive force loss and ensuring consistent and reliable results during the experimental protocols.

44. This step is critical as it allows for the normalization of force measurements. By normalizing these measurements against the muscle's cross-sectional area (CSA), it becomes possible to compare muscle strengths in a manner that is independent of size variations, ensuring a more accurate and standardized assessment of muscle functionality.

References

1. Herzog W (2000) Mechanical properties and performance in skeletal muscles. In: Biomechanics in sport: performance enhancement and injury prevention, pp 21–32
2. Andersson DC, Betzenhauser MJ, Marks AR (2012) Chapter 12 – excitation–contraction coupling in the heart. In: Hill JA, Olson EN (eds) Muscle. Academic Press, Boston/Waltham, pp 153–159
3. Kuo IY, Ehrlich BE (2015) Signaling in muscle contraction. Cold Spring Harb Perspect Biol 7(2):a006023
4. Ter Keurs H, Rijnsburger WH, Van Heuningen R et al (1980) Tension development and sarcomere length in rat cardiac trabeculae. Evidence of length-dependent activation. Circ Res 46(5):703–714
5. Starling EH (1918) The Linacre lecture on the law of the heart. Longmans, Green
6. Bowditch H (1871) Über die Eigentümlichkeiten der Reizbarkeit, welche die Muskelfasern des Herzens zeigen. Ber. K. Sächs Ges Wiss Math-phys Kl:652
7. Janssen PM, Periasamy M (2007) Determinants of frequency-dependent contraction and relaxation of mammalian myocardium. J Mol Cell Cardiol 43(5):523–531
8. Endoh M (2004) Force–frequency relationship in intact mammalian ventricular myocardium: physiological and pathophysiological relevance. Eur J Pharmacol 500(1–3):73–86
9. Janssen PM (2010) Myocardial contraction-relaxation coupling. Am J Physiol Heart Circ Physiol 299(6):H1741–H1749
10. Stull LB, Leppo MK, Marbán E et al (2002) Physiological determinants of contractile force generation and calcium handling in mouse myocardium. J Mol Cell Cardiol 34(10): 1367–1376
11. Huijing PA (1999) Muscle as a collagen fiber reinforced composite: a review of force transmission in muscle and whole limb. J Biomech 32(4):329–345
12. Moss R, Halpern W (1977) Elastic and viscous properties of resting frog skeletal muscle. Biophys J 17(3):213–228
13. Maruyama K, Kimura S, Kuroda M et al (1977) Connectin, an elastic protein of muscle: its abundance in cardiac myofibrils. J Biochem 82(2):347–350
14. Hoffman EP, Brown RH Jr, Kunkel LM (1987) Dystrophin: the protein product of the Duchenne muscular dystrophy locus. Cell 51(6):919–928
15. Petrof BJ, Shrager JB, Stedman HH et al (1993) Dystrophin protects the sarcolemma from stresses developed during muscle contraction. Proc Natl Acad Sci USA 90(8): 3710–3714
16. Pastoret C, Sebille A (1995) Mdx mice show progressive weakness and muscle deterioration with age. J Neurol Sci 129(2):97–105
17. Hakim CH, Grange RW, Duan D (2011) The passive mechanical properties of the extensor digitorum longus muscle are compromised in 2-to 20-mo-old mdx mice. J Appl Physiol 110(6):1656–1663
18. Finder JD, Birnkrant D, Carl J et al (2004) Respiratory care of the patient with Duchenne muscular dystrophy: ATS consensus statement. Am J Respir Crit Care Med 170(4):456–465
19. Ratnovsky A, Elad D, Halpern P (2008) Mechanics of respiratory muscles. Respir Physiol Neurobiol 163(1–3):82–89
20. Mead AF, Petrov M, Malik AS et al (2014) Diaphragm remodeling and compensatory respiratory mechanics in a canine model of Duchenne muscular dystrophy. J Appl Physiol 116(7):807–815
21. Pennati F, Arrigoni F, LoMauro A et al (2020) Diaphragm involvement in Duchenne

muscular dystrophy (DMD): an MRI study. J Magn Reson Imaging 51(2):461–471
22. Berlowitz DJ, Wadsworth B, Ross J (2016) Respiratory problems and management in people with spinal cord injury. Breathe 12(4): 328–340
23. Melacini P, Fanin M, Danieli G et al (1996) Myocardial involvement is very frequent among patients affected with subclinical Becker's muscular dystrophy. Circulation 94(12): 3168–3175
24. Eagle M, Baudouin SV, Chandler C et al (2002) Survival in Duchenne muscular dystrophy: improvements in life expectancy since 1967 and the impact of home nocturnal ventilation. Neuromuscul Disord 12(10):926–929
25. Muntoni F (2003) Cardiomyopathy in muscular dystrophies. Curr Opin Neurol 16(5): 577–583
26. De La Porte S, Morin S, Koenig J (1999) Characteristics of skeletal muscle in mdx mutant mice. Int Rev Cytol 191:99–148
27. Sicinski P, Geng Y, Ryder-Cook AS et al (1989) The molecular basis of muscular dystrophy in the mdx mouse: a point mutation. Science 244(4912):1578–1580
28. Blake DJ, Weir A, Newey SE et al (2002) Function and genetics of dystrophin and dystrophin-related proteins in muscle. Physiol Rev 82(2):291–329
29. Head SI, Williams DA, Stephenson DG (1992) Abnormalities in structure and function of limb skeletal muscle fibres of dystrophic mdx mice. Proc R Soc Lond B Biol Sci 248(1322): 163–169
30. Sacco P, Jones D, Dick J et al (1992) Contractile properties and susceptibility to exercise-induced damage of normal and mdx mouse tibialis anterior muscle. Clin Sci 82(2): 227–236
31. Tyml K, Budreau C (1991) A new preparation of rat extensor digitorum longus muscle for intravital investigation of the microcirculation. Int J Microcirc Clin Exp 10(4):335–343
32. Louboutin JP, Fichter-Gagnepain V, Pastoret C et al (1995) Morphological and functional study of extensor digitorum longus muscle regeneration after iterative crush lesions in mdx mouse. Neuromuscul Disord 5(6): 489–500
33. Warren GL, Hayes DA, Lowe DA et al (1994) Eccentric contraction-induced injury in normal and hindlimb-suspended mouse soleus and EDL muscles. J Appl Physiol 77(3): 1421–1430
34. Moran AL, Warren GL, Lowe DA (2005) Soleus and EDL muscle contractility across the lifespan of female C57BL/6 mice. Exp Gerontol 40(12):966–975
35. Kotake H, Ogura Y, Yamada S et al (2023) Mechanism for exercise-mediated prevention against muscle wasting on extensor digitorum longus muscle in spontaneously diabetic Torii fatty rats. J Physiol Sci 73(1):5
36. Kiriaev L, Baumann CW, Lindsay A (2023) Eccentric contraction-induced strength loss in dystrophin-deficient muscle: preparations, protocols, and mechanisms. J Gen Physiol 155(2): e202213208
37. Ritchie J (1954) The relation between force and velocity of shortening in rat muscle. J Physiol 123(3):633
38. Mashali MA, Saad NS, Canan BD et al (2021) Impact of etiology on force and kinetics of left ventricular end-stage failing human myocardium. J Mol Cell Cardiol 156:7–19
39. Camacho P, Fan H, Liu Z et al (2016) Small mammalian animal models of heart disease. Am J Cardiovasc Dis 6(3):70
40. Hasenfuss G (1998) Animal models of human cardiovascular disease, heart failure and hypertrophy. Cardiovasc Res 39(1):60–76
41. Varian KD, Xu Y, Torres CA et al (2009) A random cycle length approach for assessment of myocardial contraction in isolated rabbit myocardium. Am J Physiol Heart Circ Physiol 297(5):H1940–H1948
42. Bers DM (1997) Ca2+ sparks. Jumping the gap from the cell to cardiac muscle. Circ Res 81(4): 636–638
43. Janssen P, Lehnart SE, Prestle J et al (1998) The trabecula culture system: a novel technique to study contractile parameters over a multiday time period. Am J Physiol Heart Circ Physiol 274(5):H1481–H1488
44. De Tombe PP, Little WC (1994) Inotropic effects of ejection are myocardial properties. Am J Physiol Heart Circ Physiol 266(3): H1202–H1213
45. Lowe J, Kadakia FK, Zins JG et al (2018) Mineralocorticoid receptor antagonists in muscular dystrophy mice during aging and exercise. J Neuromuscul Dis 5(3):295–306
46. Lowe J, Floyd KT, Rastogi N et al (2016) Similar efficacy from specific and non-specific mineralocorticoid receptor antagonist treatment of muscular dystrophy mice. J Neuromuscul Dis 3(3):395–404
47. Peczkowski KK, Rastogi N, Lowe J et al (2020) Muscle twitch kinetics are dependent on muscle group, disease state, and age in Duchenne muscular dystrophy mouse models. Front Physiol 11:568909

48. Martin PT, Xu R, Rodino-Klapac LR et al (2009) Overexpression of Galgt2 in skeletal muscle prevents injury resulting from eccentric contractions in both mdx and wild-type mice. Am J Physiol Cell Physiol 296(3):C476–C488
49. Lowe J, Kolkhof P, Haupt MJ et al (2020) Mineralocorticoid receptor antagonism by finerenone is sufficient to improve function in preclinical muscular dystrophy. ESC Heart Fail 7(6):3983–3995
50. Lowe J, Wodarcyk AJ, Floyd KT et al (2015) The angiotensin converting enzyme inhibitor lisinopril improves muscle histopathology but not contractile function in a mouse model of Duchenne muscular dystrophy. J Neuromuscul Dis 2(3):257–268
51. Janssen PM, Murray JD, Schill KE et al (2014) Prednisolone attenuates improvement of cardiac and skeletal contractile function and histopathology by lisinopril and spironolactone in the mdx mouse model of Duchenne muscular dystrophy. PLoS One 9(2):e88360
52. Rafael-Fortney JA, Chimanji NS, Schill KE et al (2011) Early treatment with lisinopril and spironolactone preserves cardiac and skeletal muscle in Duchenne muscular dystrophy mice. Circulation 124(5):582–588
53. Murray JD, Canan BD, Martin CD et al (2012) The force-temperature relationship in healthy and dystrophic mouse diaphragm; implications for translational study design. Front Physiol 3: 422
54. Elnakish MT, Schultz EJ, Gearinger RL et al (2015) Differential involvement of various sources of reactive oxygen species in thyroxin-induced hemodynamic changes and contractile dysfunction of the heart and diaphragm muscles. Biophys J 108(2):292a
55. Peterson JM, Kline W, Canan BD et al (2011) Peptide-based inhibition of NF-κB rescues diaphragm muscle contractile dysfunction in a murine model of Duchenne muscular dystrophy. Mol Med 17(5):508–515
56. Acharyya S, Villalta SA, Bakkar N et al (2007) Interplay of IKK/NF-κB signaling in macrophages and myofibers promotes muscle degeneration in Duchenne muscular dystrophy. J Clin Invest 117(4):889–901
57. Janssen PM, Hiranandani N, Mays TA et al (2005) Utrophin deficiency worsens cardiac contractile dysfunction present in dystrophin-deficient mdx mice. Am J Physiol Heart Circ Physiol 289(6):H2373–H2378
58. Delfín DA, Xu Y, Peterson JM et al (2011) Improvement of cardiac contractile function by peptide-based inhibition of NF-κB in the utrophin/dystrophin-deficient murine model of muscular dystrophy. J Transl Med 9(1):1–10
59. Repas SJ, Saad NS, Janssen PM et al (2017) Memantine, an NMDA receptor antagonist, prevents thyroxin-induced hypertension, but not cardiac remodeling. J Cardiovasc Pharmacol 70(5):305–313
60. Saad NS, Repas SJ, Floyd K et al (2017) Recovery following thyroxine treatment withdrawal, but not propylthiouracil, averts in vivo and ex vivo thyroxine-provoked cardiac complications in adult FVB/N mice. Biomed Res Int 2017
61. Saad NS, Floyd K, Ahmed AA et al (2016) The effect of sorafenib, tadalafil and macitentan treatments on thyroxin-induced hemodynamic changes and cardiac abnormalities. PLoS One 11(4):e0153694
62. Saad NS, Elnakish MT, Brundage EA et al (2018) Assessment of PKA and PKC inhibitors on force and kinetics of non-failing and failing human myocardium. Life Sci 215:119–127
63. Saad NS, Mashali MA, Elnakish MT et al (2022) Effect of hypothyroidism on contractile performance of isolated end-stage failing human myocardium. PLoS One 17(4): e0265731
64. Chung J-H, Martin BL, Canan BD et al (2018) Etiology-dependent impairment of relaxation kinetics in right ventricular end-stage failing human myocardium. J Mol Cell Cardiol 121: 81–93
65. Billman GE, Nishijima Y, Belevych AE et al (2010) Effects of dietary omega–3 fatty acids on ventricular function in dogs with healed myocardial infarctions: in vivo and in vitro studies. Am J Physiol Heart Circ Physiol 298(4):H1219–H1228
66. Varian KD, Kijtawornrat A, Gupta SC et al (2009) Impairment of diastolic function by lack of frequency-dependent myofilament desensitizationin rabbit right ventricular hypertrophy. Circ Heart Fail 2(5):472–481
67. Layland J, Kentish JC (1999) Positive force-and [Ca2+] i-frequency relationships in rat ventricular trabeculae at physiological frequencies. Am J Physiol Heart Circ Physiol 276(1):H9–H18
68. Raman S, Kelley MA, Janssen PM (2006) Effect of muscle dimensions on trabecular contractile performance under physiological conditions. Pflugers Arch 451:625–630

Index

Kevin Flanigan and Stefan Nicolau (eds.), *Duchenne Muscular Dystrophy: Methods and Protocols*, Methods in Molecular Biology, vol. 2975, https://doi.org/10.1007/978-1-0716-4811-7,

MIX
Papier aus verantwortungsvollen Quellen
Paper from responsible sources
FSC® C105338

If you have any concerns about our products,
you can contact us on
ProductSafety@springernature.com

In case Publisher is established outside the EU,
the EU authorized representative is:
Springer Nature Customer Service Center GmbH
Europaplatz 3, 69115 Heidelberg, Germany

Printed by Libri Plureos GmbH
in Hamburg, Germany